Science 9
Directions

Science Directions 7
Science Directions 8
Science Directions 9

Science Directions 7 Teacher Resource Package
Science Directions 8 Teacher Resource Package
Science Directions 9 Teacher Resource Package

John Wiley & Sons Canada Limited/Arnold Publishing Ltd.

SI Metric

Our Cover Photograph

The splash of colour you see on our cover is really a tiny crystal of citric acid. In citrus fruits such as oranges, lemons, and limes, it is the citric acid that gives them their tart but pleasant acid taste. As well, citric acid is used as a flavouring agent in some foods, in carbonated beverages, and in some pills and tonics.

What is shown here is a citric acid crystal as it appears when magnified under a special light microscope that polarizes the light used. This process produces the separation of colours that you see.

Science Directions 9

PROGRAM CONSULTANT

Douglas A. Roberts
Faculty of Education, University of Calgary

AUTHOR TEAM

Mary Kay Winter • Derek Bullard
Alan J. Hirsch • Gordon R. Gore
Eric S. Grace • Barry Emerson
Linda Whitney McClelland

PUBLISHING CONSULTANT

Trudy L. Rising

John Wiley & Sons
Toronto   New York   Chichester
Brisbane   Singapore

Arnold Publishing
Edmonton

All Activities in this text have been designed to be as safe as possible, and have been reviewed by professionals specifically for that purpose. The introductory section *Safety in the Science Classroom* discusses appropriate safety procedures and specifies how they are highlighted in the text. Even though the publisher has taken these measures, safety of students remains the responsibility of the classroom teacher, the principal, and the school board.

Canadian Cataloguing in Publication Data

Main entry under title:

Science directions 9

ISBN 0-471-79581-X

1. Science — Juvenile literature.   2. Science — Problems, exercises, etc. — Juvenile literature.
I. Roberts, Douglas A., 1936–
II. Winter, Mary Kay, 1942–

Q161.2.S35 1991   500   C90-093599-5

Science Directions 9 Project Team

Project Director: Trudy Rising
Project Manager: Grace Deutsch
Project Editor: Mary Kay Winter
Developmental Editors: Bruce Bartlett, Tilly Crawley, Jane McNulty
Bias Reviewer, Copy Editor, Proofreader: Wendy Thomas
Word Processing: Diane Klim, Irene Harding
Design: Julian Cleva
Production Co-ordinator/Art Director: Francine Geraci
Production Assistant: Gail Copeland
Picture Research: Jane Affleck, Paulie Keston
Typesetting, Assembly, Film: Compeer Typographic Services Ltd.
Manufacturing: Arcata Graphics
Field Test Preparation Co-ordinator: Shona Wehm
Workshop Co-ordinator: Janet Mayfield
Field Test Design and Preparation: Bruce Campbell, Arnold Publishing
   Ltd.; Presentation Plus Desktop Publishing Inc.

Written, designed and illustrated in Canada.
Printed and bound in the United States of America.

2 3 4 5   AG   98 97 96 95 94

Contents

Acknowledgements

Science Directions 9 is the result of the efforts of a great many people. First, we would like to thank Alberta Education for reviewing this project at its various stages of development. We thank particularly the Alberta Junior High Science Advisory Committee and Bernie Galbraith, Program Manager, for their thorough analysis of the first draft, the field test (pilot) material, and the final pages.

The efforts of many other people were invaluable in developing the book. The Teacher Resource Package authors reviewed each Unit and made helpful suggestions throughout manuscript development. They are: Don Loerke, Fort McMurray; Gary Vornbrock, Camrose; Phil Maloff, Medicine Hat; Florian Borstmayer, Edmonton; Larry Hartel, Red Deer; and Bill Heinsen, Red Deer.

Each Unit of **Science Directions 9** has been field-tested. We would very much like to thank the following teachers and their students for using our materials in draft form and giving us their reactions to assist us in the development of the final text: Marion Jacobsen, John Ware Junior High School, Calgary (Units One and Two); Paul Dvorack, St. Margaret School, Calgary (Units One and Five); Steve Klein, W.D. Cuts Junior High School, St. Albert (Unit One); Steve Telfor, John Maland Junior High School, Devon (Unit One); Roman Scharabun, Sir George Simpson Junior High School, St. Albert (Units Two and Four); Phil Maloff, Medicine Hat High School, Medicine Hat (Unit Two); Gary Vornbrock, Camrose Composite High School, Camrose (Unit Two); Mary Goede–Kohn, Prairie River Junior High School, High Prairie (Units Two and Three); Oliver O'Reilly, Medicine Hat High School, Medicine Hat (Units Three and Five); Larry Hartel, Glendale School, Red Deer (Units Four and Six), John Tate, Fort McMurray Composite High School, Fort McMurray (Unit Four), Caron Cunningham, John Maland Junior High School (Unit Five); Doug Duncan, Dan Knott Junior High School, Edmonton (Unit Five); and Alberta Education field test teachers and their students (all six Units).

We are grateful to the following reviewers for assisting us in ensuring that the content was pedagogically sound and scientifically accurate: Margaret Redway (Unit One); Dr. Margaret Ann Armour (Unit One); Jean Bullard (Units One and Three); John Eix (Units One and Four); Rick Buchannan (Unit Two); Jack Fraser (Unit Three); Paul Crysler (Unit Four); Sandy Eix (Unit Four); Dr. Nancy Flood (Units Five and Six); Alex Penner (Unit Six); Dr. Ann Zimmerman (Unit Six).

Science and Technology in Society features were prepared by Valerie Wyatt (Units Two, Three, and Four), Eric S. Grace (Unit Five), and Julie E. Czerneda (Unit Six). Jane Forbes, author of material in *Science Ideas and Applications 9* on which the Unit One Science and Technology in Society feature was based, is thanked for permission to use it. We would like to credit the author team of *Science Explorations 9* for the Career Feature, Unit One. We thank Eric S. Grace and Mary Kay Winter for preparing the Unit Two Career Feature. As well, we would like to credit these specific authors of *Science Explorations 10:* Alan J. Hirsch (Career Feature, Unit Three, and Career Feature, Unit Four); Julie E. Czerneda (Career Features, Units One, Five, and Six). We also thank each of the individuals whose careers are featured for their contribution.

We would also like to thank the following for use of their material in adapted form: Jean Bullard of *Science Probe 8* (for Activity 1–2); Murray Lang of *Life Probe* (for Activity 5–3); Dr. James D. Rising (for use of his activity from *Science Explorations 10* for Activity 5–7, and for portions of his text and activities from *Science Probe 8*); Marion Gadsby of *Science Explorations 9* and E. Usha R. Finucane of *Science Ideas and Applications 9* (for portions of the Skillbuilders); Gordon Gore for his ideas for Activity 2–16 in Unit Two.

We extend thanks also to Alberta Energy and Natural Resources, from whose publication *Alberta Conservation and Hunter Education* portions of Activity 5–8 were developed.

Lastly, we thank the committed editors, production co-ordinator, designer, and artists who helped us produce this textbook.

The Authors

This is a very exciting time to be teaching science. It is also a very demanding time. Science educators all over the world are being challenged to rethink the goals, purposes, and processes of science teaching, and thus their teaching approaches. The importance of scientific literacy for all citizens has become increasingly clear. As well, the challenge to demonstrate the relationships between science, technology, and their role in the societal decision-making process is generally accepted as essential in the teaching of science.

Science Directions responds to these new challenges. This program gets students involved in the processes of science, technological problem solving, and in discussions relating science and technology to social issues. I have found it very impressive to watch teachers working with these materials. I have seen students develop and defend different viewpoints on issues. And I have seen both girls and boys equally involved in the activities.

Science Directions provides a balanced approach to science by emphasizing three important goals of science education in each book. Some Units concentrate on the nature of science and science processes. Others place their main emphasis on the relationship between science and technology. And still others expose students to issues related to science, technology, and society (STS). Together these approaches are designed to develop students' critical thinking abilities. That is, students learn to identify scientific questions, technological problems, and STS issues, and to recognize the underlying assumptions of each. They develop the ability to identify and to gather appropriate data, and to draw inferences from these data. Likewise, they are encouraged to analyze issues by identifying different viewpoints and evaluating evidence on specific issues.

It has been a pleasure for me to work with these creative, solid, and very teachable materials. With texts of this kind, science teachers can provide the kind of balanced program that meets the challenges and demands of science education for the 1990s and well into the 21st century.

Douglas A. Roberts
Program Consultant

- **Science Directions 9** has six Units. Each uses an appropriate major emphasis: nature of science; science and technology; or science, technology, and society (STS). Each is designed to develop students' critical thinking abilities.
- Many and varied *Activities* are included in each Unit: formal investigations, informal discoveries, technological challenges, among others. Some Activities invite the students to solve practical problems; others provide opportunities for them to design their own experiments.
- The Activities are followed by three levels of questioning:
  Analysis leads students to consolidate their results and observations;
  Further Analysis challenges them to reach logical conclusions and to reason beyond the immediate and obvious results they have obtained;
  Extension encourages investigations or explorations in new directions. They should be considered as optional.
- Within the text, *Probing* questions help enhance the students' understanding of textual and/or visual material by leading them to think through related problems. These questions, designed to be optional, can serve as reinforcement, a basis for independent investigations, or research.
- The *Did You Know?* heading introduces brief statements of scientific, technological, or societal interest related to the concepts under discussion.
- After every two Topics within a Unit, a *Checkpoint* set of questions provides review and reinforcement of the ideas discussed.
- Each Unit ends with a *Unit Review* consisting of three parts. The *Focus* presents the subject matter in point form. The *Backtrack* reviews for the students the general content and process skills taught in the Unit. The *Synthesizer* challenges the students to think through and often apply the knowledge, skills, and processes that have been developed in the Unit.

- Within each Unit is a two-page special feature entitled *Science and Technology in Society*. While the text at all times attempts to use concrete examples to explain abstract concepts, the special features emphasize actual situations and highlight real-life applications of the scientific ideas under consideration in the Unit. This helps the students understand that the concepts they are investigating may have an immediate impact on their society and themselves.
- Each Unit also contains a career feature, entitled *Working with . . .* or *Working by . . .* , which encourages students to consider how they might develop their own interests and capabilities into rewarding careers related to science.
- Throughout the text, new terms are presented in **bold face** type and defined in context; they also appear in the *Glossary* and in the *Index*.
- Appendices called *Skillbuilders* include text and Activities on the SI units and scientific notation, the periodic table of the elements, measurement, graphing, use of the microscope, and an introduction to cell structure. The Skillbuilders provide useful support and reference for the students. This material may be used to introduce the course, or it may be used as needed to enhance skills throughout the course.
- The text opens with extensive discussions of the safety rules to be observed at all times in science classes. The students should become thoroughly acquainted with these safety rules, and the reasons for them, at the beginning of the school year. In addition, a **CAUTION** is included within the text Activities when special care must be taken in using equipment or materials. Potentially dangerous Activities are reserved for the teacher by the designation **Demonstration**.
- The *Teacher Resource Package* provides the teacher with information on the emphases of the program and offers various teaching strategies. The package is designed to provide a wealth of information for both experienced science teachers and those new to the program.

Safety in the Science Classroom

Your school laboratory is designed so that you can perform science experiments in safety—provided you follow proper procedures. Just as you must be careful in your kitchen at home, so too must you be careful as you handle materials and use the equipment in the laboratory.

The Activities in this textbook have been tested and are safe, so long as they are done with proper care. When special attention is needed, you will see the word **CAUTION** with a note about the particular care this Activity requires.

Your teacher will give you specific information about the safety routines used in your school and will make sure that you know where all the safety equipment is.

If you follow these guidelines and general safety rules, along with your own school's rules, you will have an accident-free environment in which to enjoy science.

General Safety Precautions

- Work quietly and carefully; accidents, as well as poor results, can be caused by carelessness. Never work alone; if you have an accident, there will be no one there to help you.
- Tie back loose hair, roll back and secure loose sleeves, and make sure you are wearing shoes that cover your feet as much as possible. Don't wear scarves or ties, baggy clothing, earphones, or jewellery in the laboratory—these can catch on equipment and cause spills or damage.
- Wear safety glasses during experiments involving chemicals or breakable glass.
- Inform your teacher of any allergies, medical conditions, or other physical problems you may have.
- Never eat or drink in the laboratory.
- Do not do laboratory experiments at home unless instructed to do so by your teacher.

Safety Equipment

- Listen carefully to your teacher's instructions about when and how to use the safety equipment: safety glasses, protective aprons, fire extinguishers, fire blankets, eye-wash fountain, and showers.
- Make sure you know where the nearest fire alarm is.

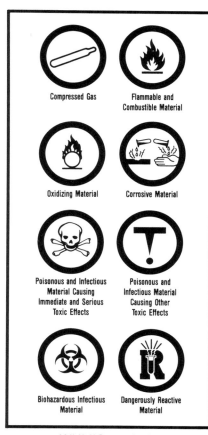

WHMIS symbols

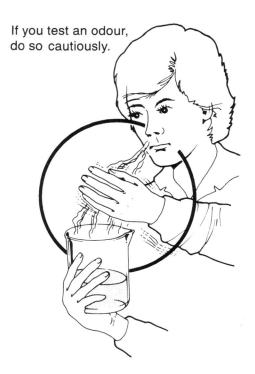

If you test an odour, do so cautiously.

Before You Begin an Activity
- Read through the entire Activity so that you know what to do.
- Make sure the work area is clean before you start. Clear away from the work area everything (books, papers, and personal belongings) except your textbook, your notebook, and a pen or pencil.
- Do not begin an Activity until you are instructed to do so.

Handling Materials
- Examine the WHMIS (Workplace Hazardous Materials Information System) safety symbols shown here. Be sure that you understand these symbols. If you see any of them on containers in your science classroom, use appropriate safety precautions as you handle the materials. (These safety symbols appear on many potentially hazardous substances. You may also find these symbols posted on walls to alert people that such materials are on the site. You should also be familiar with the consumer product safety symbols, found on many consumer products. These symbols are shown on page 3.)
- Touch substances only when told to do so. What looks harmless may, in fact, be dangerous.
- When you are instructed to smell a substance in the laboratory, ensure that you hold the container a good distance from your nose, *not* up close. Wave your hand in order to waft the fumes towards your nose.
- Never pour liquids while holding the containers close to your face. Place a test tube in a rack before pouring liquids into it.
- If any part of your body comes in contact with any harmful chemical or specimen, rinse the area immediately and thoroughly with water. If your eyes are affected, do not touch them, but rinse them with water immediately and continuously for at least 10 min.
- Wash your hands after you handle substances and before you leave the laboratory.
- Clean up any spilled substances immediately, as instructed by your teacher.
- Never pour harmful substances into the sink. Follow your teacher's instructions about how to dispose of them.

Handling a Heat Source
- Whenever possible, use electric hot plates with thermostatic controls.
- To heat a test tube using a hot plate, place the test tube in a beaker of water on the hot plate.
- Use only heat-resistant glass containers when you are heating material. Make certain the glass you use for heating is Pyrex or Kimax and is not cracked.

- Always keep the open end of a test tube pointed away from other people and from yourself.
- Never allow a container to boil dry.
- Pick up hot objects carefully by using tongs or gloves.
- Make sure that the hot plate is turned off when not in use.
- Always unplug electric cords by pulling on the plug, not the cord. Report any frayed cords to your teacher.
- Be careful how the cord from the hot plate to the electrical outlet is placed. Make sure no one will trip over it.
- If you burn yourself, immediately apply cold water or ice.

Rules for Using an Open Flame

- Before using any open flame, be sure that you know the location of the nearest fire extinguisher and/or fire blanket.
- If Bunsen burners are used in your science classroom, use them only when instructed to do so. Obtain instructions from your teacher on the proper method of lighting and using the Bunsen burner, and make sure you understand the instructions.
- Check that there are no flammable substances in the room before you light the flame.
- Use a test tube holder to hold a test tube that is being heated in the flame. Point the open end of the test tube away from yourself and from other people, and move the test tube back and forth over the flame so that heat is distributed evenly. Be ready to extinguish the flame immediately if necessary.
- Never leave a flame unattended.
- Never heat a flammable substance over an open flame.
- Follow your school's rules in case of fire.

Other Equipment

- Never use cracked or broken glassware. If glass does break while you are using it, follow your teacher's instructions in disposing of it.
- Watch for sharp or jagged edges on equipment. Report such problems to your teacher.
- After each experiment, clean all equipment and put it away. Do not leave unused equipment lying around the work area.
- Report to your teacher all accidents (no matter how minor), broken equipment, damaged or defective facilities, and suspicious-looking chemicals. In this way, you will be taking responsibility not only for your own safety, but also for the safety of those who use the laboratory after you.

It's amazing to consider just how much the materials in an apple are changed after you eat the apple. Some of it becomes part of you, while some of it may produce the carbon dioxide that you exhale. A green plant may then use a particle of this carbon dioxide to produce food and oxygen. After being released by the plant, this same oxygen might eventually assist in the blast-off of a rocket. These are only a few examples of the chemical changes that are always taking place in the world. In Unit One, you'll investigate some chemical changes in matter.

In Unit Two, you'll investigate what fluids are, how they are used in the functioning of natural and human-made devices, and how pressure influences the way they act. In doing so, you'll ask not only scientific questions, but also technological ones. **Science** is a search for how and why structures and patterns appear in the world; **technology** is a means of practical problem-solving. Though much of our present-day technology has resulted from scientific discoveries, practical applications are not the goal of scientific questions. Specific practical problems are often solved by technology, rather than science. In Unit Two, you'll investigate the science of fluids, then examine technological devices and design some of your own.

The transfer of heat warms us, cools us, and produces all sorts of practical problems that require solutions—such as how to keep warm on a cold winter's day. In Unit Three, you'll investigate the processes of heat transfer and look closely at some practical means to use them and control them.

Electrical energy is used in our world for a huge variety of functions—from running small, simple toys to lighting whole cities. In Unit Four, you'll investigate how electrical energy is generated, used, and measured. You will also design and make your own electrical device to solve a technological problem.

A study of living things includes not only knowledge of specific organisms and their relationships in natural environments, but also an understanding of the amazing diversity of organisms. How can the enormous variety in types of living things be explained? What factors affect this diversity? How can living things be grouped so that they can be compared? These are some of the scientific questions you'll explore in Unit Five.

Both scientific and technological approaches may be very helpful in dealing with environmental and scientific issues. Yet by themselves, they are sometimes not enough. To deal with such issues, a more wide-ranging approach that involves personal opinions as well as political and economic considerations may be more appropriate. This is often called a **Science, Technology, and Society** (STS) approach. In Unit Six, you'll use this approach to your investigations, putting to use the scientific and technological skills learned in earlier Units. With a case study on a specific environmental issue, you'll conclude your studies in **Science Directions**.

At the back of **Science Directions 9**, you'll find Skillbuilders. You can use them to aid you in skills for which you need review or reminders. If you need a review of graphing, for example, the Skillbuilder section is where you'll find it.

We hope that the investigations in this book help you realize how much science there is in your everyday life and how many ways there are in which science and technology affect you. You will see that a wide understanding of science and technology is important for decision making in the world today.

The Authors

Understanding Chemistry

Some of the photographs on this page are of dramatic events, some look very ordinary, and some could probably have been taken in your own home. Whatever they show, they all have something in common; they are used in this Unit to illustrate certain aspects of chemistry.

Chemistry is the science concerned with matter — properties of matter and changes in matter. **Chemists**, as well as observing properties and changes, also develop models to explain their observations. They use their knowledge to control changes and to produce new kinds of matter that have practical uses.

In this Unit, you will examine substances, observe how they change, and try to explain what you observe. You will experiment to see how you can speed up or slow down the changes. After completing the Unit, you should be able to connect all the events and objects shown here to the science of chemistry.

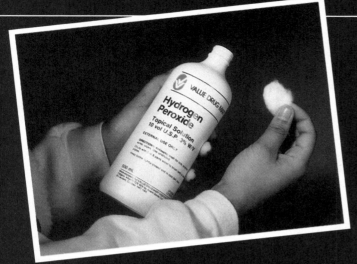

Chemicals

In a single day, you deal with thousands of substances and mixtures of substances. All of these could equally well be called chemicals and mixtures of chemicals. A **chemical** is any substance used or produced in a chemical process. For example, water is a common chemical, essential for all the chemical processes we call life. Fabrics, cosmetics, plastics, and construction materials are produced by chemical industries. "Natural" materials such as wood and cotton are also produced by chemical processes — the chemical processes in the lives of the plants that make them. These substances are all chemicals, even though we don't usually think of them in that way. The Earth itself, and all other objects in the universe, are made of chemicals. Some have chemical names as well as more familiar common names like the names shown in Table 1-1.

Table 1-1 *Some Familiar Chemicals*

COMMON NAME	CHEMICAL NAME
table salt	sodium chloride
vinegar	acetic acid
lime (a common gardening chemical)	calcium oxide
aspirin (A.S.A.)	acetylsalicylic acid
battery acid	sulphuric acid
baking soda (sodium bicarbonate)	sodium hydrogen carbonate
charcoal	carbon

In this Topic, you will investigate different materials to learn about their various properties. **Properties** are characteristics you can use to describe or identify different substances. They include what something looks like, how it feels, how it acts when it is heated or cooled, and what happens when it is mixed with other substances. Some examples of properties include:

- state (Is it a solid, liquid, or gas?)
- colour (What colour is it, or is it colourless?)
- texture (If it is a solid, is it smooth or rough?)
- lustre (If it is a solid, is it shiny or dull?)
- crystal structure (Does the solid form crystals? What shape are the crystals?)
- taste (This property should not be tested in a science classroom.)
- odour (Test for odour only if your teacher directs, and then do so carefully, using the method shown in Safety in the Science Classroom on page xii.)
- solubility (How much dissolves in water? How much dissolves in other solvents?)
- freezing or melting point (At what temperature does the liquid change into a solid or the solid change into a liquid?)
- boiling point (At what temperature does the liquid bubble and change very rapidly into a gas?)
- density (What is the mass of a specific volume of the substance?) (Information about density is supplied in Skillbuilder Three on page 372.)
- behaviour (Does it burn? Does it rot? Does it explode? Does it act as a poison? How does it change when it is heated? If it is mixed with other substances, do they change in any way?) The symbols shown in Safety in the Science Classroom on page xii are used in workplaces to warn of potentially dangerous behaviour.

Probing

The containers for many household chemicals show information, often in the form of symbols, about dangerous properties of the contents. In your own home, find and list as many examples of these symbols as you can. You are likely to find them on products such as:

- oven cleaner • paint thinner
- drain cleaner • spray paint
- contact cement • bleach
- glue •wood preservative
- gasoline • furniture polish

What does each of the four symbols mean? Why are there three different shapes?

The properties of some chemicals make them dangerous, but these same properties may also make them useful. For example, chlorine gas is poisonous. When used in small amounts in water supplies, it prevents the growth of algae and bacteria that can contaminate water. Larger amounts of chlorine would kill not only algae and bacteria, but people as well.

Whether you realize it or not, you have been gaining experience about the properties of matter all your life. In the following Activity, try using this experience to identify some common chemicals.

Chlorine is useful because it is poisonous. In swimming pools, a small amount of chlorine checks the growth of micro-organisms but does not harm humans.

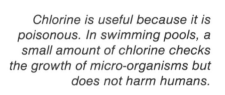

Activity 1-1

Everyday Chemistry

Problem

What common substances are these?

Procedure

1. Examine the lists of properties in Table 1-2, and identify substances A, B, and C.

Analysis

1. What properties did you use as clues to identify each substance?

Further Analysis

2. From your general knowledge, list all the properties you can for the following substances. Include behaviour as well as visible properties.

    (a) air
    (b) wood
    (c) iron
    (d) windshield washer fluid
    (e) flour
    (f) paper
    (g) apple juice

3. Pick three common substances and list several properties of each. Give your list to a friend and ask your friend to identify the substances.

Table 1-2 *Properties of Some Familiar Substances*

SUBSTANCE A	SUBSTANCE B	SUBSTANCE C
is liquid at room temperature	is solid at room temperature	is solid at room temperature
is clear and colourless	consists of white crystals	forms large colourless crystals
does not mix with oil	dissolves in water	does not dissolve in water
freezes at 0°C	melts at 146°C	when cut, its flat surfaces sparkle
	turns black when heated above	with all colours of the rainbow
	146°C	melts at 3500°C
	burns if exposed to flame	is very hard

Identifying Unknown Substances

Problem

How can you identify unknown substances on the basis of their properties?

Materials

samples of 4 unknown substances in petri dishes
information in Table 1-3
hand lens

Procedure

CAUTION: If you spill any of the solids, wash the spill immediately with lots of water. Potassium nitrate and copper(II) sulphate are poisonous. If they come in contact with your skin, wash them off thoroughly with cold water. Do not taste any of the samples.

1. Briefly examine the substances in the petri dishes and attempt to match each sample with one of the descriptions in Table 1-3.
2. Using a hand lens, examine the substances more carefully and compare them to the photographs.
3. Sketch what you see, and make any further identification.

Analysis

1. For the substances that were easy to identify, what properties allowed you to identify them?

Table 1-3

NAME	APPEARANCE
(a) sucrose (icing sugar)	powdered white solid
(b) sodium chloride (table salt)	small white solid crystals
(c) calcium carbonate (chalk)	powdered white solid
(d) sulphur	powdered yellow solid
(e) calcium hydroxide (slaked lime)	powdered white solid
(f) copper(II) sulphate (bluestone)*	blue solid crystals
(g) aluminum potassium sulphate (alum)	small white solid crystals
(h) potassium nitrate	small white solid crystals

*The complete chemical name for bluestone is really "hydrated copper(II) sulphate" or "copper(II) sulphate pentahydrate." In this book it is called simply "copper(II) sulphate" (pronounced "copper-two sulphate").

sucrose (icing sugar)

sodium chloride (salt)

calcium carbonate (chalk)

powdered sulphur

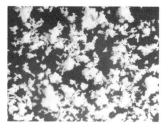
calcium hydroxide (slaked lime)

copper(II) sulphate (bluestone)

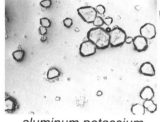
aluminum potassium sulphate (alum)

potassium nitrate

Further Analysis

2. If appearance alone did not allow you to identify the substance, what further tests could you do? (Hint: What type of behaviour could you test for?)

3. Look at the three products in the photograph. What properties could you use to distinguish among them?

Extension

4. Use a suitable reference to find out more about at least two of the substances listed in Table 1-3. Write a report and include in it a description of how they are produced and what they are used for.

5. (a) At home, find five different samples of white powders such as flour, powdered soap, cornstarch, icing sugar, baking powder, cream of tartar, citric acid, vitamin C powder, powdered milk, coffee whitener, scouring powder, lime, powdered aspirin.
   (b) Put each powder in a small bottle and label the bottles A to E.
   (c) Trade with a friend, and try to identify each other's samples.

Using Properties to Classify

Substances can be classified in many ways. For example, they may be solids, liquids, or gases. They may be green, blue, any other colour, or colourless. They may be either ''soluble in water'' or ''insoluble in water.'' Each of these classification systems is useful for different purposes.

From your previous studies, you learned that all matter can be classified as either a mixture or a pure substance. Mixtures contain at least two substances. If you can see the different parts of the mixture, it is called a **mechanical mixture**. Concrete, soil, and granola cereal are examples of mechanical mixtures. If the mixture appears to be all one substance it is called a **solution**. Vinegar, soft drinks, and clear tea are examples of solutions. The properties of mixtures vary, depending on the proportions of the parts. For example, a drink may be sour if the sugar solution is dilute, or sweet if a larger amount of sugar is used. Mortar will be hard if it contains suitable proportions of sand and cement, but weak and crumbly if the proportions are different.

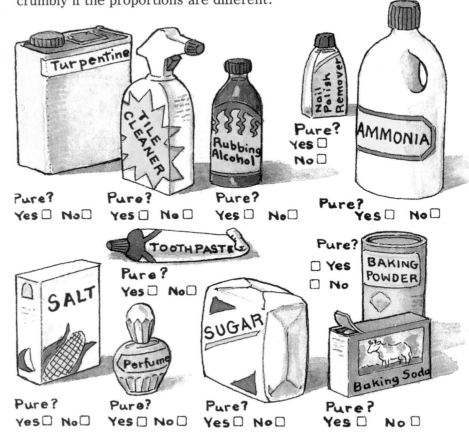

Most common products are mixtures, but a few are pure substances. What substances do the products shown here contain? (Look on page 19 to find out.)

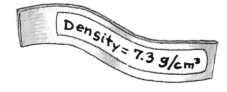

Unlike mixtures, pure substances have definite properties that are always the same. For example, values such as melting point, boiling point, and density of a pure substance can be determined accurately and recorded in standard reference tables. If you have an unknown material that you think might be a pure substance, you can experimentally determine its properties and then compare your results with the values in those tables.

In the following Activity, you will inspect samples of unknown substances and classify them as pure substances, solutions, or mechanical mixtures. For example, examine the three samples in the illustration on this page. You should be able to tell immediately which of these is a mechanical mixture. What about the other two? You can't distinguish a solution from a pure substance just by looking at them. You also need to compare the information on the labels with the values in Table 1-4 on the next page. If the melting point, boiling point, or density indicated on the label matches one of those values, the sample is likely to be a pure substance. If there is no match and you know the sample contains only substances listed in Table 1-4, it is likely to be a solution. The observation table below shows how you could classify these three. For some samples, you may be able to identify the substances as well as classify them.

Sample Number	Observed Properties	Information on label	Classification	Reasons for Classification	Substance(s)
1.	Solid pieces, some blue, some black powder		Mechanical mixture	2 parts are visible	carbon, copper (II) sulphate
2.	Shiny, silvery solid	Density = 7.3 g/cm³	Pure substance	Appearance of tin. Density is that of tin.	Tin
3.	clear, colourless liquid	Boiling point = 102°C	Solution	Appears to be one substance. Boiling point is not the same as any substance in Table 1-4.	Uncertain (could be water or ethanol with some solute(s).)

Observation table.

Using Data about Chemicals

Problem

How can you classify unknown materials as mechanical mixtures, solutions, or pure substances?

Materials

samples of at least 10 unknown materials (mechanical mixtures, solutions, and pure substances) containing only the substances listed in Table 1-4

data about properties of pure substances (Table 1-4)

Procedure

1. Make an observation table in your notebook like the one shown on page 7.

> CAUTION: Do not open the vials.

2. Examine each sample and describe its properties. Also note any information supplied on the label.

Analysis

1. Consult Table 1-4 and compare the properties of your samples with those of the pure substances listed. Classify each sample as a pure substance, a solution, or a mechanical mixture, and explain your reasons.
2. (a) For each sample that is a pure substance, identify the substance.
   (b) For any samples that are solutions or mechanical mixtures, suggest what pure substances they might contain.

Further Analysis

3. If you had no reference table of properties, what could you do to determine whether:
   (a) a clear colourless liquid is pure water or a solution of salt in water?
   (b) a clear colourless liquid is alcohol or water?
   (c) a reddish-brown metal is pure copper or a mixture of copper and tin?
   (d) white crystals are sugar or salt?

Table 1-4 *Properties of Some Pure Substances.*

PURE SUBSTANCE	MELTING POINT (°C)	BOILING POINT (°C)	DENSITY (g/cm³)	APPEARANCE
ethanol (alcohol)	−115	78	0.8	clear colourless liquid
aluminum	660	*	2.7	silvery-coloured solid
sodium bicarbonate (baking soda)	*	*	2.2	white solid
copper(II) sulphate (bluestone)	*	*	2.3	blue solid crystals
carbon (diamond)	3500	3930	3.5	colourless solid crystals
carbon (graphite)	4000	3930	2.3	grey-black solid
copper	1084	2336	9.0	shiny reddish solid
glycerol (glycerine)	18	*	1.2	colourless thick liquid
iron	1535	3000	7.9	grey solid
lead	327	1750	1.3	blue-grey solid
calcium carbonate (limestone)	*	*	2.9	grey-white solid
naphthalene	80	216	1.2	white solid
sodium chloride (table salt)	801	1465	2.2	white solid
calcium hydroxide (slaked lime)	*	*	2.2	white solid
sucrose (sugar)	170	*	1.6	white solid
sulphur	113	445	2.1	yellow solid
tin	232	2270	7.3	silvery-yellowish solid
water	0	100	1.0	clear colourless liquid

*When heated, these substances decompose (break apart) rather than change state.

Pure Substances

There are millions of kinds of pure substances. How can anyone expect to learn about all of them? Which ones would you start with? How would you find out which ones are most important? About 200 years ago, scientists asked these same questions and began to investigate substances by heating them, mixing them together, and trying various procedures. They found that there were certain "building blocks of matter"—certain simple substances—that were in all the materials they investigated. They called these substances elements. **Elements** are pure substances that cannot be broken down into any simpler substances.

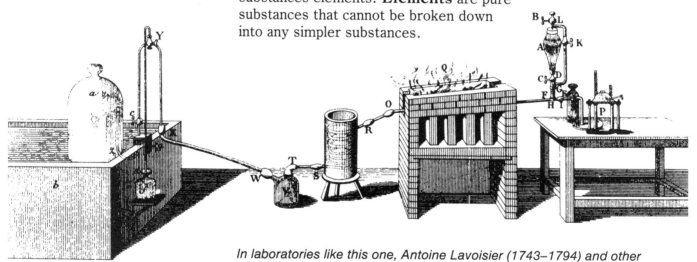

In laboratories like this one, Antoine Lavoisier (1743–1794) and other scientists conducted experiments on many elements and compounds.

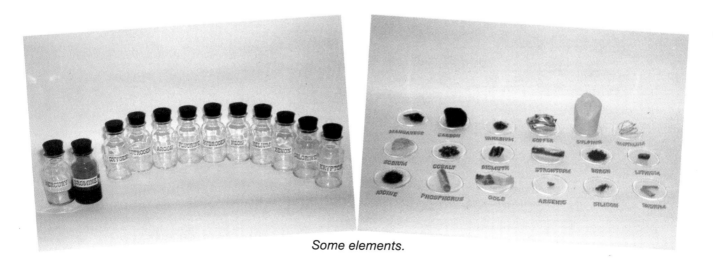

Some elements.

Carbon forms many different compounds. These are so important that there is an entire field of chemistry—*organic chemistry*—concerned with compounds of carbon.

The scientists found that other substances decomposed, or broke down. They called these substances compounds. **Compounds** are pure substances that contain two or more elements combined in a definite fixed proportion. Compounds are related to elements in the same way that words are related to letters of the alphabet. They are made by combining elements, just as words are made by combining letters. Many thousands of words can be made from the 26 letters of the English alphabet. Similarly, although there are only about 100 elements, many thousands of compounds can be made by combining them. A very few examples are shown in Table 1-5.

Table 1-5   *Elements in Some Common Compounds*

ELEMENTS	COMPOUND
hydrogen, oxygen	water
carbon, hydrogen, oxygen	any type of sugar
carbon, hydrogen, oxygen	any type of alcohol
copper, sulphur, oxygen	copper(II) sulphate
sodium, chlorine	sodium chloride .

In the entire universe, there are only about 100 elements. All of them are known to chemists, and each one has its own unique set of properties. They include several that are gases at room temperature, two that are liquids, and many that are solids. Some are common, and some are extremely rare. Some are quite safe, while others may be explosive, radioactive, or poisonous. In the following Activity, you will find out about some elements.

Table 1-6   *The Elements*

actinium, aluminum, americium, antimony, argon, arsenic, astatine, barium, berkelium, beryllium, bismuth, boron, bromine, calcium, cadmium, californium, carbon, cerium, cesium, chlorine, chromium, cobalt, copper, curium, dysprosium, einsteinium, erbium, europium, fermium, fluorine, francium, gadolinium, gallium, germanium, gold, hafnium, helium, holmium, hydrogen, indium, iodine, iridium, iron, krypton, lanthanum, lawrencium, lead, lithium, lutetium, magnesium, manganese, mendelevium, mercury, molybdenum, neodymium, neon, neptunium, nickel, niobium, nitrogen, nobelium, osmium, oxygen, palladium, phosphorus, platinum, plutonium, polonium, potassium, praseodymium, promethium, protactinium, radium, radon, rhenium, rhodium, rubidium, ruthenium, samarium, scandium, selenium, silicon, silver, sodium, strontium, sulphur, tantalum, technetium, tellurium, terbium, thallium, thorium, thulium, tin, titanium, tungsten, uranium, vanadium, xenon, ytterbium, yttrium, zinc, zirconium

The elements in your body combine into many types of compounds, such as fats, proteins, and carbohydrates. Fats are made up of carbon, hydrogen, and oxygen. *Proteins* contain carbon, hydrogen, oxygen, nitrogen, and other elements such as sulphur. *Carbohydrates* contain carbon, hydrogen, and oxygen. The most common compound is water, made up of hydrogen and oxygen. Water accounts for about 70% of your body's mass.

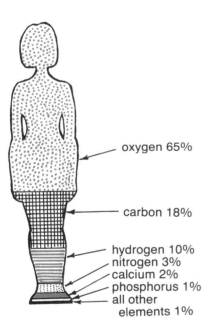

oxygen 65%

carbon 18%

hydrogen 10%
nitrogen 3%
calcium 2%
phosphorus 1%
all other
elements 1%

Cerium

Carbon

Californium

Activity 1–4

The Elements

Problem

What variety is there among elements?

Materials

index cards, each with the name of one element on it (to be distributed among the class) reference materials

Procedure

1. Use suitable references to find out some interesting facts about your element(s). Write the information on the cards. Among the things you could find out are:
   - What are its most obvious properties?
   - Where did it get its name?
   - When was it discovered?
   - What is it used for?
   - How common is it on the Earth or in living things?
   - Is it dangerous (explosive, radioactive, poisonous)?

2. Use the cards to make a file of reference material about the elements, or post cards on a bulletin board.

Analysis

1. (a) Table 1-4 on page 8 contains a list of pure substances, of which ten are compounds; Table 1-6 on page 10 contains a list of all the elements. Compare the two lists, and write the names of the ten compounds in your notebook.
   (b) Where you can, name the elements that are present in these compounds.

Further Analysis

2. Elements are often classified as *metals* or *non-metals*.
   (a) In the Glossary at the end of the book, find a meaning for "metal." List properties that metals have in common.
   (b) In your notebook, make a table with two columns, entitled "Metals" and "Non-metals." Refer to Table 1-6, then list at least ten familiar elements in each column.

Extension

3. Examine the list of elements in Table 1-6.
   (a) Name at least five elements that appear to have been named after countries or other geographical areas.
   (b) Name at least three elements that appear to have been named after people. Who were these people and why were they important?

Elements, Compounds, and the Atomic Theory

In previous studies, you have learned that many properties of matter can be explained using the *particle theory*—the idea that all matter is made up of tiny particles. We can extend this theory to explain the differences between elements and compounds. In this extended form, it is usually called the *atomic theory*.

The Atomic Theory

- All matter is made up of tiny particles, called *atoms*.

- Atoms of any one element are like one another and are different from atoms of other elements.

- Atoms may combine with other atoms to form larger particles, called *molecules*.

- Atoms are not created or destroyed by any ordinary means.

Atoms are the smallest particles of elements. Since there are about 100 elements, there are about 100 kinds of atoms. Atoms can join together in many combinations. Two or more atoms may join together, forming a larger particle called a **molecule**. Some molecules contain only identical atoms; these are molecules of an element. Other molecules contain atoms of different elements. Compounds contain atoms of at least two different elements. Examine the molecules in the illustration—the first shows a molecule of an element, and the other two show molecules of compounds. Molecules can contain as few as two atoms or can contain many thousands of atoms.

(a) In the element oxygen, atoms join together in pairs, forming oxygen molecules.

(b) In the compound methane (marsh gas), each molecule contains one carbon atom and four hydrogen atoms. (Methane is the main component of natural gas, used in home heating and in Bunsen burners.)

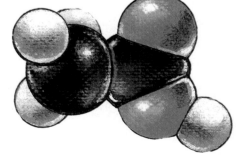

(c) In acetic acid (vinegar), each molecule has eight atoms.

Compounds and Specific Proportions

In compounds, atoms of two or more different elements combine in a specific proportion. That is, a certain number of atoms of one element join together in a *definite fixed ratio* with atoms of another element. Four examples are shown here. Each of these is a distinctly different compound, with its own properties.

Both water and hydrogen peroxide contain only atoms of hydrogen and oxygen, but the proportions are different. A water molecule never has more nor fewer than two hydrogen atoms combined with one oxygen atom; the atoms are always present in a 2:1 ratio. In a molecule of hydrogen peroxide, there are always four atoms, two of hydrogen and two of oxygen.

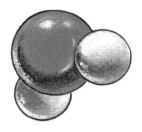

(a) A molecule of water.

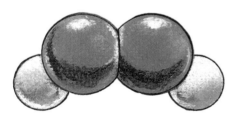

(b) A molecule of hydrogen peroxide.

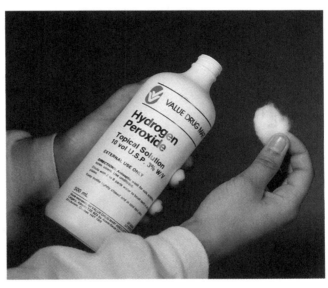

Hydrogen peroxide is a compound used as a bleach and as a disinfectant.

Both carbon dioxide and carbon monoxide contain only atoms of carbon and oxygen. They differ only in the proportion of carbon and oxygen, but these two compounds have distinctly different properties. Carbon monoxide (one of the gases present in automobile exhaust) is extremely poisonous. Carbon dioxide is present everywhere in the air and is used by plants in photosynthesis.

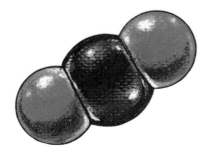

(a) A molecule of carbon dioxide.

(b) A molecule of carbon monoxide.

Pure Substances **13**

Classification of Matter

In Topic One (page 3), you learned to classify matter on the basis of properties. With the atomic theory as background, you can now complete the classification of matter into four categories: mechanical mixture, solution, compound, and element.

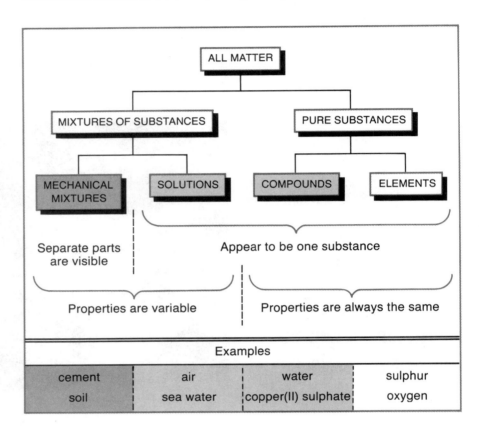

Table 1-7 *Symbols for Some Common Elements*

calcium	Ca
carbon	C
chlorine	Cl
copper	Cu
hydrogen	H
nitrogen	N
oxygen	O
sodium	Na
sulphur	S

Chemical Formulas

Chemists have developed a shorthand method of representing elements and compounds. Every element is represented by a **chemical symbol** that is either a single capital letter or a capital letter followed by a small letter (Table 1-7).

Just as single symbols represent elements, combinations of symbols represent compounds. The combination of symbols representing a particular compound is called its **chemical formula**. The chemical formula indicates which elements are in the compound, and in what proportion they are present. Compare these chemical formulas with the drawings of the molecules on page 13.

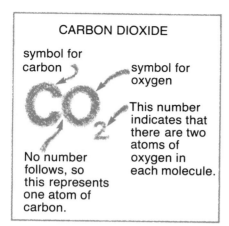

CARBON DIOXIDE

symbol for carbon

symbol for oxygen

This number indicates that there are two atoms of oxygen in each molecule.

No number follows, so this represents one atom of carbon.

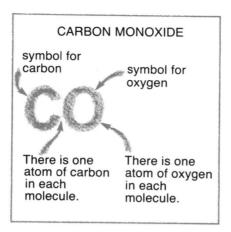

CARBON MONOXIDE

symbol for carbon

symbol for oxygen

There is one atom of carbon in each molecule.

There is one atom of oxygen in each molecule.

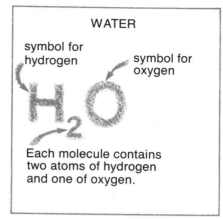

WATER

symbol for hydrogen

symbol for oxygen

Each molecule contains two atoms of hydrogen and one of oxygen.

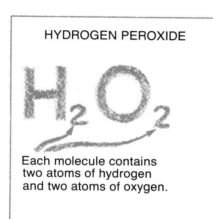

HYDROGEN PEROXIDE

Each molecule contains two atoms of hydrogen and two atoms of oxygen.

Each symbol represents an atom of an element. If only one atom is represented, no number is included. If there is more than one atom of that element, the symbol is followed by a number written below the line. This number (called a subscript) tells how many atoms there are. For example, in Table 1-8, the formula for sodium hydrogen carbonate (baking soda) tells you that the elements are present in the proportion of one atom of sodium to one atom of hydrogen to one atom of carbon to three atoms of oxygen.

Table 1-8 *Some Examples of Chemical Formulas*

sodium hydrogen carbonate (baking soda)	$NaHCO_3$
calcium carbonate (chalk)	$CaCO_3$
copper(II) sulphate	$CuSO_4$
sodium chloride (table salt)	$NaCl$
acetylsalicylic acid (A.S.A. or "aspirin")	$C_9H_8O_4$

Probing

Atoms of oxygen form molecules containing two atoms each, represented by the formula O_2. This is the form of oxygen that makes up about 20% of the air. Under certain conditions, such as in intense ultraviolet light, oxygen atoms can form larger molecules containing three atoms. This form of oxygen is called *ozone* and has the formula O_3. Find out about the importance of the ozone layer of our atmosphere, and why people are concerned about it.

Activity 1–5

Chemical Formulas

1. Explain the meaning of each letter and number in the chemical formulas in Table 1-8.
2. Refer to the symbols in Table 1-7 and write a chemical formula for each of the following. For each, state whether it is an element or a compound.

(a) a molecule of hydrogen that is made up of two atoms of hydrogen
(b) sodium nitrate, which contains atoms of sodium, nitrogen, and oxygen in the ratio 1:1:3
(c) a molecule of propane gas that is made up of three atoms of carbon and eight atoms of hydrogen

Inside the Atom

In order to understand
why elements like gold and carbon (shown here in the form of
diamond) have different properties, scientists have studied the structure of atoms.

If you could divide an element into smaller and smaller pieces, you would eventually end up with single atoms. More than 150 years ago, scientists observed that each element behaved differently from the others. This observation made them suspect that the atoms of each element must be unique and led them to propose the atomic theory. At the time, they didn't know what was different about the atoms. Since then, we have gathered a lot of evidence about the structure of atoms, and we can explain many of their different properties.

All our evidence to date supports the idea that atoms are made up of three kinds of *subatomic particles*. These are *protons*, *neutrons*, and *electrons*

Table 1-9 *The Three Subatomic Particles*

TYPE OF PARTICLE	WHERE FOUND
proton	in the nucleus
neutron	in the nucleus
electron	in a "cloud" outside the nucleus

(Table 1-9). The protons and neutrons are held tightly together, forming the *nucleus*, or core, of the atom. The electrons move very rapidly around it, as if in a cloud. Because the nucleus is very tiny and the electrons are quite far away from it, most of the atom is actually empty space. Only the protons and neutrons have been shown to have a significant amount of mass, so almost all of the mass of the atom is made up by the nucleus.

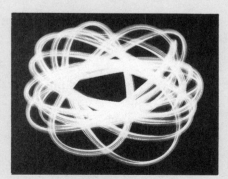

This photograph is of a beam of light from a rapidly moving flashlight. It can serve as a model of the motion of electrons around the nucleus of an atom.

ATOMIC ANATOMY!

HI I'M ADAM ATOM!

ATOM MEANS INDIVISIBLE... BUT DON'T BE FOOLED. LIKE YOUR OWN BODY. I'M MADE UP OF SMALLER PARTS.

ATOM

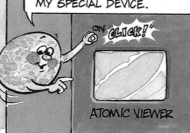

LET ME INTRODUCE YOU TO SOME OF THE PARTS ON MY SPECIAL DEVICE.

ON CLICK!

ATOMIC VIEWER

"MOST OF ME IS EMPTY SPACE. MY HEAVIEST PART IS A TINY DENSE NUCLEUS IN THE CENTRE. IF I WERE THE SIZE OF A FOOTBALL FIELD, MY NUCLEUS WOULD BE A GRAIN OF SAND IN THE CENTRE."

NUCLEUS

ATOMIC VIEW!

LET'S TAKE A CLOSE LOOK AT MY TIGHTLY PACKED NUCLEUS. THERE ARE TWO KINDS OF PARTICLES.

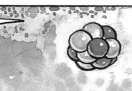

*1 THE PROTON

THIS PARTICLE HAS A POSITIVE ELECTRIC CHARGE. REMEMBER... "PRO" MEANS POSITIVE.

*2 THE NEUTRON

IT HAS NO CHARGE. REMEMBER. "NEU" MEANS NEUTRAL. (NONE)

ZZZZzzz

HEY, WAKE UP! THEY'RE WATCHING US.

IN THE EMPTY SPACE AROUND THE NUCLEUS THERE ARE LITTLE BITS OF MATTER MOVING AT HIGH SPEED.

THESE MOVE SO FAST THAT THEY SEEM LIKE A CLOUD.

NUCLEUS

THESE LITTLE HIGH SPEED PARTICLES ARE CALLED ELECTRONS.

AN ELECTRON IS 2000 TIMES LIGHTER THAN A PROTON, BUT IT HAS A NEGATIVE CHARGE THAT IS AS STRONG AS THE POSITIVE CHARGE OF A PROTON.

THAT'S RIGHT. ELECTRICALLY SPEAKING, ONE PROTON IS EQUAL TO ONE ELECTRON.

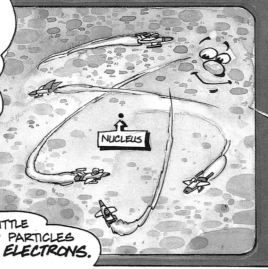

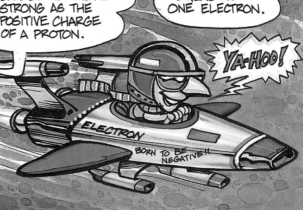

YA-HOO!

ELECTRON

BORN TO BE NEGATIVE!!

"IN ALL REGULAR ATOMS - ELECTRICAL CHARGES BALANCE OUT. WE HAVE EQUAL NUMBERS OF PROTONS AND ELECTRONS."

Atoms of each element have a different number of protons in the nucleus. For example, evidence shows that the nuclei of all hydrogen atoms contain just one proton; all oxygen nuclei contain eight protons; and all copper nuclei contain 29 protons. For every whole number up to about 100, there is a different element with that number of protons. Very large nuclei (larger than uranium, which has 92 protons) tend to be unstable and break apart; that is why there are only about 100 naturally occurring elements. (For more information about the protons in atoms, see Skillbuilder Two, *The Periodic Table of the Elements*, on page 346.)

The number of neutrons in the nuclei of atoms varies; many elements have roughly equivalent numbers of neutrons and protons.

Protons have an electrical charge that we call "positive." Neutrons have no electrical charge. Because the protons are in the nucleus, the nuclei of all atoms are positively charged. In hydrogen, for example, the nucleus has a charge of $1+$, while the nucleus of an oxygen atom has a charge of $8+$. Electrons have an electrical charge opposite to that of the protons, a charge we call "negative." In every atom, the number of electrons is equal to the number of protons, so the charges cancel each other out. The net charge is 0; that is, the atom itself has no electrical charge. Some examples are shown in Table 1-10.

Scientists have discovered that various arrangements of electrons cause the different properties of

Table 1-10

ELEMENT	NUMBER OF PROTONS	CHARGE IN NUCLEUS	NUMBER OF ELECTRONS	TOTAL NEGATIVE CHARGE	NET CHARGE ON ATOM
hydrogen	1	$1+$	1	$1-$	0
oxygen	8	$8+$	8	$8-$	0
copper	29	$29+$	29	$29-$	0
uranium	92	$92+$	92	$92-$	0

elements and compounds. Certain arrangements of electrons are more stable than others; in order to become more stable, atoms give up, gain, or share electrons. Atoms of many elements, such as sodium, chlorine, oxygen, and hydrogen, are particularly likely to join with other atoms and are unlikely to exist on their own. Atoms of some elements, however, may be more stable on their own and much less likely to form molecules. Helium atoms, for example, always exist independently; no compounds of helium have ever been discovered in nature. *In general, atoms join together if that makes the arrangement of their electrons more stable.*

What's the Evidence?

Many scientists have contributed to our present theories of atomic structure. From the following list, pick one name and find out when that person lived and what contribution he or she made.

- Henri Becquerel
- Neils Bohr
- Sir James Chadwick
- Marie Curie
- Dmitri Mendeleev
- Sir Ernest Rutherford
- J. J. Thomson

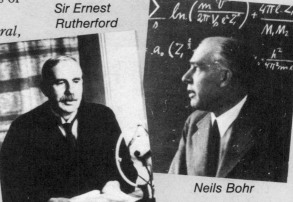

Sir Ernest Rutherford

Neils Bohr

Marie Curie

Dmitri Mendeleev

Henri Becquerel

Checkpoint

1. (a) What is meant by "property of a substance"?
   (b) List three examples of properties that you can learn about just by looking at a substance.
   (c) List three examples of properties that involve the behaviour of a substance.
   (d) List all the properties you can think of for silver, gold, air, and cooking oil.

2. Refer to Table 1-4 on page 8 and describe how you could distinguish between
   (a) iron and tin;
   (b) sugar and salt;
   (c) alcohol and water.

3. All types of matter can be classified into four categories, as shown on page 14. In the following lists, all except one of the materials belong to the same category. For each, indicate which category this is and which material does not belong in that category.
   (a) salt water, concrete, chocolate chip cookie batter, granola cereal
   (b) water, baking soda, chalk, salt water
   (c) carbon dioxide, oxygen, hydrogen peroxide, methane
   (d) carbon, hydrogen, oxygen, water

4. Use the atomic theory of matter to explain the difference between
   (a) an element and a compound;
   (b) an atom and a molecule.

5. Use a sketch to show how there can be two different compounds, each containing only the elements carbon and oxygen.

6. "Compounds have properties that are always the same but mixtures have variable properties." Explain this statement, using the atomic theory of matter in your explanation.

7. (a) The chemical formula for water is H_2O. Explain the meaning of each letter and number in the formula.
   (b) Write the chemical formula for nitrogen dioxide, whose molecules contain one atom of nitrogen and two atoms of oxygen.

8. Suppose someone tells you that a green object contains copper. You are not convinced, as you have seen that copper wires and copper jewellery are reddish-brown metal.
   (a) Is it possible that this green substance really does contain copper? Explain.
   (b) With the assistance of a chemist, what tests might you carry out to settle the question?

9. The pie graph shows the elements that are present in the crust of the Earth. Compare this with the elements in the human body (page 11).
   (a) What similarity can you see?
   (b) What element makes up a large part of the crust of the Earth, but less than 1% of the human body?
   (c) What element makes up a large part of the human body, but less than 1% of the crust of the Earth?

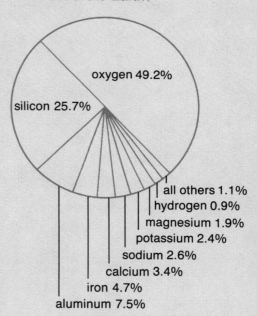

oxygen 49.2%
silicon 25.7%
all others 1.1%
hydrogen 0.9%
magnesium 1.9%
potassium 2.4%
sodium 2.6%
calcium 3.4%
iron 4.7%
aluminum 7.5%

Substances in the products shown on page 6:	
Turpentine	pinene, diterpene, and other substances
Tile cleaner	bleach, detergent, and sodium hydroxide
Rubbing alcohol	isopropanol and water
Nail polish remover	acetone and methanol
Ammonia	ammonium hydroxide and water
Toothpaste	active ingredients, flavourings, and fillers
Baking powder	sodium hydrogen carbonate, a dry acid, and starch
Salt	sodium chloride, with a little sodium iodide
Perfume	a solvent, and many dissolved substances
Sugar	sucrose (a pure substance)
Baking soda	sodium hydrogen carbonate (a pure substance)

Science and Technology in Society

How Are Electric Guitars Made?

The next time you listen to your favourite electric guitar music, you might think about what you learned in science class. The sound of an electric guitar depends on the properties of the materials used to make the guitar. These properties contribute to the sounds the guitar makes.

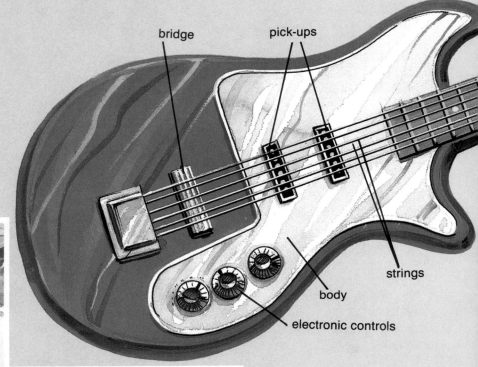

bridge pick-ups

strings

body

electronic controls

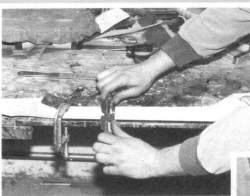

Maple is often used for the neck because it is a strong wood. The neck must be strong enough to withstand the pull of the strings without bending. The best necks use one piece of wood that runs the length of the guitar. A metal rod inside the neck helps to keep it from bending.

Ebony, a very hard wood, is a good choice for the finger board of a guitar.

The body is made of wood. The wood must be of good quality and prepared properly. The moisture content of the air (humidity) must be controlled while the guitar is being made. The type of wood affects the quality of sound. A guitar made of dense maple will have a harsh, metallic sound. Red alder and butternut, which are less dense woods, produce a more mellow sound.

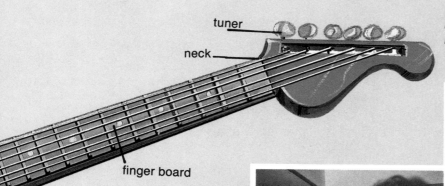

tuner
neck
finger board

The final product is unique. It produces a sound affected by the properties of all its parts: wood, metal, acrylic—and, of course, the talent and skill of the player!

The strings contain the metals chromium, nickel, and steel. Stainless steel is an important part of the strings. Steel has two properties that are needed. It is magnetic, so it can affect the pick-ups. Also, it does not readily rust.

Even the finish applied to the guitar affects the sound. Some finishes deaden the sound. Other finishes change colour or crack with time. The last finish sprayed on the guitar must be long lasting and attractive. The best choice is acrylic lacquer.

The pick-ups on the guitar detect the vibrations of the strings. The pick-up closest to the neck detects sounds of low pitch (bass response). The pick-up near the bridge detects higher pitched sounds (treble response). Controls allow the player to modify the sound.

Think About It

For any other musical instrument that you choose, find out what materials it is made of. List as many of them as you can, and describe what properties affect the quality of the sound.

Changes

In this Topic, you will be observing the different ways matter can *change*. There are two types of change—chemical change and physical change.

In any **chemical change**, one or more new substances are formed—substance(s) with properties different from those of the starting material(s). As an example, think of the changes that occur when food cooks. When an egg is heated, it changes colour and becomes solid. The cooked egg has properties quite different from the properties of the uncooked egg. As another example, think of the changes that happen to the body of a car as it becomes older. Where the steel was once hard and shiny, a crumbly reddish brown substance forms; the steel rusts. You can wipe away the soft rust and polish the steel again, but you cannot keep doing that forever. The steel has changed into rust. Eventually, if the rusting continues, holes will appear. Cooking and rusting are chemical changes.

Steel can rust away completely.

When food is cooked, chemical changes occur.

You are already familiar with physical changes as well. You have seen water change to ice, but you have also seen ice melt, and you know that the water is still there. It hasn't been used up. The water molecules haven't been broken apart. Ice is not a new type of matter, it is simply frozen water. As well, you know that when you dissolve sugar in water it may become invisible, but it is still there. You can tell by tasting it or evaporating the water and collecting the sugar. This kind of change, in which no new substance is produced, is called a **physical change**. Freezing, melting, boiling, and dissolving are examples of physical changes, as are cutting things into pieces, ploughing a field, and moulding a sculpture.

A physical change. When liquid boils it changes to a gas, but the gas can be changed back into a liquid having identical properties.

Certain clues, shown in Table 1-11, may indicate that a chemical change has occurred. But don't jump to conclusions too quickly. Any of these could also accompany a physical change. *You need to consider several clues in order to determine what type of change has taken place.* In the following Activity, you will have an opportunity to observe examples of both types of change and to use these clues to identify which is which.

Table 1-11   *Clues That May Indicate a Chemical Change*

New colour appears.
Heat or light is given off.
Bubbles of gas are formed.
Solid material (called a precipitate) forms in liquid.
The change is difficult to reverse.

Physical and Chemical Changes

Problem

Of the following five changes, which are physical and which are chemical?

NOTE: Follow your teacher's instructions for safe treatment and disposal of substances produced in each Part of this and all other Activities.

You will be observing changes in five investigations.

PART A

Materials

safety glasses
test tube in rack
stopper for test tube
copper(II) sulphate
water
2 mL measuring scoop

Procedure

1. Make a table for your observations similar to the one shown. In the table, describe the water and copper(II) sulphate.
2. Pour water into the test tube to a depth of about 4 cm.
3. Add one scoop of copper(II) sulphate, put the stopper in the tube, and mix by turning the tube upside down several times.
4. Remove the stopper and record your observations in the table.

PART B

Materials

safety glasses
test tube in rack
copper(II) sulphate solution
stirring rod
steel wool (approximately
    1 cm × 1 cm × 2 cm)

Procedure

1. Pour copper(II) sulphate solution into the test tube to a depth of about 4 cm.
2. In your table, describe the steel wool and the solution.
3. Using the stirring rod, push the piece of steel wool into the solution; record your observations.

> CAUTION: Copper(II) sulphate is poisonous.

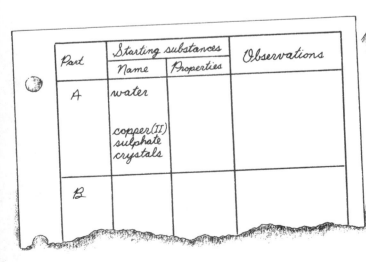

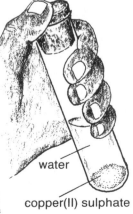

steel wool
stirring rod
copper(II) sulphate
water
copper(II) sulphate

Part	Starting substances		Observations
	Name	Properties	
A	water		
	copper(II) sulphate crystals		
B			

PART C

Materials

safety glasses
test tube in rack
magnesium ribbon (2 cm strip)
dilute hydrochloric acid

Procedure

CAUTION: Dilute
hydrochloric acid is
corrosive. Any spills on the
skin, in the eyes, or on
clothing should be washed
immediately with cold water.

1. Pour the acid into the test
   tube to a depth of about
   4 cm.
2. In your table, describe the
   dilute hydrochloric acid and
   the magnesium.
3. Add the magnesium ribbon to
   the test tube; observe and
   record any changes.

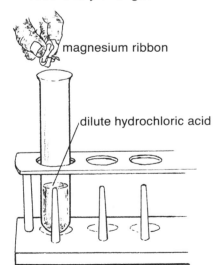

magnesium ribbon

dilute hydrochloric acid

PART D

Materials

safety glasses
2 test tubes in rack
potassium iodide solution
lead(II) nitrate solution

Procedure

CAUTION: Any compound
containing the element
lead is poisonous. Wash
your hands after using the
lead(II) nitrate solution.
Dispose of your chemicals
as your teacher directs.

1. Pour potassium iodide
   solution into one test tube
   and lead(II) nitrate solution
   into the other, both to a depth
   of about 2 cm.
2. In your table, describe the
   two solutions.
3. Pour one solution into the
   other; observe and record
   any changes.

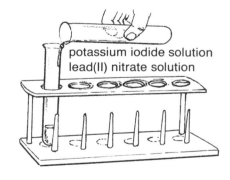

potassium iodide solution
lead(II) nitrate solution

PART E (DEMONSTRATION)

CAUTION: Your teacher
will conduct Part E if the
classroom has good
ventilation.

Materials

safety glasses
Pyrex test tube
test tube clamp
2 mL measuring scoop
ammonium chloride
glass wool
heat source: open flame

Procedure

1. Place one scoop of
   ammonium chloride in the
   test tube.
2. Plug the tube with glass wool
   to inhibit the escape of any
   gas.
3. In your table, describe
   ammonium chloride.
4. Heat the bottom end of the
   tube gently over an open
   flame for 3 s *only*; observe
   and record your observations
   as the test tube cools.

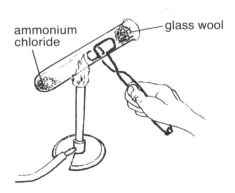

ammonium
chloride

glass wool

Analysis

1. In Part A, was there a physical or a chemical change? What evidence do you have for this inference?
2. In Part B, what type of change occurred? Give a reason for your answer.
3. In Part C, what appeared to change in the test tube? Was the change physical or chemical?
4. (a) In Part D, what evidence did you use to decide whether there was a physical or a chemical change?
   (b) What *two* properties of the new substance were different from the properties of lead(II) nitrate and potassium iodide?
5. In Part E, did the ammonium chloride undergo a physical or a chemical change? How do you know?

Further Analysis

6. If you wanted to test more properties of the new substance formed in Part B, how could you recover it from the test tube?
7. For each of the following changes, tell whether the change is physical or chemical. Include two reasons for your inference in each case.
   (a) Frost forms on windows.
   (b) Dynamite explodes.
   (c) Detergent removes grease from a dirty pot.
   (d) Concrete becomes hard after it is poured.
   (e) A candle burns.
   (f) The burner on an electric stove glows red.
   (g) Iodine turns blue-black when added to starch.
   (h) Coffee changes colour when cream is added.

Extension

8. Collect some or all of the following: vinegar, sugar, lemon juice, baking soda, steel wool. Carefully mix them together in pairs and watch for evidence of a chemical change. List the pairs of substances that underwent chemical changes.

Reactants and Products

Another term for chemical change is **chemical reaction**. In every chemical reaction, there is something used up and something produced. Any substance that is used up in a reaction is a **reactant**, and any substance that is produced is a **product**. There may be one or more reactants and one or more products.

Chemical reactions vary greatly. They may be spectacular or dull, explosive or very slow, common or rare. A complete description of a reaction would include all these aspects, but for convenience, you can use a short way of representing a chemical reaction, a way that tells you simply what is used up and what is produced. A *word equation* gives the names of all the reactants (separated by "plus" signs). Following these, an arrow points to the names of all the products (separated by "plus" signs).

The three chemical reactions that you observed in Activity 1-6 are represented in the following word equations. Read each equation aloud, and identify the reactants and the products. (Remember that "(II)" is read as "two," so, for example, iron(II) sulphate sounds like "iron-two sulphate.")

copper(II) sulphate + iron → iron(II) sulphate + copper

magnesium + hydrochloric acid → hydrogen + magnesium chloride

lead(II) nitrate + potassium iodide →

lead(II) iodide + potassium nitrate

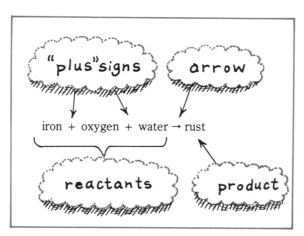

Read this equation as "Iron plus oxygen plus water produces rust."

The beauty of some chemical reactions cannot be described in equations.

Mass and Chemical Change

What exactly happens in a chemical reaction? Matter changes, but does the *amount* of matter also change? The earliest chemists, who were called "natural scientists," carried out many experiments in attempts to answer these questions. They learned a lot, but they were limited in two ways. First, their instruments did not allow them to make accurate measurements. Second, until the atomic theory of matter was developed (in 1808), they couldn't explain all their observations.

You aren't limited as they were. You have both modern technology and the atomic theory at your disposal. Your classroom balances can make far more accurate measurements than could the crude instruments of the pioneering chemists. As well, you can use the atomic theory to predict what could happen in a reaction. In the next Activity, you will have the opportunity to use both the technology and the theory.

Probing

The gas produced when magnesium reacts with hydrochloric acid is hydrogen (Activity 1-6). Use a suitable reference to find out what properties hydrogen has in common with another gas, helium, and in what ways these two elements are different.

The airship Hindenburg *was filled with hydrogen. In 1927, when it exploded and caught fire, 36 passengers and crew died.*

Mass of Reactants and Products

Problem

How is total mass affected by a chemical reaction?

IF A GAS IS PRODUCED, WILL THE PRODUCTS FLOAT AWAY?

WOULD A SOLID PRODUCT HAVE MORE MASS?

WHAT HAPPENS TO THE ATOMS?

Materials

sodium carbonate solution
calcium chloride solution
250 mL Erlenmyer flask
small test tube
stopper for flask
balance

Procedure

1. As a class, discuss and predict how a chemical reaction might affect the total mass of reactants and products.
2. Check that your test tube is the correct size for your flask. (See the illustration. The test tube should be small enough to fit inside, but large enough that it cannot lie flat.)
3. Pour the sodium carbonate solution into the flask to a depth of about 1 cm.
4. Pour calcium chloride solution into the test tube until it is about three-quarters full. Carefully dry off the outside of the test tube and gently put it in the flask.
5. Put a stopper firmly in the flask. Check that the outside of the flask is dry.

6. Record your description of the reactants, then determine the total mass of the reactants and their containers. Record the mass.

7. Invert the flask while firmly holding the stopper.

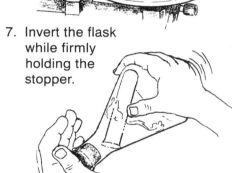

8. Observe the reaction. Then measure the mass of the flask and its contents.

Analysis

1. (a) What evidence is there that a chemical reaction has occurred?
   (b) What were the reactants?

Further Analysis

2. Was the total mass before the reaction different from that after the reaction? Explain in terms of the atomic theory.
3. The products of this reaction are calcium carbonate and sodium chloride. Write a word equation for this reaction.
4. There was another chemical present in both the flask and the tube, but it did not take part in the reaction. Suggest what substance this was.
5. Which product(s) do you think you could see in the flask after the reaction? Explain your answer.

Missing Mass?

Your observations in Activity 1-7 probably led you to a very logical inference. Observations from your everyday life may point towards other inferences. If you trust your senses, you *know* that burning reduces a pile of wood to a mere handful of ashes. Following a forest fire, a large tree can become a small pile of rubble. In burning, matter appears to be destroyed. In the next Activity, you'll observe another reaction in which matter appears to be destroyed. Observe carefully, and try to explain the results.

A Change in Mass

Problem

How can you explain an apparent difference in mass?

Materials

safety glasses
sodium hydrogen carbonate (baking soda)
dilute hydrochloric acid
250 mL beaker
test tube
2 mL measuring scoop
balance

Procedure

> CAUTION: Wash any spills of hydrochloric acid thoroughly with plenty of cold water.

1. Be sure that the test tube and beaker are clean and dry. Pour dilute hydrochloric acid into the test tube to a depth of about 5 cm. Put one scoop of sodium hydrogen carbonate into the beaker.
2. Determine and record the total mass of the beaker, the test tube, and the substances in them.
3. Remove the beaker from the balance. *Slowly* pour the acid into the beaker. Observe and record what happens.

4. Put the test tube back in the beaker, and again measure and record the total mass.

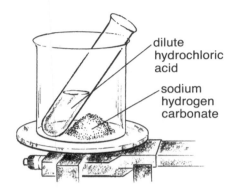

dilute hydrochloric acid

sodium hydrogen carbonate

Analysis

1. What evidence is there that a chemical reaction occurred?
2. (a) What were the reactants?
   (b) What was the difference in mass between the reactants and the products?

Further Analysis

3. (a) What might account for the difference in mass?
   (b) If you were to make a fair comparison between this Activity and Activity 1-7, how would you modify the procedure?
   (c) How could this modification be dangerous?
   (d) What property of one of the products causes the potential danger?
4. (a) What do you think the products of this reaction are?
   (b) What could you do to test your hypothesis? (You will know how to identify one of the products after you complete Topic Four.)

5. Suggest a reason why you were told to remove the beaker from the balance before mixing the two reactants.
6. In her science class, Ms Cardinal carried out a demonstration. She carefully determined the mass of a piece of magnesium ribbon. Then she burned it, being careful to collect all the pieces of white ash. Finally, she determined the mass of the ash. Look at her results, and try to explain them. (Hint: First write a word equation for the reaction. There are *two* reactants and *one* product.)

Burning magnesium.

Mass of magnesium ribbon = 3.0
Mass of ash
= 4.8
Difference =

Probing

Explain why only a small pile of ashes remains when a large pile of wood is burned. Name as many of the products and reactants in this reaction as you can.

Conservation of Mass

When you are determining the mass of reactants and products, you must consider the mass of any gases as well as the mass of the solids and liquids. For more than 200 years, scientists tried to devise methods to trap the gases that are used or produced in reactions and to find ways to measure their masses. After years of experimenting, in which the masses of *all* the reactants and *all* the products were determined, scientists agreed that mass is neither gained nor lost in any chemical reaction. This conclusion is stated as a *law*. A scientific **law** is a general statement that sums up the conclusions of many experiments.

The Law of Conservation of Mass
In a chemical reaction, the total mass of the reactants is always equal to the total mass of the products.

This law ties in well with the atomic theory of matter. According to the theory, atoms are never created or destroyed. In chemical reactions, the atoms and molecules are simply rearranged. Molecules may be broken apart and new ones may be formed, but the atoms in the products are the same ones that were in the reactants.

Did You Know?

The law of conservation of mass does not apply to the *nuclear reactions* that occur in the interior of the Sun, in atomic bombs, and in nuclear reactors. In nuclear reactions, there *is* some loss of mass: the mass is *changed into* energy. The idea that mass and energy are interchangeable was first suggested by Albert Einstein.

Einstein stated the relationship between mass and energy in his famous equation

$$E = mc^2$$

In this equation, E represents energy, m represents mass, and c^2 represents an extremely large number (the speed of light multiplied by the speed of light). A *very tiny* amount of mass is equivalent to a *very large* amount of energy.

Albert Einstein (1879–1955).

Chemical Reactions

In Topic One, you learned to identify substances on the basis of their properties. Substances such as sulphur, aluminum, copper, or copper(II) sulphate have a distinctive appearance and are easy to tell apart. Substances that look similar, such as salt and sugar, may be difficult to distinguish by appearance, but can be distinguished by their behaviour. For example, when salt is heated to a high enough temperature (801°C), it undergoes a *physical change*—it melts. Sugar, on the other hand, undergoes a *chemical change* when it is heated—it turns brown, then black. In this Topic, you will investigate various reactions and identify their products.

Their different properties make it easy to distinguish between these white crystalline solids.

Reactions as Identification Tests

When you observe a chemical reaction, you may be able to identify what is produced. If each product has a distinctive appearance, this may be easy to do. But some substances are not so easy to tell apart. For example, clear colourless gases such as carbon dioxide, hydrogen, and oxygen all look alike. Other properties, such as melting point and boiling point, may be difficult to determine. Table 1-12 shows that all these substances melt and boil at extremely low temperatures. Instead of using physical differences to distinguish these substances, you can use chemical tests. A **chemical test** is a distinctive chemical reaction that allows you to identify positively an unknown substance.

Table 1-12 *Three Clear Colourless Gases*

	MELTING POINT (°C)	BOILING POINT (°C)
Oxygen (O_2)	−218	−183
Hydrogen (H_2)	−259	−253
Carbon dioxide (CO_2)	−57	*

*Solid carbon dioxide may change directly from a solid to a gas at −79°C, without changing first to a liquid.

Oxygen

One common type of reaction is combustion, another name for burning. **Combustion** is a chemical reaction in which oxygen is one of the reactants and in which heat is produced. Many substances, such as wood, kerosene, natural gas, and diesel fuel, can undergo combustion. These substances burn readily in air, which is only about 20% oxygen. In pure oxygen, they burn much more intensely. It is this chemical property of oxygen—it promotes combustion—that allows you to identify it. If you suspect that a clear colourless gas may be oxygen, you can carry out the following test.

- Light a wooden splint.
- Blow out the flame but leave the splint glowing.
- Hold the glowing splint in a small amount of the unknown gas.
- *If the splint bursts into flame, the gas is oxygen.*

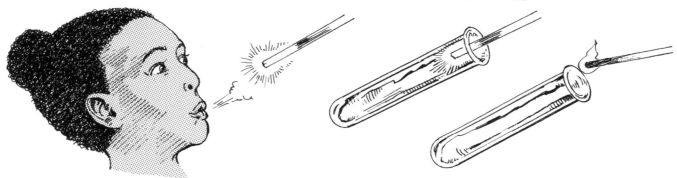

Positive test for oxygen.

Hydrogen

Hydrogen burns explosively. This chemical property of hydrogen allows you to identify it. If you have a clear colourless gas that is lighter than air, and you suspect that it may be hydrogen, you can carry out the following test. (This test should be attempted only on a *small* amount of gas in an open-mouthed, shatter-proof container. It should be done only under teacher supervision.)

- Light a wooden splint.
- Hold the burning splint in a small amount of the unknown gas.
- *If you hear a loud "pop," the gas is hydrogen.*

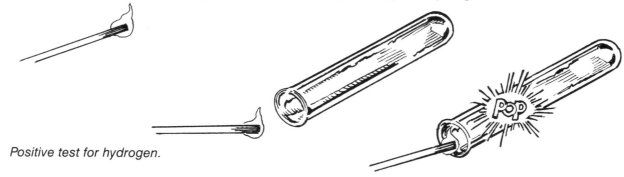

Positive test for hydrogen.

unknown gas bubbled through tubing

beaker of limewater

or

a few drops of limewater added to gas

Positive test for carbon dioxide.

Carbon Dioxide

Carbon dioxide does not burn and does not allow other materials to burn. If you put a burning splint into carbon dioxide, the flame goes out. This test would demonstrate that the gas is neither hydrogen nor oxygen, but it would not prove that the gas *is* carbon dioxide. (There are other clear colourless gases that would behave in the same way.)

The test for carbon dioxide uses a liquid called *limewater*, a clear colourless solution of calcium hydroxide (slaked lime) in water. Carbon dioxide reacts with the dissolved calcium hydroxide, producing an insoluble white product that causes the mixture to appear cloudy. Suppose you suspect that a clear colourless gas may be carbon dioxide. You can carry out this test.

- Bubble the unknown gas through the limewater solution.

Or

- Add a few drops of the limewater solution to the gas and swirl it around.
- *If the limewater turns milky, the gas is carbon dioxide.*

Activity 1–9

Testing for Carbon Dioxide

Problem

How is carbon dioxide commonly produced?

Materials

limewater
new straw
seltzer tablet
baking soda
vinegar
fizzy soft drink
various beakers, flasks, or large-sized test tubes

> CAUTION: Dispose of the straw after you have used it.

Procedure

1. Use a straw and some limewater in a test tube to test the air you breathe out. Does it contain carbon dioxide?
2. With your group, plan a way to test air for the presence of carbon dioxide. Try your method.
3. Plan methods to test whether carbon dioxide is present when
   - a seltzer tablet is put in water,
   - a can of pop is opened,
   - vinegar is added to baking soda.

   Try your methods.

Analysis

1. List some ways in which carbon dioxide is produced.

Further Analysis

2. In the test for carbon dioxide, a chemical reaction produces calcium carbonate and water. Write a word equation for this reaction.
3. Carbon dioxide is produced in many ways, but how is it used up? Recall your work with green plants; describe one reaction in which carbon dioxide is a *reactant*.

Water

There are many clear colourless liquids. One easy way to distinguish water from any other liquid is to test it with a cobalt(II) chloride test paper. This is simply paper that has solid cobalt(II) chloride within it. When dry, the paper is blue. If you suspect that a clear colourless liquid may be water, you can do this test.

- Touch the cobalt(II) chloride test paper to the liquid.
- *If the paper turns pink, the liquid is water.*

Cobalt(II) chloride test paper.

Combustion

What happens when something burns? You have learned that oxygen is always one of the reactants. In the next Activity, you will try some chemical tests to identify some of the products of one common combustion reaction.

If air can be prevented from reaching the fire, combustion is stopped.

A Common Reaction

Problem

What are the products when wax burns?

Materials

candle
glass plate (or small candle holder)
2 large jars
match
limewater
cobalt(II) chloride paper

Procedure

> CAUTION: See Safety in the Classroom on page xii. Follow these rules when using an open flame.

1. Light the candle. Fasten it to the glass plate by dripping some wax onto the plate, then placing the candle in the pool of molten wax.
2. Place a clean dry jar upside down over the candle.

3. Observe what happens to the candle. Then remove the jar and observe its inside surface. Test the surface with blue cobalt(II) chloride paper.
4. Set the second clean dry jar right side up. Place the candle and its stand into it.

5. Light the candle again. Allow it to burn for about 2 min, then blow it out.
6. Immediately add a few drops of limewater to the jar and gently swirl it around. Observe the drops.

Analysis

1. What evidence do you have that a chemical reaction occurred?
2. (a) Describe what you observed on the walls of the upside-down jar. What substance was present?
   (b) Describe what you observed in the second jar. What substance was present?

3. (a) What are two of the products of combustion of candle wax? Write a word equation to represent this reaction.
   (b) You may have found evidence of a third product. If you did, describe this product. What do you think it is?

Further Analysis

4. Explain why you used an upside-down jar to detect one product and a right-side up jar to detect the other. (Hint: Refer to the data about density in Table 1-12 on page 32.)
5. When sulphur burns, the product is sulphur dioxide. Write a word equation for this reaction.
6. Ordinary glass is a solution of sodium silicate and calcium silicate. It is made by heating together sand, limestone, and sodium carbonate. During this process, the following two reactions occur. Write a word equation for each of them.
   (a) Calcium carbonate (limestone) and silicon dioxide (sand) react to form calcium silicate and carbon dioxide.
   (b) Sodium carbonate and silicon dioxide react to form sodium silicate and carbon dioxide.

Energy in Chemical Reactions

In science **energy** means the ability to do work. There are many forms of energy. In your home you may use *electrical energy* to wash your clothes, run a fan, and operate a tape deck. *Gravitational energy* causes water to fall over a waterfall. Plants use *light energy* in photosynthesis. A photographic film also uses light energy to produce an image. *Elastic energy* can send an elastic band across a room. Anything that moves is said to have *energy of motion* (also known as *kinetic energy*). A pencil sharpener uses *mechanical energy* to shave wood off a pencil. When you talk on your telephone, *sound energy* is converted to electrical energy and converted back to sound energy at the other end of the telephone line. As you have previously studied, all forms of energy can be converted from one form to another.

Energy can also be stored. Water held back behind a dam contains *potential energy* because of its position high above the river bed. The potential energy of chemical compounds is called *chemical energy*. The chemical compounds that you eat, such as proteins, fats, and carbohydrates, have potential energy. As these substances take part in chemical reactions in your body, energy is gradually released—energy that you use to carry on your daily life. A flashlight battery also uses chemical compounds as a source of energy. As the stored chemical energy is released, it is converted to useful electrical energy.

Reactions such as these, in which energy is released, are called **exothermic reactions**. Combustion reactions are exothermic—they release energy in the forms of both heat and light. You do need to supply a small amount of heat to get the reaction started (by holding a lighted match to a candle, for example). But once the reaction has started, energy is produced. A dynamite explosion is another example of an exothermic reaction, one that releases not only heat and light, but also sound and mechanical energy.

Large amounts of energy can be released during chemical reactions.

In exothermic reactions, the products contain less stored energy than the reactants did. There are also reactions in which the products contain *more* energy than the reactants. These need to have energy *supplied* to them. These are **endothermic reactions**—reactions that absorb energy from their surroundings. The most familiar endothermic reactions are those that occur during cooking. As you fry an egg, for example, the white begins to become solid. If you were to remove the egg from the heat, the reaction would stop. The egg would remain half cooked until you heated it again. In the following Activity, you will have a chance to observe a less familiar endothermic reaction.

Activity 1–11

An Endothermic Reaction (DEMONSTRATION)

Problem

How can you identify an endothermic reaction?

Materials

electrolysis apparatus
water
sodium sulphate
power supply (6V DC)
electrical leads
wooden splints

Procedure

1. Assemble the apparatus as shown. (Sodium sulphate or an equivalent solute is essential, in order to allow the electric current to pass through the water.) Turn on the power.
2. Observe and record what happens. Keep watching until one of the test tubes is full of gas. Be sure to record the effect of turning the power on and off.

3. Test each gas by putting a flaming or glowing splint into the test tubes. Record your observations.

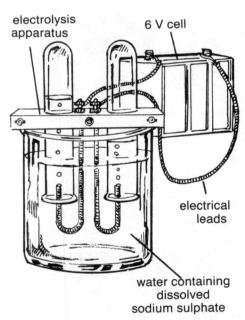

Apparatus used to show the effect of electricity on water.

Analysis

1. (a) Describe the gases produced.
   (b) Describe the results of your identification tests.
   (c) Which gas was more abundant?
2. (a) What evidence do you have that a chemical change occurred in the water?
   (b) What observation leads you to infer that the reaction was endothermic?
   (c) What was the source of energy?

Further Analysis

3. In this reaction, water *decomposed*, or broke down into its parts. (In other investigations, scientists have shown that the sodium sulphate does not take part in the reaction.) Write a word equation for the decomposition of water.

1. Suggest five clues you would consider before deciding whether a change is physical or chemical.

2. Which of the following are physical changes? Which are chemical changes? Defend your answers.
   (a) leaving garbage to rot
   (b) burying garbage
   (c) cutting up carrots
   (d) burning wood
   (e) giving a silver spoon a black coating by dipping it in a soft-boiled egg
   (f) making a pot of tea
   (g) bleaching a stain
   (h) cutting the lawn

3. When a candle burns, the solid wax undergoes a physical change as it melts. Then the melted wax undergoes a chemical change as it burns.
   (a) What are the products of the chemical change?
   (b) How you can identify each of the products?

4. In the test for water, the blue substance is *anhydrous cobalt(II) chloride* (anhydrous means ''without water'') and the pink substance is *cobalt(II) chloride hydrate*. Write a word equation for this change.

5. In the test for carbon dioxide, the white insoluble product is calcium carbonate. Water is also produced. Write a word equation for this reaction.

6. Write word equations for the following reactions.
   (a) Zinc and sulphuric acid react, producing zinc sulphate and hydrogen gas.
   (b) Sodium chloride is produced from the reaction of metallic sodium with chlorine gas.

7. (a) State the Law of Conservation of Mass.
   (b) State the law in another way, using the atomic theory.

8. (a) In the photograph, an endothermic reaction is about to take place. Where will this reaction occur, and what form of energy will cause it?
   (b) Give an example of an exothermic reaction that produces
   • heat and light;
   • electricity;
   • several forms of energy at once.

9. (a) Write a word equation to represent the breakdown of water into its two elements.
   (b) What form of energy can cause this reaction to take place?
   (c) Compare this reaction to the burning of hydrogen. How are the two reactions similar? How are they different?

10. Ace Transport Company in Edmonton receives a container of expensive wool from New Zealand. The container, still in perfect condition with its seal unbroken, is labelled ''105 kg'' and ''Contains mothballs to prevent insect damage.'' The Ace Transport agent in Edmonton determines that the package has a mass of only 102 kg. How could you explain the apparent disappearance of some of the cargo? Try to suggest two or three explanations.

Acids and Bases

THE TEACHER, STILL ANGRY, MANAGED TO PRODUCE AN ACID SMILE.

SHE WAS PROGRAMMING IN BASIC.

"OOOOOH...I'VE GOT ACID INDIGESTION!"

"LET'S MAKE THIS OUR BASE."

"THE BASIC PROBLEM IS THAT..."

HE SLID INTO THIRD BASE.

"THE NEXT TRYOUT WILL BE THE ACID TEST."

You are already familiar with various substances that are acids or bases. You have used many at home, and some you have used in the science classroom. They are simply groups of compounds that have several properties in common. **Acids** taste sour, are soluble in water, and undergo similar chemical reactions. **Bases** taste bitter, are soluble in water, feel slippery, and react with acids. Substances that are neither acidic nor basic, such as water, are said to be **neutral**. In this Topic, you will learn about acids and bases, how acidity is measured, and what happens when acids and bases are mixed. Finally, you will design and carry out an experiment to assess how much base there is in a common medicine.

How can you tell, without tasting it, whether a substance is an acid or a base? You could try mixing it with an indicator solution. An **indicator** is a substance that is a different colour in an acid than it is in a base. One of the most useful indicators is litmus, but many other common substances can be used as well, as you will see in the following Activity.

Investigating Acidity

Problem

What can you find out about acids, bases, and indicators?

Materials

safety glasses
water
dilute hydrochloric acid
dilute sodium hydroxide solution
  (a base)
test tube rack
test tubes
indicators (e.g., red and blue
  litmus paper, phenolphthalein,
  bromthymol blue, grape juice,
  tea, etc.)
samples of substances to be
  tested (e.g., salt water,
  swimming pool water, lake or
  river water, cola drink, soft
  drinks, vinegar, various fruit
  juices, oven cleaner, drain
  cleaner, household ammonia,
  baking soda solution, cream
  of tartar solution, various
  detergents, shampoo, milk,
  sour milk, solution of aspirin
  tablets or antacid tablets,
  bleach, egg white, water from
  boiled vegetables)

Procedure

1. Add a sample of each indicator to water, acid, and base. Record the colours in a table in your notebook.
2. Test each of your samples with one or two indicators. Keep a careful record of your results.

> CAUTION: Wash any spills of hydrochloric acid, sodium hydroxide, or cleaning agents immediately with cold water. Never mix cleaners, as they may react together.

Analysis

1. List any materials that were not distinctly acidic or basic.
2. List the materials that were distinctly acidic or basic in one of four columns in a table like Table 1-13.
3. Examine your four lists. What general observations can you make? (For example, do acidic foods have any other characteristics in common? Are cleaning materials generally acids or bases?)

Further Analysis

4. (a) In Step 1, which liquid was the control?
   (b) What is the purpose of a control in an investigation?
5. Which indicator(s) do you think are most useful? Why?
6. Suppose your teacher forgot to order red litmus paper for your class. The storeroom is overstocked with blue litmus paper, but the students need both kinds. What would you advise your teacher to do?
7. Blueberries in rolls made with yeast remain blue, but those in muffins made with baking soda may turn slightly green. Suggest an explanation for this.

Litmus is produced from lichen, a plant-like organism.

Table 1-13 *Foods*

FOODS		NON-FOODS	
ACIDIC	BASIC	ACIDIC	BASIC
lemon juice			

How Acidic? How Basic?

Both vinegar and battery acid are acidic; both baking soda and household ammonia are basic. Obviously there are differences among various acids, and differences among various bases. Both strong acids and strong bases are dangerous. Some can cause serious skin damage or produce fumes that can harm your lungs. Weak acids and weak bases may be harmless.

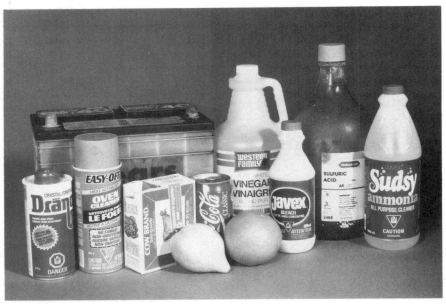

Which of these acids and bases are dangerous, and which are relatively harmless?

Rainwater is naturally slightly acidic. The "acid rain" that is falling in heavily industrialized parts of the world has an unusually low pH because it contains additional acids, formed from pollutants in the air. You will learn more about the effects of acid rain in Unit Six.

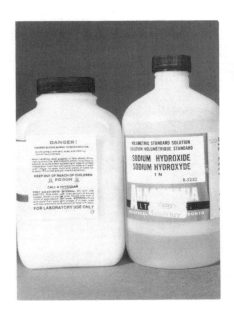

Even a dilute solution of sodium hydroxide, such as you used in Activity 1-12, should be used carefully. Solid sodium hydroxide is likely to be labelled with the WHMIS symbols shown above, and must be used with extreme caution.

Acidity is measured on the *pH scale*, a scale of numbers running from 0 to 14. If a substance is neither an acid nor a base, it is neutral and has a pH of 7.0. Acids have pH values below 7; the more acidic the solution, the lower the pH value. Anything with a pH between 0 and about 3 is a strong acid. Bases have pH values above 7; the more basic the solution, the higher the pH value. Very strong bases, such as drain cleaners, have a pH close to 14.

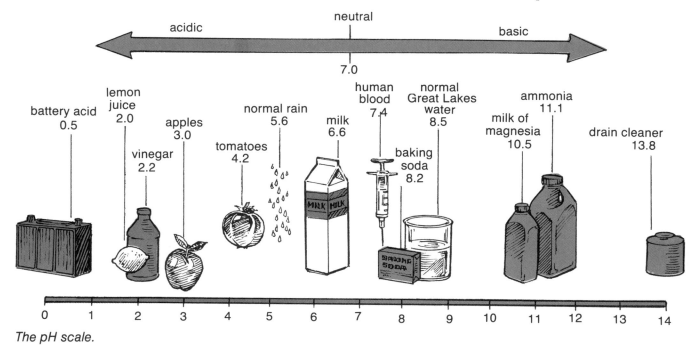

The pH scale.

Universal Indicator

You have seen several acid-base indicators. Not all indicators change colour at exactly pH 7. Some may change at a lower or higher pH. **Universal indicator** is a mixture of several of these indicators. Because of all the different substances in it, universal indicator turns a different colour for each number on the pH scale. Tests with universal indicator are much more precise than tests with litmus or other simple indicators.

The colours of universal indicator.

Mixing Acids and Bases

Acids and bases, whether they are strong or weak, react together. In this type of reaction, called **neutralization**, both the acid and the base are used up. Water and a type of compound called a *salt* are produced. For example, when hydrochloric acid reacts with sodium hydroxide, table salt (sodium chloride) is produced.

hydrochloric acid + sodium hydroxide → sodium chloride + water

Another salt you may have used in science classes is potassium nitrate, which can be produced in a similar reaction.

nitric acid + potassium hydroxide → potassium nitrate + water

Since weak acids and bases do not readily harm us, we use them to neutralize spills of stronger, more dangerous bases or acids. Vinegar and a solution of baking soda, weak and harmless in themselves, can be kept as a safety precaution against more dangerous bases and acids.

When ammonia (a base) spilled from this tank truck, nearby residents were evacuated.

Antacids

The acid in your stomach normally has a pH of about 2. It aids in digestion of food and helps to kill off any bacteria that enter your body along with the food. If you eat too quickly or eat when you are under stress, your stomach may produce too much acid, making you feel unwell. For relief, some people swallow an antacid preparation to neutralize the excess acid. The antacid tablet contains a mild base. In the following Activity, you will use baking soda to neutralize acid and then use your results to assess the amount of base in an antacid tablet.

Many over-the-counter medicines are bases.

Assessing an Antacid

Problem
How does an antacid tablet compare with baking soda?

Materials
safety glasses
dilute hydrochloric acid in a
  dropper bottle
indicator solution (e.g.,
  bromthymol blue)
100 mL beaker
stirring rod
baking soda
water
balance
petri dish
antacid tablet

Procedure
1. Measure 2 g of baking soda into the beaker. (Refer to Skillbuilder Three, *New Ideas on Measuring*, page 348.) Add 2 mL water, then add a few drops of indicator solution.

CAUTION: Wash any spills of hydrochloric acid thoroughly with plenty of cold water.

---

2. While stirring the baking soda solution with the stirring rod, add hydrochloric acid slowly, counting the number of drops added.

3. As soon as the indicator has changed colour, stop adding acid, and record the total number of drops added. Record this number as the "drops to neutralize 2 mL of baking soda."
4. Plan the steps you will take to compare the tablet to baking soda. (Hint: Do not plan to use the whole antacid tablet. Use only small pieces in the investigation.) Obtain your teacher's approval, and carry out your investigation.
5. Answer the questions here and on the following page.

Analysis
1. (a) Why did the colour of the indicator solution change when acid was added?
   (b) Suppose you added more baking soda to the beaker after the original baking soda had already been neutralized. What colour would the indicator turn? (If you have your teacher's permission, you might test your prediction.)
2. (a) Calculate the number of drops of hydrochloric acid required to neutralize the entire antacid tablet.

$$\frac{\text{mass of tablet}}{\text{mass used}} \times \text{drops acid used} = \underline{\quad} \text{ drops of hydrochloric acid}$$

   (b) Calculate the mass of baking soda that is equivalent to one antacid tablet.

$$\frac{\text{drops to neutralize tablet}}{\text{drops to neutralize 2 g baking soda}} \times 2\text{ g} = \underline{\quad} \text{ g of baking soda}$$

Further Analysis
3. (a) Would you want your antacid tablet to dissolve in your stomach instantly or over a period of time? Explain your answer.
   (b) What is one possible advantage of
   • liquid medicines over solid ones?
   • solid medicines over liquid ones?
4. (a) What other substances, besides bases, might there be in antacid tablets?
   (b) Why do you think these substances are included?

Extension

5. Repeat this Activity with different brands of antacid tablets. When you have compared the effectiveness of the brands, compare the prices as well. Is one a better bargain than the others?

6. Look up "neutralize" in a dictionary. In one or two paragraphs, write a short story in which "neutralize" is used with a non-scientific meaning.

7. Vinegar contains acetic acid. You can compare the amount of acetic acid in different brands of vinegar by comparing the amount of weak base needed to neutralize a certain volume. Test several types and brands of vinegar and prepare a chart illustrating your results.

Did You Know?

Many fish oils are bases that have a distinctive odour. Lemon juice neutralizes the bases and eliminates the "fishy" smell.

Controlling Chemical Reactions

You can cook an egg in only three minutes. When you're waiting for a hamburger to cook, the reactions may seem to take forever! Other reactions take considerable time as well. You may have painted a room using paint that becomes "dry to the touch" in two hours. But the instructions tell you not to put on a second coat for at least 24 hours. The extra time is required for further chemical reactions that make the paint harder.

A reaction that proceeds too quickly can be dangerous.

Factors Affecting Rates of Reaction

The speed of a chemical reaction is called the **reaction rate**. Some reactions have a naturally fast rate; others are naturally slow. We can influence the rate in many ways. For example, we try to slow the burning of a building, stop the rusting of iron, and hasten the cooking of foods. In industry, chemists work to find just the right conditions, so that certain reactions will occur more quickly and others will be slowed or stopped. In this Topic, you will discover several ways to slow down and speed up reactions.

Hot or Cold?

Problem

How does temperature affect reaction rate?

Materials

4 Alka-Seltzer tablets
4 large glasses or beakers (250 mL each or larger)
ice
hot water (e.g., from a kettle)
thermometer
watch or timer

Procedure

1. Prepare a table in which you can record your results for each of the four glasses. The two columns should be headed "Temperature" and "Time."

> CAUTION: Use boiling water only in heat resistant glassware (e.g., Pyrex). For ordinary glass use hot, not boiling, water.

2. Fill one glass with ice and water, and the others with cool, hot, and very hot water.
3. Use the thermometer and record the temperature of the water in each of the four glasses.

4. Before the water has a chance to cool, drop one tablet into each of the four glasses *at the same time*. Record the time taken to complete the reaction in each glass.

Analysis

1. What happened to the Alka-Seltzer tablets in water?
2. What evidence do you have that a chemical reaction occurred?
3. How did temperature affect the rate of the reaction?

Further Analysis

4. (a) Make a graph of time vs. temperature. Plot your four points on the graph and join them with a smooth curve. Give your graph an appropriate title and select a scale suitable to your results. (For a review of graphing, see Skillbuilder Four, *Graphing*, on page 354.)
   (b) Use your graph to predict the amount of time that would be necessary for a similar reaction at 40°C and at 60°C.
5. Suppose you obtained data for this reaction at temperatures of 0°C, 20°C, 50°C, and 95°C. Your friend's data were for 0°C, 30°C, 70°C, and 80°C. How would your two graphs be different? How would you expect them to be the same?
6. Recall the theory that particles in matter are constantly moving and that they move more quickly at higher temperatures. Use this theory to explain your results in this investigation.

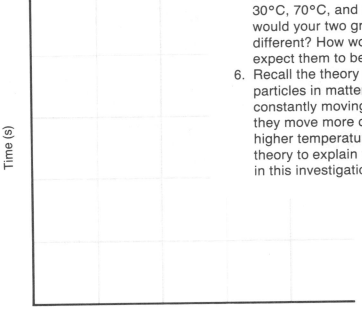

Temperature (°C)

Temperature

Temperature affects the rate of all reactions. For some, the effect may be hardly noticeable or even impossible to measure. For others, the effect may be great. For example, with just the right temperature, a hamburger cooks to perfection. But if you try to hurry it up with too much heat, other reactions occur as well. You could end up with an inedible blackened lump!

Surface Area

For two substances to react together, they must come into close contact. The closer the contact, the faster the reaction will be. When one of the reactants is a solid and the other is a liquid or a gas, the surface area of the solid will have an effect on the rate of reaction. When you have completed this short Activity, you should be able to predict what effect it will have.

Activity 1 – 15

Predicting

1. Examine the illustration. Both (a) and (b) represent a volume of 8 cm³.
   (a) Calculate the total surface area of the large block.
   (b) Suppose the large block is cut into eight smaller blocks of 1 cm³ each. What is the total surface area?

2. When a large piece of a solid is broken into smaller pieces, how does this affect the surface area?

3. Predict how the size of pieces might affect the rate of a chemical reaction. Use your result from Question 1 to explain your prediction.

4. Which burns more quickly, one large log or several small sticks? Explain this in terms of the surface area of the wood.

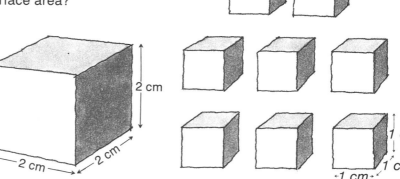

(a)

2 cm

2 cm   2 cm

surface area =
2 cm × 2 cm × 6 = _____ cm²

(b)

1 cm

1 cm

1 cm

surface area =
1 cm × 1 cm × 6 × 8 = _____ cm²

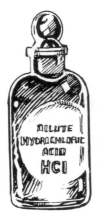

Concentration

As you learned in your earlier study of solutions, **concentration** refers to the amount of solute present in a specific amount of solution. You have been using *dilute* hydrochloric acid, which contains relatively little hydrochloric acid in water. (*Concentrated* hydrochloric acid contains much more acid in the same volume and is too dangerous to use in a classroom.) How do you think the rate of a reaction would be affected if the acid were even more diluted? In the next Activity, you will have a chance to make and test predictions about the reaction between calcium carbonate and hydrochloric acid. The word equation for this reaction is:

These two substances react together.

calcium carbonate + hydrochloric acid →

carbon dioxide + water + calcium chloride

Activity 1–16

Reaction Rate

Problem

How does surface area affect reaction rate? How does concentration affect reaction rate?

CAUTION: Wear safety glasses at all times. Wash any spills of hydrochloric acid thoroughly with plenty of cold water.

CAUTION: To mix, make gentle circular motions. Do not shake the tube up and down. Direct the test tube opening away from yourself and others.

Procedure

1. Decide which variable your group will investigate. Design a controlled experiment.
2. Write a step-by-step method, including a list of the materials you will need. Obtain your teacher's permission, and carry out your investigation.

Use a mortar and pestle to make powdered chalk from larger pieces.

Analysis

1. Describe the reactants and products in this reaction.
2. What evidence do you have that the same reaction occurred in each test tube?
3. Did any observation suggest that the reaction was endothermic or exothermic? Explain your answer.

Further Analysis

4. Explain your results.
5. Explain the results of a group who investigated the other variable.
6. If you wanted to use the same materials but make the reaction proceed as quickly as possible, what materials would you mix together?
7. How might the rate of this reaction be affected if the reactants became warm?

Metals and Corrosion

Metals are elements that are generally hard and shiny. As well, they can be pulled into wires or pressed into sheets. Many metals have another property in common—they *corrode*. **Corrosion** is a chemical reaction, the "eating away" of a metal as it reacts with other substances in the environment. Various metals corrode in different reactions at different rates. The corrosion of iron, called "rusting," can eventually use up all the metal. Gold, however, does not corrode at all. Aluminum and copper do corrode, but only at the surface of the metal; the corrosion product forms a continuous layer that protects the metal underneath. Some metal **alloys** (solid solutions of metals), such as brass, do resist corrosion, but these are too expensive to use in large-scale construction.

Iron and steel are used extensively in construction. For both of these substances, corrosion is a serious problem. The rust that forms is a crumbly, flaky compound that does not protect the metal underneath from further corrosion. In the next Activity, you will investigate conditions that affect the rate of this reaction.

Did You Know?

The roofs of the Parliament buildings in Ottawa are covered with sheets of copper that are green. The green colour is due to the product of corrosion. Over a period of years, the red-brown colour of copper is gradually replaced with a green layer as the surface of the copper reacts with gases in the atmosphere. The green layer is a mixture of several compounds of copper, including copper(II) carbonate and copper(II) sulphate.

The Parliament Buildings in Ottawa.

Assessing Corrosion

Problem

What factors affect the rate of rusting?

Materials

safety glasses
small iron objects
250 mL beakers
plastic wrap
cotton balls or paper towels
various substances, depending on factors to be investigated (e.g., salt, dilute acid, copper, zinc, etc.)
rusted objects (or photographs of rusted objects)

> **CAUTION: Wash any spills of hydrochloric acid thoroughly with plenty of cold water.**

Procedure

1. Observe several rusted objects or photographs of rusted objects. Some should be only a little rusty and some should be almost completely rusted away. As a class, decide on a "rating scale" for rusting.

2. Decide what variable(s) you will investigate. Table 1-14 lists some possibilities.
3. Plan your procedure, obtain your teacher's approval, and set up the equipment needed to make your comparison.
4. For at least one week, carry out your investigation and record your results in a table.

Compare the rate of rusting under different conditions. (If you want to keep the nails damp, put moist cotton balls or paper towels in the beakers.)

Table 1-14 *Suggestions of Variables to Investigate*

COMPARE THIS		WITH THIS
Dry conditions	vs.	Damp conditions
In water with no air	vs.	In water with air above
Damp conditions	vs.	Damp conditions with acid
Damp conditions	vs.	Damp conditions with salt
Damp conditions	vs.	Damp, touching copper
Damp conditions	vs.	Damp, touching zinc
Damp, in freezer	vs.	Damp, at room temperature

Analysis

1. (a) List conditions that favour the formation of rust.
   (b) List any condition in which the formation of rust is slowed or prevented.
2. Rate the rusting of all the objects used by members of your class, according to the scale agreed upon in Step 1.

Rust Protection

Because corrosion is a chemical reaction, all the factors that affect reaction rates apply to it. For example, rusting of iron or steel occurs more slowly if the temperature is low, if little or no surface area is exposed to the air, if the surroundings are dry, and if no salt or acid are present. Rusting cannot occur at all if the iron or steel is totally protected from air and moisture.

Corrosion poses problems in many aspects of life, from bridge construction, to car manufacturing, to the production of computer chips. Chemists, engineers, and technologists are continually working to understand and prevent corrosion damage.

Further Analysis

3. Think back to Activities 1-14 and 1-16, in which you investigated reaction rates. Use those results in making the following predictions.
(a) Which do you think would rust away more quickly, steel wool or steel nails? Why?
(b) Which might become more badly rusted, a bicycle left outside all year in Edmonton or one left outside all year in Vancouver? Explain your answer.

4. If you or your classmates used copper and/or zinc in this investigation, you may be able to make some inferences about these two metals.
(a) Which metal do you think reacts most readily? Explain what observations led you to make this inference.
(b) Why do people who use metal boats in the ocean sometimes attach a zinc bar to the boat?

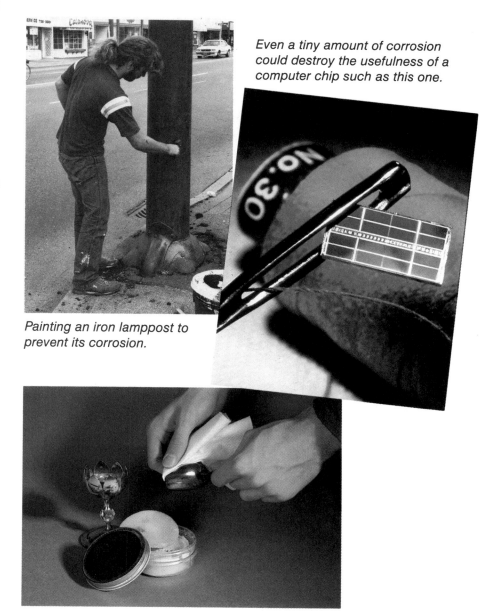

Even a tiny amount of corrosion could destroy the usefulness of a computer chip such as this one.

Painting an iron lamppost to prevent its corrosion.

When black silver sulphide forms on silver objects, you can remove it by polishing.

Working with Chemistry

A Toxicologist

Margherita Conti is a biologist who specializes in the field of toxicology—the study of poisons. In her job with the federal government, she works with applications to license new pesticides. "My responsibility is to review the toxicological information submitted by the company. We assess the hazards, considering effects on reproduction, interference with normal development, ability to produce permanent changes in cells, and increased risk of cancer. After reviewing the data, the government provides advice on the manufacture, handling, and application of pesticides."

In high school, Margherita took all the mathematics and science courses she could. While she was in university, she had a summer job at the Pesticides Division of Health and Welfare Canada. That, along with her courses in chemistry, genetics, and physiology, led to her interest in toxicology.

A Forensic Scientist

As a forensic scientist, Douglas Lucas applies his knowledge of science to legal problems. The courts may call upon forensic scientists to test for drugs in post-mortem samples of blood, to sift through ash looking for clues to the cause of a fire, or to examine an almost limitless range of substances. Forensic scientists need a wide general knowledge of many sciences in order to obtain and interpret information.

"When I was in high school I worked part time in a drugstore, and that led me to consider a career in pharmacy. Later, in university, I became more interested in the chemistry of drugs. My background in analytical chemistry, pharmaceutical chemistry, and pharmacology led to a job in forensic science. I developed expertise in the field of breath tests for alcohol and helped to organize training courses for police officers in the use of the Breathalyser." After ten years, Douglas Lucas became director of his laboratory. He expects that scientific investigations will become more complex in the future and that most analytical equipment will be computer controlled. "But although computer training is increasingly important, we should never overlook the fundamentals of science."

Checkpoint

1. In your notebook, complete the comparison of acids and bases as in Table 1-15.

2. (a) What substances are produced when an acid reacts with a base?
   (b) What type of reaction is this?
   (c) Write a word equation that represents an example of this type of reaction.

3. Sodium hydroxide is a strong base. When it reacts with fat, soap and water are produced.
   (a) Write a word equation for this reaction.
   (b) Suggest why sodium hydroxide is an effective ingredient in oven cleaners.
   (c) After using an oven cleaner, some people rinse the oven with a dilute solution of vinegar in water. What is the purpose of the vinegar rinse?

4. (a) What would be the advantage in using brass or stainless steel for the body of a car?
   (b) Why do you think cars are not made of these materials?

Table 1-15

	ACIDS	BASES
Characteristics	sour-tasting ▄▄▄▄ ▄▄▄▄	▄▄▄▄ ▄▄▄▄ ▄▄▄▄
Examples	▄▄▄▄ ▄▄▄▄ ▄▄▄▄	baking soda ▄▄▄▄ ▄▄▄▄

5. The following types of care could have slowed the rusting of the car shown on the left below. Use principles of chemistry to explain the effect of each one.
   (a) parking it outside, rather than in a heated garage
   (b) washing it frequently, especially in winter
   (c) putting rubber mats over the carpet
   (d) repairing chipped paint promptly

6. Imagine a metal with properties that make it perfect for every application. Write a paragraph or two describing this ideal material and telling how you would use it to improve the quality of your life.

Focus

- Properties can be used in the identification and classification of substances.
- Pure substances can be either elements or compounds.
- The atomic theory is useful in explaining properties of elements and compounds.
- Changes in matter may be physical changes or chemical changes.
- Chemical reactions can be represented by word equations.
- In any chemical reaction, the total mass of the reactants always equals the total mass of the products.
- Many kinds of energy can be produced or used in chemical reactions.
- Acids and bases, and the reactions between them, are common in everyday life.
- The rate of a particular reaction can be affected by the temperature, the surface area of a solid, and the concentration of a solution.
- Principles of chemistry can be applied to the prevention of corrosion.

Backtrack

1. (a) Define the following terms: pure substance, element, compound.
   (b) Give two examples of elements and two examples of compounds.

2. A black material is heated. A gas is released and a white substance remains. Was the original material an element or a compound? Explain your answer.

3. (a) What is an atom?
   (b) How are molecules related to atoms?

4. For each situation illustrated below, decide whether the change is physical or chemical. Give your reasons in each case.

(a) Wood rots.

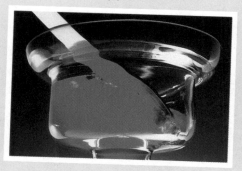

(c) Jelly changes from a liquid to a solid.

(e) Paint dries.

5. (a) What do all chemical reactions have in common?
   (b) Give four examples of differences among chemical reactions.

6. Look around your home and select three situations where chemical reactions are occurring. For each, explain how you know it is a chemical reaction.

(b) A large amount of energy is released during blast-off.

(d) An ice cube "disappears."

7. The following equations represent three different chemical reactions.

Reaction X
calcium carbonate + hydrochloric acid →
calcium chloride + carbon dioxide + water

Reaction Y
lead nitrate + sodium iodide → lead iodide + sodium nitrate

Reaction Z
potassium chlorate → potassium chloride + oxygen

(a) Which has only one reactant?
(b) Which has (have) a product that is a gas?
(c) In which reaction is there a substance that is poisonous?

8. Identify the reactants and products in the following reactions, then write a word equation for each.
(a) Hydrochloric acid reacts with sodium hydroxide to produce sodium chloride and water.
(b) Zinc sulphide is produced by the reaction of zinc with sulphur.
(c) Hydrogen peroxide gradually decomposes, forming water and oxygen.
(d) Carbon dioxide and water are produced in our bodies. These are the end products of a series of reactions involving sugar and oxygen.

9. Turn back to page 1 and examine the photographs. In your notebook write a title for each one, connecting the picture to the science of chemistry.

10. When it is heated, limestone decomposes into calcium oxide (a solid) and carbon dioxide (a gas).
(a) Write a word equation for this reaction.
(b) If you heated some limestone, how would you expect the mass of the limestone to compare with the mass of calcium oxide? Explain your answer.

Synthesizer

11. Examine this diagram of molecules.
(a) What substances do the drawings represent?
(b) What reaction is shown?
(c) Write a word equation for this reaction.

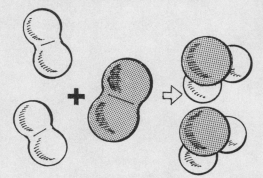

12. When a welding torch is used, the following reaction takes place.

acetylene + oxygen →
carbon dioxide + water

(a) What forms of energy are produced in this reaction?
(b) Name the reactants and products.
(c) Classify the pure substances shown in the word equation as either elements or compounds.
(d) If a welding torch is improperly adjusted, it produces a smoky yellow flame. If the flame touches a metal object, it leaves black marks on the metal. What might the black material be? Explain your answer.

Using a welding torch.

Fluids and Pressure

There would be no life on Earth without the two most common fluids, air and water. Although these two are essential for all living things, many other fluids are also important in our everyday lives. For example, fluids help heat our homes, cool our food, extinguish fires, and move goods and people around.

The student riding her bicycle enjoys a smooth ride because of the pressure of a fluid in the tires. A fork-lift can raise skids loaded with boxes, and a "cherry-picker" can lift a passenger high in the air because of the behaviour of a fluid under pressure. A runner's muscles receive oxygen and nutrients because his heart exerts enough pressure to send an important fluid to all the cells of his body. Weather systems move across the country because colder, "heavier" fluids exert more pressure than warmer, "lighter" ones. Delivering water to crops or to city homes involves complex systems with many more parts than are shown in these photographs.

To design and build technological devices, to predict weather patterns, or to understand the effects of high blood pressure, you need some knowledge of fluids. What is there to know about fluids? What instruments can supply the information? How do fluids help us do work? In what other ways do you use fluids? In this Unit, you'll discover some of the properties of fluids that make them so useful. As well, you'll try out some of the technology that makes use of fluids and pressure.

Fluids and Pressure   59

The Nature of Fluids

A **fluid** is anything that flows. You can probably think of many situations in which liquids flow. For example, each of these photographs shows flowing water. Have you ever seen a gas flowing? How can you demonstrate that gases are fluids? In this Activity, try designing a demonstration to convince your classmates that gases really are fluids.

(a)

(b)

(c)

Flowing Gases

Problem

How can you demonstrate that gases flow?

Materials

a variety of easily obtainable materials, such as:
sheet of paper, straw, balloon, chalk dust, pencil, beaker, slide projector, flashlight

Procedure

1. Before you design your demonstration, discuss with your group the following questions.

- The people in photograph (a) on page 60 cannot see or feel the flowing water. What sense tells them that the water is falling close by? How could you use this sense to show that a gas is moving?
- The child in photograph (b) is not looking at the water, yet he knows it is flowing. What sense is telling him? How could you use this sense to show that air (or some other gas) is moving?
- What indicates to the boys in photograph (c) that the water is flowing? How could you use this idea to show that a gas is flowing?

2. Use an idea from one of the photographs, or an idea of your own, as you design a demonstration to show that gases flow.
3. Present your demonstration to the class, and explain it.

Analysis

1. How successful was your demonstration? How could you have improved it? If your classmates are still unconvinced, explain why they are.
2. Are you convinced? Explain why or why not.
3. In your own words, describe what the word "flow" means.
4. What kinds of matter flow?

Solids that Seem to Flow

All gases and liquids can flow, so all gases and liquids are fluids. This photograph shows a material that is not a gas or a liquid, but appears to flow. Flour, sand, and wheat may also appear to flow. Are they considered to be fluids?

When a solid consists of a very large number of tiny pieces, it may appear to behave like a fluid. A fine powder can be poured from one container into another. But have you ever seen water form a heap, as flour does when you pour it? Can you make a pile of milk, as you can of sand or wheat? This is one characteristic of fluids that solids never have. The surface of a fluid at rest is flat, but when a solid like sand is poured, the surface of the solid will not be flat. Therefore, only liquids and gases are considered to be fluids.

When tiny pieces of a solid are poured, they pile up into a heap.

Fluids and the Particle Theory

For an explanation of the behaviour of fluids, recall the particle theory of matter. This theory states that all matter is made up of tiny particles. These particles are so tiny that even a single grain of sand, wheat, or flour would be made up of billions upon billions of particles. In solids, the particles are held in fixed positions, so when a grain of sand moves, all of its billions of particles move together. The particles vibrate, but they never change places. Because the particles cannot move freely past each other, solids do not flow.

In liquids, the particles are still packed closely together. However, they can move around more freely than they can in solids and are able to change places with other particles. When you pour a liquid, the particles do not stay locked together in clumps as they do in sand grains. They move independently past one another, so the liquid is able to flow.

The particles in gases are much farther apart than they are in either solids or liquids. There is a lot of space between them, so particles can easily move past one another. The gas has no difficulty flowing.

Particles in a solid, a liquid, and a gas.

How Fluids Flow

All fluids do not flow equally easily. For example, pancake syrup and molasses flow more slowly than water does. Are there other liquids that flow very slowly? Is there a fluid that flows faster than water? Can you speed up or slow down the rate of flow? In this Topic, you'll investigate the flow rate of several fluids.

Engineers sometimes need to know about flow rates of fluids when they are designing machinery or distribution systems. For design purposes, *qualitative* descriptions like ''slow'' or ''fast'' are not enough. Engineers need to know *how* slow or *how* fast; they need *quantitative* descriptions. In this Topic, you'll make quantitative measurements of the flow rates of several fluids. These measurements will enable you to provide quantitative information about the fluids.

Using a Syringe to Experiment with Fluids

Throughout this Unit, you'll be using syringes in your investigations. These syringes are standard medical or laboratory syringes that have been modified in two ways, as shown in the diagram. The needle has been removed to provide an unobstructed opening, and a wooden platform has been attached to the top of the piston. This platform provides a surface upon which you can place various masses, as required by the Activities.

Although volumes of liquids are usually stated in millilitres (mL) or litres (L), the volume in syringes has traditionally been indicated in cubic centimetres (cm³). You may find that your syringe is marked in ''cc''s. This is not a *Système International* symbol, but it is still in use in some parts of the world. (The ''cc'' stands for ''cubic centimetre.'')

$$1 \text{ mL} = 1 \text{ cm}^3 = 1 \text{ cc}$$

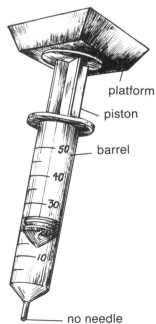

The modified syringes you will use throughout this Unit are like the one shown. They have no needle and have a platform attached to the top of the piston.

Flow Rate of Liquids and Gases

Problem

Which flows more quickly—air or water?

Materials

support stand
2 burette clamps
50 mL modified syringe
250 mL beaker
1 kg mass (or mass that causes appropriate flow rate)
timing device

Procedure

1. Set up the apparatus as shown.

2. Adjust the piston until the syringe contains exactly 50 mL of air.
3. Working in pairs, one partner should place the 1 kg mass on the top platform, as the other starts timing.
4. Record the number of seconds required for the piston to be pushed in completely.
5. Practise filling a syringe with water, using the technique in the illustration. Ensure that the syringe contains 50 mL of water, but no air. Clamp the syringe in place as before.

6. Repeat Steps 3 and 4, using water instead of air.

Filling a Syringe with 50 mL of Water

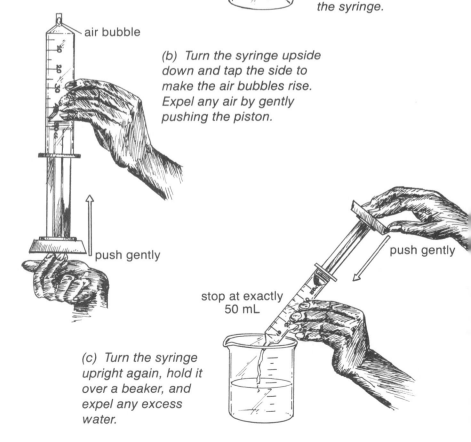

(a) Draw more than 50 mL of water into the syringe.

(b) Turn the syringe upside down and tap the side to make the air bubbles rise. Expel any air by gently pushing the piston.

(c) Turn the syringe upright again, hold it over a beaker, and expel any excess water.

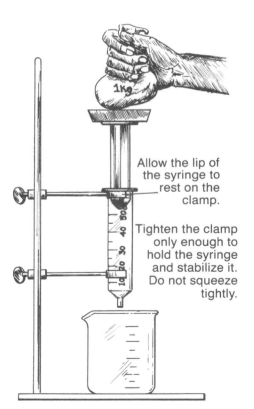

Allow the lip of the syringe to rest on the clamp.

Tighten the clamp only enough to hold the syringe and stabilize it. Do not squeeze tightly.

Apparatus used to compare flow rate.

Air and Water

Some fluids flow much more easily than others. Because air flows freely, objects can travel through air more quickly and easily than they can through water. Boat designers make use of this fact. Since air can flow around a boat more easily than water can, boats designed like the one in the photograph are able to reach higher speeds than boats that ride lower in the water.

Analysis

1. Which fluid flows more slowly, air or water? How many seconds longer did it take for the slower fluid?

Further Analysis

2. (a) According to the particle theory of matter, how are liquids different from gases? (b) Use the particle theory to suggest a possible reason for your results in this Activity.
3. Do you think it likely that your results hold true for all liquids and gases? Use the particle theory to explain your answer.
4. Running along the beach is much easier than running in waist-deep water. Use your results in this Activity to explain why.

At slow speeds, hydrofoil craft operate like ordinary boats. At high speeds, they "take off" and "fly" over the surface of the water, supported by winglike surfaces (hydrofoils).

Viscosity

There are many different kinds of liquids in the world. How do the rates of flow of other liquids compare with that of water? A liquid's "thickness," or resistance to flow, is called its **viscosity**. For example, molasses is said to be more viscous than water because it does not flow as readily. In other words, molasses has a slower *flow rate* than water does. Flow rate is an important property of liquids that are transported by pipeline. The pipeline must be designed to be suitable for the fluids that are to flow through it.

In some industries, liquids such as lubricants must be produced with a specific viscosity. If motor oil, for example, is not viscous enough, it will not protect the engine parts from friction; but if it is too viscous, it will not flow to all the parts of the engine that need protection. In the next Activity, you'll see how you can make quantitative comparisons of the vicosity of liquids.

Cars and airplanes are designed to reduce "drag" so they will be able to attain higher speeds and use less fuel. In a reference book, find a meaning for "aerodynamics" and "turbulence." Choose a car, and explain where principles of aerodynamics have been applied in its design.

Comparing the Flow Rate of Different Liquids

Problem

How can you compare the rate of flow of four liquids?

Materials (for each group)

support stand
2 burette clamps
50 mL modified syringe
250 mL beaker
2 masses (1 kg each)
timing device
liquid to be used in your group's
    apparatus: isopropanol,
    glycerol, vegetable oil, or
    water

Procedure

1. In your notebook, prepare a data table like the one shown.

> CAUTION: Isopropanol is poisonous.

2. When you have been assigned to a group, set up the apparatus as in Activity 2-2. Fill the syringe with the liquid assigned to you. Instead of a single 1 kg mass, place two 1 kg masses on the top platform.
3. In your data table, record the time required for the liquid to be pushed completely from the syringe.
4. Move in turn to other tables, and repeat Steps 2 and 3 for the other liquids.
5. After everyone has tested the four liquids, return them to the containers provided, wash the dirty syringes in hot soapy water, and rinse thoroughly.

Table 2-1

| | INDIVIDUAL RESULTS | | | CLASS AVERAGE |
LIQUID	VOLUME (mL)	TIME (s)	FLOW RATE (mL/s)	FLOW RATE (mL/s)
glycerol	50			
isopropanol	50			
vegetable oil	50			
water	50			

Analysis

1. Which liquid was easiest to draw up into the syringe? Which was most difficult?
2. Calculate the flow rate in mL/s for each liquid.

$$\text{flow rate} = \frac{\text{volume of liquid (mL)}}{\text{time (s)}}$$

3. Collect all the class results in a table on the chalkboard. For each liquid, calculate the class average value for the flow rate. Compare this answer to your individual results. Explain any differences.
4. (a) In your notebook, list the four substances in order from most viscous (slowest flowing) to least viscous (fastest flowing).
   (b) If you included air in your list, where would it fit? (Refer to your data for Activity 2-2.)

Further Analysis

5. Draw a bar graph illustrating the flow rates of the four liquids tested in this procedure.
6. (a) According to the particle theory of matter, how can you describe the motion of particles in a liquid?
   (b) Use the particle theory to suggest possible reasons why some liquids flow more easily than others.
7. The kind of straw normally served with a milk shake is different than the kind normally served with a soft drink. Why is this?

8. Suppose you are asked to design a pipeline to carry molasses from one part of a factory to another. How should the pipe diameter compare with the diameter of the water pipes in the factory? Explain your answer.

Extension

9. Use a dropper to place single, separate drops of isopropanol, glycerol, vegetable oil, and water on a clean, dry glass plate. Note any differences in height among the four drops. With your eye at the level of the tabletop, observe the differences in the shapes of these drops.
   (a) Sketch and label each drop.
   (b) Use the particle theory to suggest reasons for any differences in height.
   (c) Does there appear to be any relationship between the heights of the four drops and the flow rates you measured for the liquids? If there is, suggest an explanation.

Did You Know?

More than a dozen different fluids are moved by pipeline in Canada today. In addition to the most common ones — water, natural gas, and crude oil — milk, ammonia, and carbon dioxide are also moved by pipeline.

Activity 2-4

Changing the Viscosity of a Liquid

Problem

How is the viscosity of a liquid affected by temperature?

Materials

support stand
2 burette clamps
50 mL modified syringe
4 beakers (250 mL each)
2 masses (1 kg each)
timing device
vegetable oil
ice
thermometer
3 containers for water baths (e.g., 600 mL beakers)
stirring rod
(hot plate, if tap water at 60°C is not available)

CAUTION: Use water baths to heat the oil. Do not heat it directly over a hot plate.

Procedure

1. Prepare three water baths. One, containing ice and water, should be at approximately 0°C. The others should be warmer than room temperature, one at about 40°C, and one at about 60°C. (You may be able to use water from a hot tap for both of these.)
2. Pour more than 50 mL of vegetable oil into each of the four 250 mL beakers. Place one in each of the water baths, and leave one at room temperature. Stir them occasionally with the stirring rod, and check the temperature with the thermometer.
3. In your notebook, prepare a data table like Table 2-2.
4. Record the temperature of the oil at room temperature. Using the method you used in Activity 2-3 (with two 1 kg masses), determine the time 50 mL of the oil takes to flow from the syringe.

Table 2-2 *Flow Rate of Vegetable Oil*

TEMPERATURE (°C)	VOLUME (mL)	TIME (s)	FLOW RATE (mL/s)
	50		
	50		
	50		
	50		

5. Repeat Step 4 using the other three samples, which are at different temperatures. For each, record the temperature and the time the oil takes to flow from the syringe.
6. Calculate the flow rate at each of the four temperatures.

Analysis

1. (a) At what temperature did vegetable oil flow most quickly? What does that tell you about its viscosity at this temperature?
   (b) At what temperature did vegetable oil have the slowest flow rate? What does that tell you about its viscosity at this temperature?
2. (a) Draw a line graph of your results. Put the manipulated variable, temperature, on the horizontal axis, and the responding variable, flow rate, on the vertical axis.
   (b) Write a statement about the relationship between viscosity and temperature in liquids.

Further Analysis

3. (a) Extend the lines on your graph to temperatures above and below those you measured. (For a review of extrapolation, see Skillbuilder Four, *Graphing*, on page 354.)

(b) Using your graph, predict the temperature at which vegetable oil would flow as easily as water. (Refer to your results in Activity 2-3.)
(c) At what temperature might vegetable oil stop flowing entirely?
4. (a) According to the particle theory, how does an increase in temperature affect the particles of a liquid?
   (b) Use the particle theory to explain why temperature affects viscosity.
5. Motor oil is used in car engines to lubricate moving parts. It is sold in several grades, such as SAE 5, SAE 10, SAE 30, and SAE 40. The numbers (assigned by the Society of Automotive Engineers) indicate viscosity —the lower the number, the less viscous the oil.
   (a) Which grade of oil would you use in your car in summer? Explain your answer.
   (b) Which grade would you use in winter? Explain why.

Extension

6. Today, most motor oil is referred to as *multigrade* and is identified by numbers like 5 W 30, 10 W 30, or 10 W 40.
   (a) Consult a local service station operator to find out the difference between multigrade and single grade motor oil, and how to safely dispose of used motor oil.

(b) What environmental problems can be caused by the incorrect disposal of used oil?
7. Repeat this experiment using a different liquid, such as glycerol or molasses. You may find that you need to run the test at lower or higher temperatures.
8. (a) In cold Canadian winters, the oil in a car engine may become so viscous that the motor cannot start. To overcome this problem, the driver may install a device in the vehicle. Find out what device this is, and how it solves the problem.
   (b) In the far north, where winter temperatures are still more severe, even the device mentioned in part (a) won't help. Contact a northern oil exploration company or a northern mining company, and find out how they overcome the problem.

The viscosity of bubble gum can be a problem.

Changing Viscosity

Bubble gum can be thought of as a liquid with a high viscosity. What would you do if you had to clean a thick wad of bubble gum off your shoe? One way that works well is to place the shoe in the freezer, to make the gum even more viscous. It becomes hard and can then be removed easily because it does not change shape, or flow, when you lift it off the shoe.

The ability to change a liquid's viscosity is important in the petroleum industry. For example, the petroleum in Alberta's oil sands is more viscous than molasses and is too viscous to pump to the surface. To overcome this problem, high-pressure steam is injected for several weeks into the oil-sand formation to heat it and reduce the viscosity of the petroleum. It is left for several weeks to allow time for the oil to percolate to the extraction site and the oil is then pumped to the surface.

Oil pipelines have heaters at intervals along their length, to maintain the temperature of the oil at all seasons of the year. Without heaters, cooled oil could become too viscous to flow easily through the pipeline.

Probing

Use a reference book to find out what different kinds of lava are produced in volcanoes. How are the shapes of the volcanoes affected by different viscosities of the lava? Why are the shapes affected in this way?

Checkpoint

1. (a) Write the meaning of the word "fluid." Explain how fluids are different from other materials.
   (b) What is meant by the word "viscosity"? What factor can cause the viscosity of a fluid to change?
2. Think of the fluids (both liquids and gases) that are used in your home. List at least ten of them.
3. (a) Name two fluids that have a higher viscosity than water.
   (b) Name two fluids that have a relatively low viscosity.
   (c) In a quantitative comparison of viscosity, what property of liquids can you measure? What units are used in this measurement?

4. (a) To make maple syrup, tree sap is boiled in open pots until it is reduced to about 5% of its original volume. How do you think this affects its viscosity?
   (b) Sometimes a spoonful of the hot syrup is thrown on the snow, producing maple candy. What change happens to the sap as it is made into candy?
5. (a) Asphalt is the black, sticky material that binds gravel in the pavement of streets and highways. Paving is almost always done in the summer. What advantages are there to doing this work while the weather is hot?

(b) Why is it unwise to walk on a tar and gravel roof on a hot sunny summer day? What damage to the roof could result?
6. (a) Examine the illustration below. How many fluids are shown or suggested? List as many as you can.
   (b) For each of the fluids you named in part (a), suggest at least one way it could be transported, and describe any particular difficulties or safety concerns.
   (c) List all the fluids named in part (a) that could damage the environment if they were handled carelessly. Describe how this could happen and what damage might result.

Many fluids are shown or suggested in this scene.

Buoyancy

Balloons float in air and ships float in water; this tendency to float is called *buoyancy*. Buoyant objects take up space in the fluid, so they push some of the fluid away, or **displace** the fluid. As they push the fluid away, it pushes back, causing them to float. But many objects, such as stones, cars, and most everyday items, don't float —they fall in air and sink in water. These objects also take up space in the fluid; they must push away the fluid around them. As they displace the fluid, does it push back? And what about fluids themselves? Can one fluid displace another? Or are solids the only things that can displace fluids? In this Topic, you'll investigate more about the behaviour of fluids.

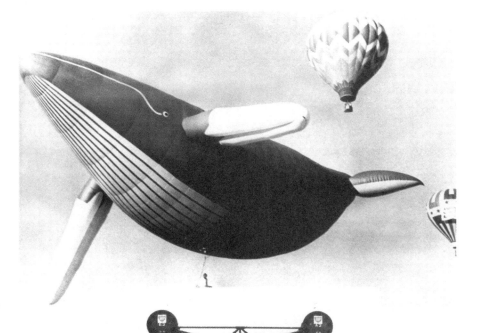

A floating "whale balloon."

The world's largest floating oil platform, in the North Sea near Scotland.

Floating One Gas on Another

Problem

How can one gas displace another?

Materials

600 mL beaker
250 mL beaker
graduated cylinder
5 mL measuring spoon
cardboard (15 cm × 15 cm)
short piece of candle (3 cm)
wooden splint
vinegar
baking soda

Procedure

1. Pour 100 mL of vinegar into the 600 mL beaker. Add 5 mL of baking soda. Place the cardboard on the beaker, leaving a small opening for air to escape.
2. Stand the candle in the 250 mL beaker and light it with the wooden splint. Watch the candle for several seconds to assure yourself that it burns steadily.
3. When the vinegar and baking soda have stopped fizzing, slowly pour the gas down the side of the 250 mL beaker as shown in the illustration. Do not pour out any of the liquid.

Analysis

1. Describe how the candle burned:
   (a) before you poured the gas into the beaker;
   (b) as you first started pouring the gas into the beaker;
   (c) when you continued to pour the gas into the beaker.
2. (a) When vinegar and baking soda react, carbon dioxide gas is produced. How do you know that the carbon dioxide flowed? (Explain, using your experience with combustion in Unit One.)
   (b) When carbon dioxide and air are put in the same container, where does the carbon dioxide go? Where does the air go?

Pour the gas down the side of the beaker, not directly onto the candle.

Further Analysis

3. (a) What happened to the air in the 250 mL beaker?
   (b) When you added baking soda to the vinegar in the larger beaker, what do you think happened to the air that was originally in the beaker? How do you know this?
4. (a) Design a fire extinguisher that uses baking soda and vinegar to generate carbon dioxide gas. Consider that:
   • the vinegar and the baking soda must be kept separate until the fire extinguisher is used;
   • vinegar is an acid and will corrode most metals.
   (b) Make a labelled drawing of your fire extinguisher.
   (c) List the instructions for using your fire extinguisher.

Extension

5. Consult a reference book or your local fire department to find out the difference between Class A, Class B, and Class C fires. For which class(es) of fire(s) may carbon dioxide extinguishers be used?
6. Check the fire extinguisher in your science classroom. What type is it? Why do you think this is a suitable type for this situation?

Buoyant Force

Recall that a **force** is a push or a pull and that forces are measured in newtons (N). (For example, the weight of an object is a measure of the pull of gravity on it; weight is therefore measured in newtons.) When something floats, it does so because the fluid exerts a force on it. This upward force of a fluid is called its **buoyant force**. Buoyant force also is measured in newtons. A gas can exert a buoyant force and cause another gas to float, as you saw in Activity 2-5. A liquid as well can exert a buoyant force and make an object float. Does a liquid exert a buoyant force on objects that sink —objects like the anchor in the illustration? You can investigate this problem in the next Activities.

Activity 2-6

Buoyant Force of a Liquid on a Solid

The student in the photographs is using a rope to hold up a bleach jug full of sand. Examine the pictures and discuss with your group the following questions.

1. By how much do you think the weight appeared to change when the jug was immersed in the water?
2. Estimate, in litres, how much water was displaced when the jug was completely submerged.
3. (a) What apparently caused the jug to weigh less when it was in the pail of water?
(b) By how much do you think its weight appeared to change? Explain your answer.

The jug full of sand is too heavy to hold comfortably for long.

When it's lowered into the pail of water, it's a lot easier to support.

Buoyant Forces of Various Liquids

Problem

What difference is there in the buoyant force exerted by different liquids?

Materials

3 beakers (600 mL each)
500 g mass
5 N spring scale
glycerol
vegetable oil
water

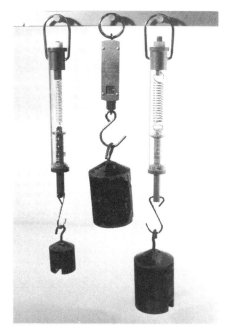

Spring scales.

Procedure

1. Find the weight of the 500 g mass in air. Record this measurement to the nearest 0.1 N.
2. *As accurately as possible*, fill a clean beaker to the 400 mL mark with water.

3. Lower the 500 g mass into the liquid. Submerge it completely, but do not let it touch the bottom or sides of the beaker. Record its apparent weight to the nearest 0.1 N while it is submerged in the liquid.
4. Read the combined volume of the liquid and the 500 g mass. Record the information in your notebook. Thoroughly clean and dry the 500 g mass.
5. Repeat Steps 2 to 4 using glycerol instead of water.
6. Repeat Steps 2 to 4 using vegetable oil instead of water.
7. Wash the 500 g mass in hot soapy water, rinse it, and dry it. Store the liquids.
8. Answer the questions here and on the following page.

Analysis

1. When the 500 g mass was immersed in the liquids, did its weight appear to increase or decrease?
2. How much did the weight of the 500 g mass appear to change when you immersed it in
   (a) water;
   (b) glycerol;
   (c) vegetable oil?
3. (a) When you submerged the 500 g mass in the liquids, what was the total volume?
   (b) What is the volume of the 500 g mass? (You may want to refer to Skillbuilder Three, *New Ideas on Measuring*, page 348.)
4. The buoyant force exerted on an object by a liquid is the difference between the weight of the object in air and its apparent weight in the liquid.

   $$\begin{array}{ccc} \text{buoyant} & \text{weight} & \text{weight} \\ \text{force} & = \text{in air} & - \text{in liquid} \\ \text{(N)} & \text{(N)} & \text{(N)} \end{array}$$

   Calculate the buoyant force exerted on the 500 g mass by water, glycerol, and vegetable oil.
5. In your notebook, make a table like Table 2-3. Fill in the first column by listing the three liquids in order of increasing buoyant force exerted on the 500 g mass. Keep your table, to complete the remaining columns in Activities 2-8 and 2-9.

Table 2-3 *A Comparison of Glycerol, Vegetable Oil, and Water*

	LIQUIDS, IN ORDER OF		
	BUOYANT FORCE EXERTED ON SOLID (ACTIVITY 2-7)	BUOYANT FORCE EXERTED ON LIQUID (ACTIVITY 2-8)	MASS OF 1 mL (ACTIVITY 2-9)
Least ↓			
Greatest ▼			

Further Analysis

6. Draw a bar graph to show the buoyant force exerted by each of the liquids. Show the liquids in order of increasing buoyant force.

7. Think of a way to completely submerge a balloon. Examine the illustration below and answer the following questions.
   (a) Which of the three balloons will be the hardest to push under the water?
   (b) On which balloon will water exert the greatest buoyant force?
   (c) How is the buoyant force related to the volume of the balloon?
   (d) Describe how you can estimate how much buoyant force water will exert on an object.

8. Suggest what causes buoyant force to be different for different liquids.

Extension

9. You can cause a "bottle imp," or Cartesian diver, to float or sink at will. Make a Cartesian diver using a 2 L plastic pop container, water, and a medicine dropper. Experiment with your design until you can control the movement of the diver. Explain what causes it to dive when the bottle is squeezed.

What causes the Cartesian diver to float or sink?

The fact that oil floats on water is important in the oil industry. By simply pumping from oil wells, companies would be able to extract only about 25% of the oil that is there. But when water is pumped *into* the oil-bearing formation, the oil floats and moves upwards towards the site of extraction. Often more than 75% of the oil can be removed from the well.

When an oil tanker is involved in an accident, large amounts of oil may be released, to float on the water. In recent years, there have been several major oil spills on the world's oceans. For any one major spill, find out what caused the spill, what damage was done to the environment, and what was done to minimize the damage.

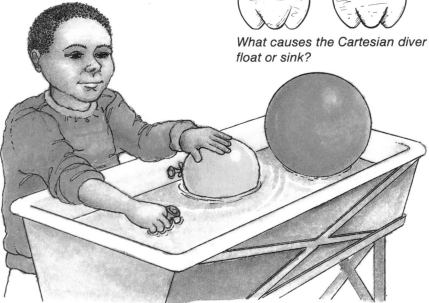

Buoyant Force Exerted by Liquids on Other Liquids

Problem

Can liquids exert a buoyant force on other liquids?

Materials

4 test tubes
test tube rack
glycerol
vegetable oil
food colouring
water

Procedure

1. Before you begin, refer to your bar graph from Activity 2-7 illustrating the buoyant force exerted by the three liquids. Based on those results, predict whether water, glycerol, or vegetable oil will float or sink when the three liquids are mixed. Record your prediction.

2. Pour about 5 mL of water into one test tube, about 5 mL of glycerol into another, and about 5 mL of vegetable oil into a third. To the water, add two or three drops of food colouring so you will be able to distinguish it from the other liquids.

3. Examine the illustration showing the correct method for gently pouring the liquids. Starting with the liquid you think will sink to the bottom, *slowly* pour the three liquids into the fourth test tube.

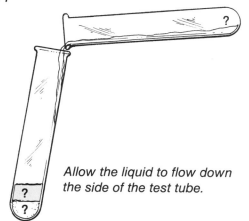

Allow the liquid to flow down the side of the test tube.

Analysis

1. (a) Make a labelled sketch of the test tube with the three liquids lying in layers.
   (b) How does the order in your sketch compare with the order you predicted in Step 1?
2. What fluid was floating above all three liquids in the test tube?
3. Compare the order of the liquids in your bar graph from Activity 2-7 with the order of the liquid layers in the test tube. Is there any similarity? If so, try to suggest a reason that would explain both sets of data.
4. (a) Which liquid exerted the most buoyant force? Explain your answer.
   (b) List the three liquids in the second column of your table from Activity 2-7 (page 75).

Heavier? Lighter?

You have seen that when a fluid is displaced, it exerts a force upwards on whatever displaces it. This buoyant force is different for different fluids. The buoyant force of a particular fluid causes some, but not all, substances to float. Most people would say that oil floats on water because oil is "lighter" than water, and that a metal sinks because it is "heavier" than water. These are not the scientific words. Scientists use a different term for this property of "heaviness." In the next Topic, you'll take a closer look at this important property.

Density of Fluids

People were building ships long before they understood buoyant force. They built boats of wood, which floats easily. Today's ships are made of metal, which normally sinks in water. In addition to their own mass, huge ferries like the ones shown here carry thousands of tonnes of cargo. How can these huge vessels float?

When anything floats, a buoyant force is acting on it. But what causes buoyancy? Is there any liquid on which a chunk of lead could float? Is there any way to predict which kinds of matter will float and which will sink? How are floating and sinking important in everyday life? Answers to these questions will appear in this Topic.

Why Things Float

Problem

What property determines whether a liquid will float or sink in another liquid?

Materials

balance
graduated cylinder
3 beakers (250 mL each)
dropper
glycerol
vegetable oil
water

Procedure

1. In your notebook, prepare a table like Table 2-4 for your data.
2. Use the balance to find the mass of the graduated cylinder. Record its mass in all three spaces in your table.
3. Pour almost 50 mL of glycerol into the cylinder, then use the dropper to measure *exactly* 50 mL.
4. Find the total mass of the graduated cylinder and glycerol, and record it in your table.
5. Return the glycerol to the container provided. Wash the dropper and graduated cylinder in hot, soapy water. Rinse and dry.
6. Repeat Steps 3 to 5 using vegetable oil instead of glycerol.
7. Repeat Steps 3 to 5 using water instead of glycerol.

What causes some logs to sink in water and others to float?

Table 2-4

LIQUID	MASS OF GRADUATED CYLINDER +50 mL LIQUID (g)	MASS OF GRADUATED CYLINDER (g)	MASS OF 50 mL LIQUID (g)	MASS OF 1 mL LIQUID (g)
glycerol				
vegetable oil				
water				

Analysis

1. Do the following calculations for each of the three liquids, and enter the results in your table.
   (a) Find the mass of 50 mL of liquid, by subtracting the mass of the empty cylinder from the mass of the cylinder and liquid.
   (b) Divide by 50 to find the mass of 1 mL of liquid.
2. In your table from Activity 2-7 (page 75), complete the third column by listing the three liquids in order from least mass to most mass of 1 mL.

Further Analysis

3. Suggest a hypothesis that will allow you to predict which kinds of matter will float on a particular fluid and which kinds will sink.

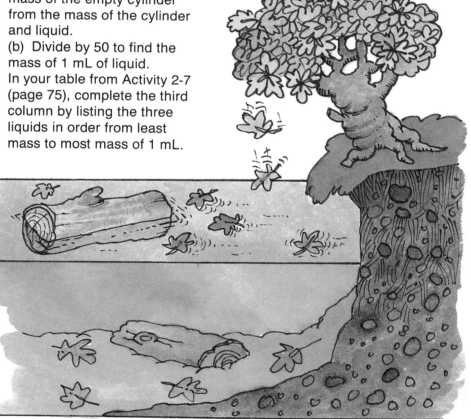

The Meaning of Density

In Activity 2-9, to have a fair comparison of the "heaviness" of the three liquids, you needed to compare equal volumes. When you compare the masses of equal volumes of different kinds of matter, you are comparing their densities. **Density** is the amount of mass in a certain volume of a substance.

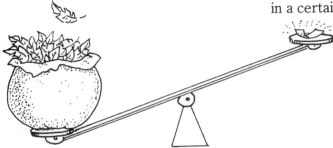

If you have enough feathers, they can have more mass than a piece of gold.

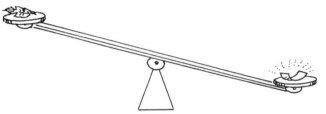

In an equal volume, the gold has more mass.

To find the density of any material, you need to know both the volume and the mass of the sample. In the laboratory, the density of fluids is measured in g/mL; the density of solids is measured in g/cm³. (Remember, 1 mL is the same volume as 1 cm³.) In Activity 2-9, you calculated the mass of 1 mL of water, glycerol, and vegetable oil. Expressed as grams per millilitre, you found the densities of these liquids. Scientists have recorded the densities of many substances, as you can see in Table 2-5. Compare your values for the densities of the three liquids with the values given here.

One fluid will float on another whose density is greater. For example, air can float above carbon dioxide (as you found in Activity 2-5), and both these gases float over water. A solid's density also determines whether or not it will float on a liquid. Both of the kinds of wood listed in the table will float on water, but none of the metals will. Some of them, though, do float on liquid mercury.

Table 2-5 *Approximate Densities of Common Materials*

FLUID	DENSITY (g/mL)
helium	0.0002
nitrogen	0.00125
air	0.0013
oxygen	0.0014
carbon dioxide	0.002
isopropanol	0.79
vegetable oil (varies)	0.9
water	1.00
glycerol	1.26
mercury (a metal)	13.55

SOLID	DENSITY (g/cm³)
wood (western red cedar)	0.37
wood (birch)	0.66
sugar	1.59
salt	2.16
aluminum	2.70
iron	7.87
nickel	8.90
copper	8.92
lead	11.34
gold	19.32

The low density of air is the reason that modern ships float. All ships—early wooden sailing ships and today's huge metal freighters—are built so that they contain a large volume of air. Even though the metal used in the ship's construction and the substances in the cargo may be more dense than water, the overall density of the ship containing a large volume of air is less than the density of water.

Metal ships enclose a large volume of air.

Activity 2-10

Density and Buoyancy

Use the densities given in Table 2-5 to answer the questions.

1. (a) What is the most dense substance listed in the table?
   (b) What is the least dense substance listed?
2. (a) Name two gases that are less dense than air.
   (b) Name two gases that are more dense than air.
3. (a) Name all the solids listed in Table 2-5 that will float on liquid mercury.
   (b) Name all the solids listed in Table 2-5 that will sink if placed on liquid mercury.
4. If the three liquids, glycerol, mercury, and water were placed in the same container, in which order would you expect to find them (from top to bottom)?
5. A block of an unknown metal is 5 cm × 3 cm × 2 cm. The block has a mass of 235 g. Of what metal do you think the block is made? Give reasons to support your answer.

Extension

6. Design and conduct an experiment to determine the density of 25-cent coins. Based on the results of your experiment, what metal do you think is used in these coins? (Refer to Table 2-5 on page 80.)

Measuring Density with Hydrometers

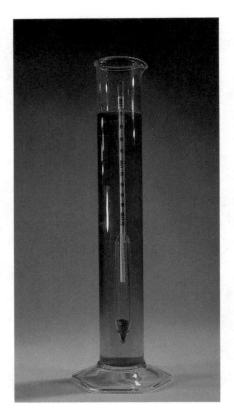

A commercial hydrometer.

You can easily determine the density of a liquid if you have an appropriate hydrometer. A **hydrometer** is a device that uses buoyancy to measure density directly. How does it give you this measurement? As the hydrometer floats on a liquid, it remains partly above and partly below the surface. In a relatively dense liquid, the instrument floats higher above the surface; in a less dense liquid, it is more submerged. By reading the hydrometer's scale at the liquid's surface, you can find out the density. There are several different kinds of hydrometers, for use with different liquids. In Activity 2-11, you'll assess two simple hydrometers, then design and make your own improved model.

Using a hydrometer to measure density is useful in industries as different as candy factories and oil refineries. When a specific liquid is being prepared in a manufacturing process, its readiness may be related to its density. Density depends in part on the concentration of the solution (the amount of solute dissolved in a specific amount of the solvent). How does a hydrometer measure the concentration of a solution? In Activity 2-12 you'll find out, using your own hydrometer.

Did You Know?

More than 2000 years ago, Archimedes, a famous Greek mathematician, discovered how buoyant force could help him solve a practical problem. He had been asked to find out if the king's crown was really pure gold—without melting it down. Archimedes solved the problem by weighing the crown in air and weighing it again when it was submerged in water. He repeated this procedure with an *equal mass* of pure gold. When he found that the buoyant force acting on the crown was more than the buoyant force acting on the pure gold, he knew that the crown must displace a larger volume of water than the pure

gold did. This meant that the metal in the crown had a lower density than that of pure gold. Archimedes was able to tell the king that the crown was not pure gold.

The idea that allowed him to solve the problem is still called *Archimedes' Principle: The buoyant force on an object in a fluid is equal to the weight of the fluid displaced by the object.*

According to legend, the idea occurred to Archimedes when he was in his bath. He jumped out and ran through the streets shouting "Eureka!", which means "I've found it!"

Homemade Hydrometers

PART A

Problem

How effective are these student-made hydrometers?

Materials

3 beakers to contain liquids
empty plastic, single-serving
  coffee creamer
sand
wooden dowel
short wood screw
vegetable oil
glycerol
water

Procedure

1. Make hydrometers like the ones shown.

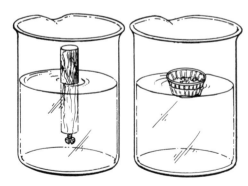

In Part A, you'll use hydrometers like these. In Part B, you'll design and make your own.

2. Test each hydrometer in water, glycerol, and vegetable oil. As you test each one, consider the following questions:
   (a) Does it float on all three liquids?
   (b) Is there any difficulty in placing a scale on the hydrometer?
   (c) Could the hydrometer identify two liquids that were of similar densities?
   (d) Will the hydrometer still work accurately after being used many times?
   (e) How could the hydrometer be adapted for use with liquids that are less dense than vegetable oil or more dense than glycerol?
3. Make marks on the side of each hydrometer to show the level at which it floats in each of the three liquids. Use the density data in Table 2-5 (page 80) and the marks to help you calibrate a scale showing density in grams per millilitre on the side of each hydrometer.

PART B

Problem

How accurate a hydrometer can you make?

Materials

various, from home or classroom

Procedure

1. Design and make a hydrometer, improving on the two preliminary designs. Use as many of their strengths and eliminate as many of their weaknesses as possible.
2. Put the three test liquids in unlabelled containers and have a classmate attempt to identify them. Is the hydrometer able to distinguish between these liquids?
3. Mix glycerol and water (half and half) to produce a fourth liquid. Does your hydrometer clearly show that this new liquid is not one of the original three test liquids?
4. Make any improvements that will make your hydrometer more accurate, more sensitive, or more durable.

Analysis

1. (a) Did your hydrometer work effectively in both Steps 2 and 3? If so, what design features made this possible?
   (b) If not, explain why you think your hydrometer did not work as well for these tasks.
   (c) Was your hydrometer easy to use? Was it accurate? Was it reliable? Explain its good points.

Extension

2. Try mixing isopropanol and water in various ratios (1:1, 1:2, 1:3) and testing them with your hydrometer. Is it sensitive enough to distinguish among these mixtures?

CAUTION: Isopropanol is poisonous.

Testing Your Hydrometer

Problem

How effective is your hydrometer in determining the concentration of a solution?

Materials

hydrometer (from Activity 2-11)
balance
1000 mL graduated cylinder
dropper (for measuring exact volume of water)
large jar or beaker (1500 mL capacity)
600 mL beaker
scoop
long stirring rod
deep, narrow container (e.g., a graduated cylinder)
sugar
sugar solution of unknown concentration

Procedure

1. Carefully measure exactly 1 L (1000 mL) of water and pour it into your 1500 mL beaker. Your group will be asked to prepare one of the following solutions:

   Solution 1 – 0 g of sugar
                  per 1 L of water
   Solution 2 – 50 g of sugar
                  per 1 L of water
   Solution 3 – 100 g of sugar
                  per 1 L of water
   Solution 4 – 200 g of sugar
                  per 1 L of water
   Solution 5 – 400 g of sugar
                  per 1 L of water

2. Use the 600 mL beaker and the balance to measure the correct mass of sugar for your group's solution. (Refer to Skillbuilder Three, *Measuring Accurately*, on page 348.) Add this sugar to the water and stir until it is completely dissolved.

3. Pour your solution into the deep narrow container. Attach a small sign to indicate the solution's concentration (e.g., 100 g sugar per 1 L water).

4. When all the sugar solutions have been prepared, test your hydrometer in each one. Make a scale on it, showing the level at which it floats in the different solutions. For example, a line marked "0" could indicate a sugar concentration of 0 g/L, a line marked "50" could indicate 50 g/L, etc.

5. Place your hydrometer in the unknown solution, and read its concentration on the scale on the hydrometer.

Analysis

1. How is density of a sugar solution related to its concentration? Explain how you know.

2. Are the scale markings on your hydrometer far enough apart to permit accurate readings? Suggest ways in which you could improve the design of your hydrometer.

Further Analysis

3. (a) In your notebook, make a two-column table. In the first column, list the concentrations of the solutions prepared for this investigation.
   (b) Measure the distance (in millimetres) from the 0 g/L mark to each of the other marks on your hydrometer scale, and list these in the second column.
   (c) Make a line graph, plotting concentration of the sugar solution (the manipulated variable) on the horizontal axis, and distance (the responding variable) on the vertical axis. On the horizontal axis, include points from 0 g/L to 600 g/L.
   (d) Plot the points from the table on your graph.
   (e) Draw a smooth, continuous line that best fits the points you have plotted. Describe the shape of this line.
   (f) Use your graph to predict where the 300 g/L and the 600 g/L marks should be located on your hydrometer.

4. (a) According to the particle theory, what makes up a solution?
   (b) Use the particle theory to explain how density is related to concentration of a solution.

Extension

5. Prepare sugar solutions with concentrations of 300 g/L and 600 g/L and test your predictions from question 3(f).

Temperature and Density of Fluids

The density of a fluid can be affected by heat. For example, the hot air inside the balloon in the photograph is less dense than the cooler air outside it. The buoyant force exerted by the cooler, denser air keeps the balloon and its passengers afloat. But if the air inside the balloon cools, the overall density of the balloon increases, and the balloon drifts downwards.

Temperature also affects the density of liquids. You may have noticed patches of warm water on the surface of a lake. Have you ever let yourself sink down below one of these warm patches? You don't go very far before you notice that the water is cooler. Because the cool water is more dense, it exerts a buoyant force that keeps the warm water floating above it.

Most fluids continue to become denser as they get cooler. But water, unlike any other substance, reaches its maximum density at 4°C. When it is cooled below 4°C, it becomes slightly less dense. In winter, the water at 4°C sinks to the bottom of lakes. Colder water floats on top of this dense bottom layer. At the surface, where it is exposed to the frigid winter air, ice forms. The warmer, lower layers remain liquid. If water were a more normal fluid, lakes would freeze from the bottom up and not from the top down!

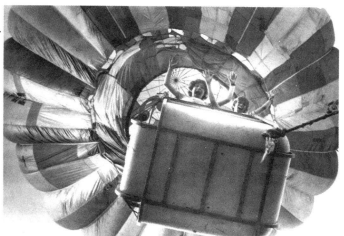

A hot-air balloon.

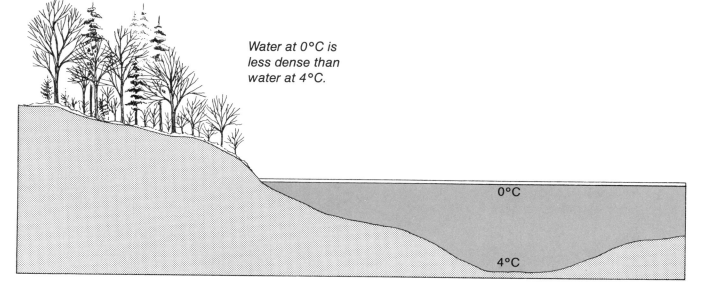

Water at 0°C is less dense than water at 4°C.

0°C

4°C

Checkpoint

1. (a) Explain the meaning of: force, buoyant force, displace.
   (b) How does buoyant force make the work easier when you are clearing rocks from a swimming area?

GET THE BUOYANT FORCE TO HELP!

2. (a) Which has more mass, a kilogram of feathers or a kilogram of bricks? Explain your answer.
   (b) Which has more volume, a kilogram of feathers or a kilogram of bricks?
   (c) Combine the ideas from both (a) and (b) into a single sentence.

3. Which of the solids listed in Table 2-5 on page 80 will float:
   (a) on water?
   (b) on air?
   (c) on mercury?

4. (a) You can float on water, but you can also make yourself sink. Estimate the approximate density of your body.
   (b) Why do you think it is easier to swim in the ocean than in a freshwater lake?

5. A fishing weight has a mass of 100 g and a volume of 9 cm³. Calculate the mass of 1 cm³ of the metal in the fishing weight.

6. To make maple syrup, tree sap is boiled in open pots until most of the water has evaporated.
   (a) What effect would this have on the density of the product?
   (b) Why would the density change?
   (c) How might a hydrometer help maple-syrup producers?

7. (a) Why could you not use the same hydrometer to measure the density of all liquids?
   (b) Describe several differences you might find between two hydrometers that are used in different industries.

8. The block of wood shown in the illustration has a mass of 1.1 kg.
   (a) Calculate its density in grams per cubic centimetre.
   (b) Would this kind of wood float or sink in water? Explain your answer.

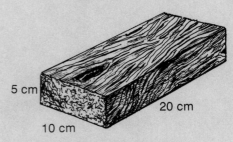

5 cm
10 cm
20 cm

9. In Activities 2-9 and 2-10 you probably found that the buoyancy and density rankings for glycerol, vegetable oil, and water were similar. However, their ranking for flow rate (Activity 2-3) was different. Use the particle theory to suggest an explanation for this.

10. Many fish can adjust the density of their bodies by controlling the amount of gas in a sack called a swim bladder. How would the amount of gas in the swim bladder have to change in order for a fish to swim deeper underwater? How would it change so the fish could swim nearer to the surface? Explain your answers.

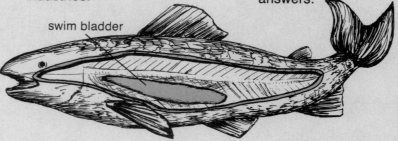

swim bladder

Fluid Pressure

The devices shown here make fluids shoot out at high speed. The shooting results when a force is applied that pushes the fluid through a small opening in the container. Fire hoses, syringes, and garden sprinklers also function because the fluid in them exerts pressure. **Pressure** measures the amount of force applied to a certain area. The amount of pressure a fluid exerts depends on two factors: the size of the force, and the total area on which the force acts. As you might expect, the larger the force, the greater the pressure of the fluid. But the same does not hold true for area: the pressure is *greater* when the force acts on a *smaller* area.

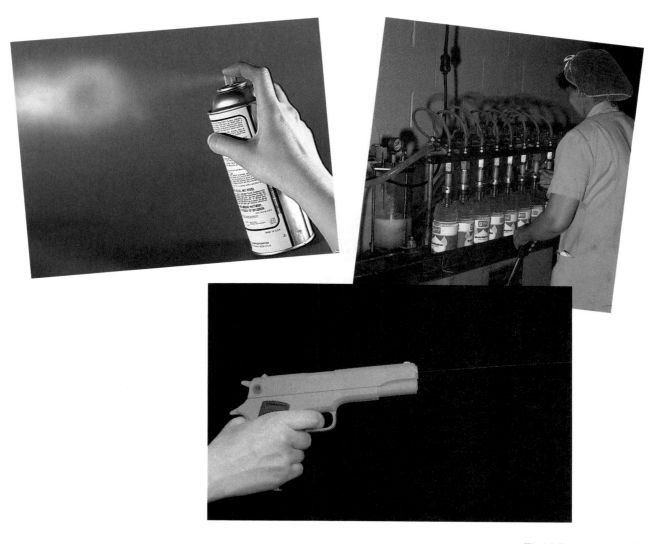

For example, water in a garden hose exerts very little pressure if the hose has no nozzle. It has just enough pressure to move upwards against gravity, but it doesn't move very fast or very far. For a more effective hose, gardeners and firefighters need a stream of water that travels a long distance. There are two ways they could produce this. One is simply to adjust the tap or fire hydrant to increase the rate of flow. The other way to increase the pressure is to reduce the area of the opening. If you place your thumb on the opening, you make it much smaller. This results in a great build-up of water pressure! Gardeners and firefighters use adjustable nozzles in their hoses, to change the area of the opening and thereby adjust the pressure at the mouth of the hose. You can test this for yourself the next time you take a shower. Try adjusting the nozzle, or placing a fingertip over part of it, and observe what happens.

Do liquids and gases behave in a similar way when under pressure? In what direction(s) does the pressure act? How are air pressure and water pressure measured, and why are they important? In this Topic, you'll find answers to these questions. In the next Activity, observe how **confined fluids** (fluids in a completely enclosed container) behave when under pressure.

The Effects of Force on Confined Fluids

Problem

What effect does a force have on a confined fluid?

Materials

support stand
2 burette clamps
50 mL modified syringe
solid rubber stopper with snug
   half-hole for tip of syringe
250 mL beaker
4 masses (1 kg each)

Procedure

1. In your notebook draw a table like Table 2-6.
2. Calculate the force with which gravity will pull on each of the masses listed in Table 2-6, and note these in the appropriate column. (Recall that force is measured in newtons. The force of gravity on a 1 kg mass is 10 N.)
3. Set up the apparatus as shown. Withdraw the piston until there is exactly 50 mL of air in the syringe. Carefully insert the open end of the syringe into the hole in the stopper. It should fit closely, making an airtight seal.
4. Place one 1 kg mass on the platform, and record, in the appropriate column of the table, the volume occupied by the fluid. Repeat with two 1 kg masses on the platform, then with three masses, then with four.

Table 2-6

MASS ON PISTON (kg)	FORCE ACTING ON CONFINED FLUID (N)	VOLUME OF AIR (mL)	VOLUME OF WATER (mL)
0	0	50	50
1			
2			
3			
4			
0			

5. Remove all the masses and record the volume of fluid in the syringe.
6. Repeat Steps 3 to 5, starting with 50 mL of water in the syringe. Be sure to remove *all* the air from the syringe, using the method shown on page 64.

Analysis

1. (a) How did the force affect the gas?
   (b) How did the force affect the liquid?
2. Did the volume of the gas change by the same amount each time the force was increased?
3. (a) What happened when the force was removed from the gas?
   (b) What happened when the force was removed from the liquid?

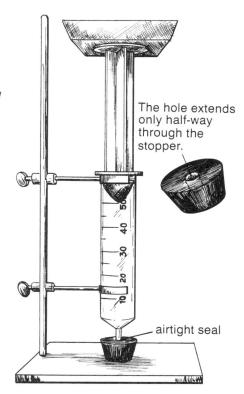

The hole extends only half-way through the stopper.

airtight seal

The tip of the syringe should fit tightly into the half-hole in the stopper.

Further Analysis

4. (a) According to the particle theory, how are gases different from liquids?
   (b) Why do you think the confined gas behaved as it did when a force acted on it?
   (c) Why did the confined liquid behave as it did?
5. If there had been an opening in the syringe (that is, if the fluids had not been completely confined), how would they have behaved?
6. Draw a line graph of your data to show how a force affects the volume of a confined fluid. Place the manipulated variable (the force) on the horizontal axis and the responding variable (volume) on the vertical axis. Show water with one colour and air with another colour on the same graph.
7. (a) What effect do you think a force has on the density of a confined gas? Explain your answer.
   (b) What effect do you think a force has on the density of a confined liquid? Explain your answer.
8. (a) What evidence was there that the confined fluids were pushing on the piston? In what direction did they push?
   (b) Use the particle theory to explain your answer.

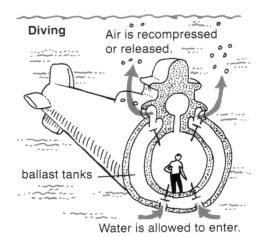

Diving

Air is recompressed or released.

ballast tanks

Water is allowed to enter.

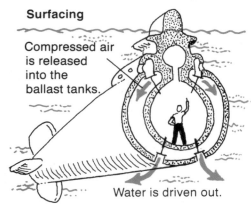

Surfacing

Compressed air is released into the ballast tanks.

Water is driven out.

Water and compressed air are used to increase or decrease the submarine's density.

Forces Applied to Fluids

Hoses, toy water pistols, and syringes depend on the fact that liquids are nearly **incompressible**. That is, the volume doesn't decrease when pressure is applied. Apparently the particles in a liquid are already almost as close together as possible. Instead of making the volume smaller, the force is transmitted throughout the liquid. The liquid exerts *fluid pressure* everywhere on the container's inner surface. If the container is not completely enclosed, the fluid pressure moves the liquid out the opening.

As you have seen, gases are highly **compressible**. When a force is exerted on a confined gas, its volume shrinks. As more and more force is applied, the volume of the gas continues to shrink. The particles making up the gas are forced more closely together, and the force is transmitted throughout the gas. The trapped gas exerts considerable fluid pressure on its container. If the container of a compressed gas is suddenly opened, the pressure will push the gas out rapidly.

Because of this, compressed air is used in submarines. Submarines submerge when their ballast tanks are filled with sea water. The sea water adds to the overall density of the submarine, so it sinks below the surface. To make the submarine rise, compressed air is released into the ballast tanks. The air spreads out and drives the sea water out of the tanks. The air-filled tanks reduce the overall density of the submarine, enabling it to float to the surface. When the ship is to submerge again, pumps re-compress the air for storage, and sea water is allowed to re-enter the tanks.

Another useful property of gases is their ability to push back when they are compressed. You observed this in Activity 2-13. The air confined in the syringe pushed back on the masses and prevented the piston from moving all the way to the bottom. This ability is often used to cushion shocks. The air in an automobile tire pushes back against the force applied by the weight of the car. If the car hits a bump, the extra force compresses the air in the tires. In this way, the effect of the force is spread out over the whole tire, instead of being transmitted directly to the body of the car and its passengers. When the extra force is removed, the air returns to its original volume, pushing the tire back to its original shape. Because gases are compressible, they are more suitable than liquids for this use.

Large tires enable this "Snowbus" to drive over the surface of the Columbia Ice Fields.

Units for Measuring Pressure

As you know, the amount of pressure in a tire is important. If your bicycle tires are underinflated, pedalling requires a lot of energy. If they are overinflated, you get a very bumpy ride. To check that tires are inflated correctly, you need to measure the air pressure inside them.

Recall that the amount of pressure a fluid can exert depends on two factors: the size of the force, and the total area on which the force acts. This is expressed in the following formula.

$$P = \frac{F}{A}$$

In this formula, P represents pressure, F represents force, and A represents area. Thus, if the force is stated in newtons (N) and the area is stated in square metres (m^2), then the units of pressure are newtons per square metre (N/m^2).

Refer to the cube shown in diagram (a), which represents a cube of water measuring 1 m × 1 m × 1 m. The area at the bottom of the cube is its length times its width, or 1 m². The mass of this volume of water (1 m³) is 1000 kg. The force of gravity pulling on 1000 kg of water is 10 000 N. From these values, we can calculate the pressure exerted at the bottom of the cube.

$$P = \frac{F}{A} = \frac{10\ 000\ \text{N}}{1\ \text{m}^2} = 10\ 000\ \text{N/m}^2$$

In diagram (b), imaginary cubic metres of water are stacked one on top of the other. Under the top cube, the pressure is 10 000 N/m², the same as it is in diagram (a). Under both cubes, however, the pressure is 20 000 N/m², because twice as much water is stacked over the same area.

In diagram (c), which shows 10 cubes of water, the force at the bottom of the pile is 10 × 10 000 N, or 100 000 N. The area, however, is still only 1 m². Thus, the pressure is 100 000 N/m².

$$P = \frac{F}{A} = \frac{100\ 000\ \text{N}}{1\ \text{m}^2} = 100\ 000\ \text{N/m}^2$$

However, pressure is not usually measured in N/m², as there is a standard SI unit for pressure — the pascal (Pa).

$$1\ \text{Pa} = 1\ \text{N/m}^2$$

A pressure of 1 Pa is so small that you would barely be able to feel it. If you poured a layer of water only 0.1 mm thick over an area of one square metre, the pressure that the water would apply to that surface would be 1 Pa. Because 1 Pa represents such a small amount of pressure, the kilopascal (1 kPa = 1000 Pa) is a more common unit.

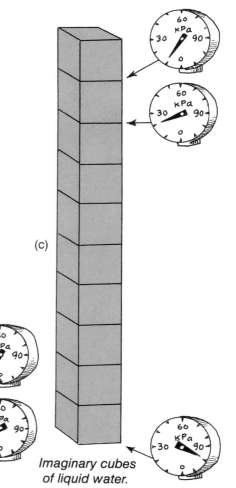

Imaginary cubes of liquid water.

Therefore, the pressure at the bottom of a 1 m cube of water is 10 000 Pa, or 10 kPa. At the bottom of two cubes, the pressure is 20 kPa. There are 100 kPa of liquid pressure acting at the bottom of the 10-cube stack. It is clear that a greater depth of fluid has a greater mass. This results in greater force of gravity pulling on the fluid. With this greater force, the fluid exerts a higher pressure.

The pressure exerted by a liquid is affected by its density as well as its depth. If the cubes in the illustration contained sea water, which is more dense than pure water, the pressure at each level would be greater.

Pressure in Liquids

Check that you understand the units for pressure by answering the following questions.

1. In your notebook, make a table like Table 2-7. Refer to the blocks shown below, which represent quantities of water, and fill in the table.

2. If block (a) were mercury instead of water, what would be the pressure, in kilopascals, at the bottom? Add a third column to your table, and follow the same steps as in Question 1. (The density of mercury is 13.55 g/mL.)

3. Refer to the stack of 10 cubes in diagram (c) on page 92. What would be the pressure (in kilopascals) at the bottom of the fifth cube from the top? What would it be at the bottom of the seventh cube?

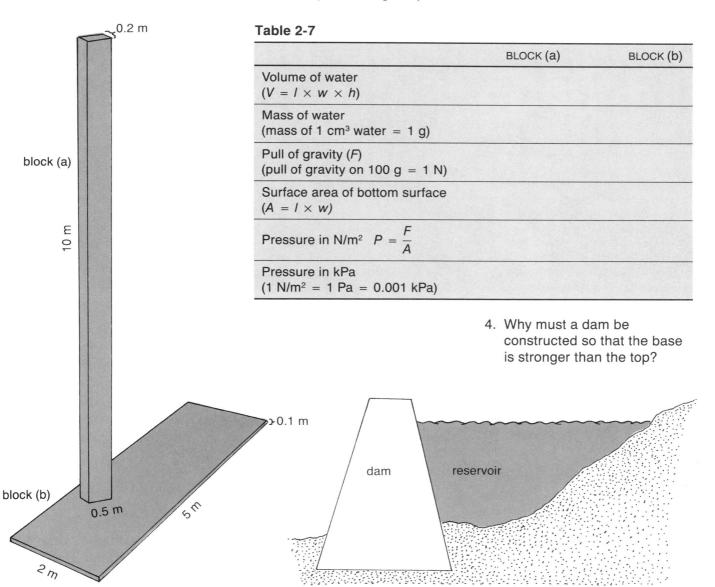

Table 2-7

	BLOCK (a)	BLOCK (b)
Volume of water ($V = l \times w \times h$)		
Mass of water (mass of 1 cm³ water = 1 g)		
Pull of gravity (F) (pull of gravity on 100 g = 1 N)		
Surface area of bottom surface ($A = l \times w$)		
Pressure in N/m² $P = \dfrac{F}{A}$		
Pressure in kPa (1 N/m² = 1 Pa = 0.001 kPa)		

4. Why must a dam be constructed so that the base is stronger than the top?

0.2 m

block (a)

10 m

block (b)

0.1 m

0.5 m

5 m

2 m

dam

reservoir

The Direction of Fluid Pressure

In the ocean, behind a dam, or in a swimming pool, gravity pulls the water towards the centre of the Earth. This downward pull is what causes the water at the bottom to exert pressure. But when a fluid is under pressure, does it just push downwards?

The container shown in the illustration has eight small holes in its side. When it is filled with water, water pours out through these holes. In the top view, you can see that each stream of water travels the same distance horizontally, and all go in different directions. Therefore, the force pushing the water out must be the same size at all the holes. The amount of pressure must be the same and must act equally in all directions. Pressure acts downwards, sideways, and upwards. (The action of pressure upwards is the cause of buoyant force.)

From each hole, there is a stream of coloured water.

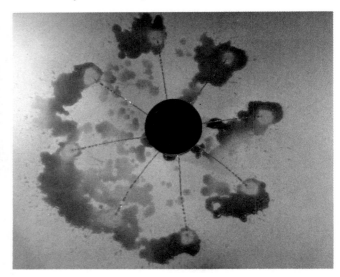

Top view of the same container.

Atmospheric Pressure

The air we breathe is a fluid under pressure. Earth's atmosphere extends outwards into space for a distance of more than 700 km. Even though air has a low density, the column of air above each square metre of Earth's surface is so tall that it has a very large mass—about 10 130 kg. The Earth's gravity pulls on this column with a force of 101 300 N. Thus, the air pressure exerted by the atmosphere at sea level is 101 300 N/m², or 101 300 Pa, or 101.3 kPa. On mountain tops, the column of air is not as tall, and the air itself is less dense. Therefore, the atmospheric pressure is considerably less. For example, the top of Mount Everest (the highest mountain on Earth) is 8848 m above sea level. The average atmospheric pressure at the top of Mount Everest is only about 36 kPa.

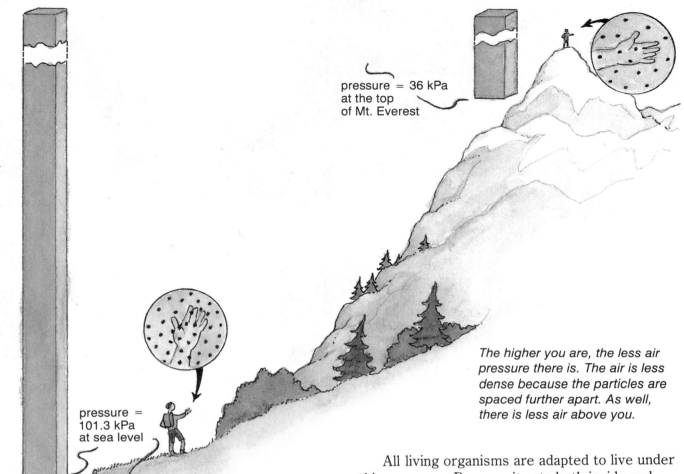

pressure = 36 kPa
at the top
of Mt. Everest

pressure =
101.3 kPa
at sea level

The higher you are, the less air pressure there is. The air is less dense because the particles are spaced further apart. As well, there is less air above you.

All living organisms are adapted to live under this pressure. Because it acts both inside and outside your body, pushing in all directions, you don't even notice it. However, when the outside pressure changes, your body must adjust to it. For example, when you take off in an airplane, the outside air pressure decreases as you get higher. If the air pressure inside your ears does not adjust quickly enough, you may feel an uncomfortable sensation. If you swallow, air can get into the middle ear from the back of your throat. Your ear then feels normal because the pressure is again equal on both sides of the eardrum.

Atmospheric pressure changes slightly from day to day, because of changes in the temperature and moisture content of the air. Meteorologists construct weather maps by plotting air pressure readings from locations all over the globe. These data help forecasters to predict future weather patterns.

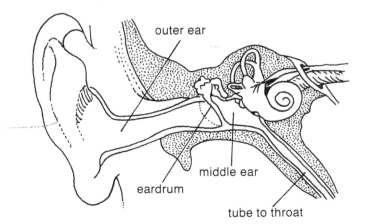

outer ear

middle ear

eardrum

tube to throat

The eardrum separates the outer ear from the middle ear.

Fluid Pressure **95**

A diver who returns too quickly from a deep dive to the surface may suffer from "the bends," or decompression sickness. Under the high pressure during a dive, nitrogen dissolves in fluids in the diver's body. Under the much lower pressure of the atmosphere on the surface, the nitrogen comes out of solution, forming bubbles. The bubbles in the muscles cause intense pain; in blood vessels or in the brain, they can be fatal. To prevent the bends, divers must rise to the surface very slowly. They must stay for periods of time at intermediate depths, eliminating excess nitrogen from the body as they breathe.

Water Pressure

A column of water, being more dense than air, exerts more pressure than the same depth of air. If you dive to the bottom of a swimming pool, you will feel in your ears the effect of the extra pressure. Your body doesn't notice the normal 101 kPa of air pressure, but it does feel the additional 30 kPa at the bottom of the pool. There is a limit to the amount of additional pressure that the human body can tolerate.

Using special diving equipment and breathing from tanks of compressed air, divers can go under water to depths of 50 m, where the pressure is about 500 kPa. If they go deeper, or if they stay too long at this depth, they lose the ability to think clearly. This is because, at the higher pressure, gases from the air become more soluble in the divers' bodies. More nitrogen and oxygen dissolve in the blood, and the higher concentration of nitrogen and oxygen affects the brain. For safety, divers strictly regulate the depth of their dives and the amount of time spent under water.

Living night and day in underwater stations, people can work under water for a week or more, swimming outside the station to work. Maintenance to offshore oil-drilling rigs may be carried out in this way. Workers must return to the surface only very slowly, allowing their bodies plenty of time to re-adjust to normal pressure.

Canada's research submersible, Pisces IV, was launched in 1972. Because of its spherical shape and its thick steel hull, this vessel can dive to 2000 m, where the pressure (20 000 kPa) would crush any ordinary submarine. Inside, the pressure is about 100 kPa, and the air is kept clean and fresh, so the occupants need not wear any special equipment.

This octopus and rat-tail fish are adapted for living under a pressure of more than 15 000 kPa.

Devices for Measuring Pressure

Mercury Barometers

A device that measures air pressure is called a **barometer**. Many weather stations use **mercury barometers**, like the one shown in the diagram. The height of the column of mercury indicates how hard the air pressure is pushing. The greater the air pressure, the higher the column. A normal air pressure of 101 kPa will support a column of mercury 76 cm tall.

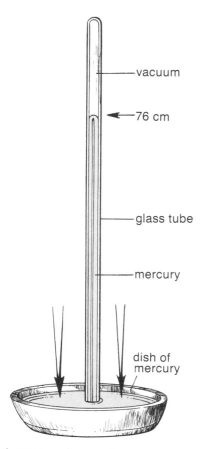

A mercury barometer.

A mercury barometer consists of a sealed glass tube that is completely filled with liquid mercury, then inverted in a dish of mercury. Gravity acting on the column of liquid in the tube pulls it down towards the bowl, leaving a vacuum at the top of the sealed tube. Because there is no air in this space at the top of the tube, gravity is the only force pulling the mercury downwards; there is no air pressure in the tube. Air pressure, pushing down on the mercury in the bowl, pushes mercury back up into the tube. The mercury level stabilizes at a point where those two forces are equal. When the air pressure changes, the height of the column of mercury changes as well.

Aneroid Barometers

A convenient barometer for home use is the **aneroid** barometer. Because it contains no liquid whatever, it avoids the mess and danger of mercury spills. The photographs show two types of aneroid barometer. Similar aneroid barometers are used in airplanes, to indicate altitude.

Both of these are aneroid barometers. Aneroid barometers contain a sealed container from which some of the air has been removed. As outside air pressure changes, the lid moves in and out. A series of levers magnifies the movement and transfers this movement to the pointer. (In the photograph on the left, you can just see the top of the container.)

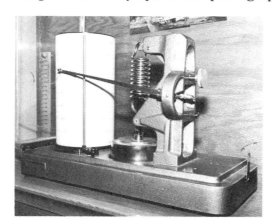

Construct a modified manometer like the one shown in diagram (c) below. (A piece of rubber from a balloon, stretched tightly over the thistle tube, is an effective diaphragm.) Observe what happens to the liquid levels when you hold the thistle tube at different depths in a fish tank or large pail of water. Try holding it sideways and upside down. Fill a tall cylinder with water, and hold the thistle tube at similar depths in it. Do you predict the pressure will be the same, or different? If you can, replace the water in the tank with isopropanol (less dense than water) or glycerol (more dense than water), and repeat your investigation.

Manometers

A device similar to a barometer can be used to measure the pressure of other gases. This device, called a **manometer**, consists of a U-shaped tube partly filled with liquid, usually mercury. The diagram shows a manometer that could be put together using materials normally found in a school science laboratory (for safety reasons, coloured water is used in place of mercury).

Both manometers and aneroid barometers can be readily constructed from materials found at home or in your science classroom. In the next Activity, you'll construct a barometer, make a scale for it, and use the barometer to observe and record changes in atmospheric pressure.

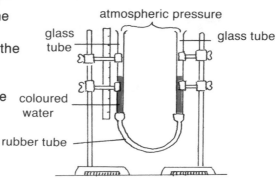

(a) In a manometer, the difference in the height of the liquid in the two glass tubes indicates the difference in the pressure in the two sides.

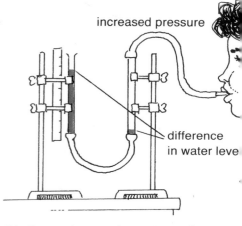

(b) If you place a clean piece of tubing on one side of a manometer and blow into it, you increase the pressure there. The liquid level is pushed down on that side and rises on the other.

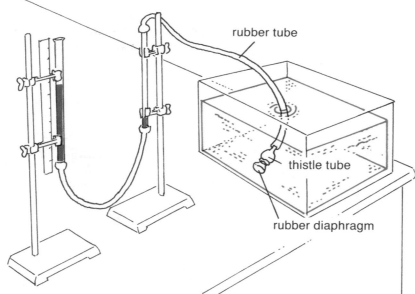

(c) This manometer has been modified so that it can be used to measure pressure in a liquid. As the pressure on the diaphragm changes, the liquid moves up or down in the U-shaped tube.

Making a Barometer

Problem

How effective is a homemade barometer?

Procedure

1. Examine the student-made barometer shown on this page and the photograph of the barometer on page 97. Plan the design of your instrument, and collect all the materials.

2. Build your barometer, and place it indoors, in a location where the temperature will remain constant. (*This is important, as temperature changes would affect the accuracy of your barometer readings*.)

3. You will need about a week to make a scale. Consult the television or radio weather reports for air pressure readings at least twice each day (morning and evening). Mark each reading on your scale at the level the barometer is pointing to at the time. As well, record in a table in your notebook the date, time, pressure, and weather conditions at the time of each reading.

4. When the scale is complete, bring your barometer to school.

Analysis

1. What were the highest and lowest atmospheric pressures during your test period? Is your barometer sensitive enough to show these differences clearly?

2. Compare the air pressure readings on your scale with those on barometers of different designs. Assess which designs appear to be most effective. Discuss possible reasons for this.

3. Compare the design of your homemade barometer with the design of the commercial barometer in the illustration.
   (a) In what way(s) are they alike?
   (b) How do they differ?
   (c) What single part of the commercial barometer is most likely responsible for its greater accuracy? Give reasons to support your answer.

4. Compare the readings of atmospheric pressure you recorded on your barometer with the weather conditions at the time of the reading. Is there any relationship between atmospheric pressure and daily weather? Describe the evidence you found that supports your answer.

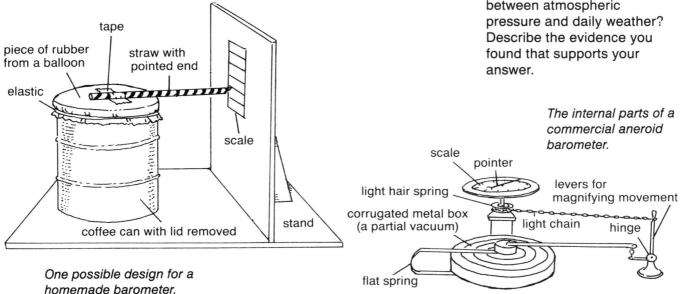

One possible design for a homemade barometer.

The internal parts of a commercial aneroid barometer.

Science and Technology in Society

SCUBA Science

"Don't hold your breath!" cautions your instructor. It's the last thing you hear before the water closes over your head. You're sitting on the bottom of a swimming pool, and every instinct tells you not to breathe. Then, you remember your instructor's words and nervously inhale.

Air fills your lungs. You can hardly believe it! You are breathing underwater. With a little more training, you'll be nearly as free as a fish to explore the world under the waves. What makes your new freedom possible? A Self-Contained Underwater Breathing Apparatus (or SCUBA) supplies air that flows into your lungs through a tube you clench between your teeth.

Without a large supply of air, early divers were severely limited in how long they could stay underwater. Some divers tried to take their air supply with them, in sheepskin bags for example, but their efforts were in vain. More recently, people on the surface supplied air to the diver through a long tube, but this method was cumbersome and not very effective. The Self-Contained Underwater Breathing Apparatus revolutionized diving. An average tank contains a closet-sized volume of air (about 3 m^3) compressed into a space the size of two shoeboxes. This air allows you plenty of time to enjoy your freedom in the underwater world.

A SCUBA is more than just a tank filled with compressed air. If you had only that, you would have two problems. First of all, you wouldn't be able to breathe directly from the tank. Its high pressure would push you back with considerable force, and you wouldn't be able to hold on to the mouthpiece. Second, water pressure would prevent you from breathing in. The pressure isn't enough to crush bones or muscles, but it would squeeze the hollow cavities that make up your lungs. Without a SCUBA, your lungs would be squeezed smaller and smaller the deeper you went.

In 1943 Jacques-Yves Cousteau and Emile Gagnan solved both problems with their invention of the breathing regulator, now a part of the SCUBA. The regulator has two parts. The first stage, attached directly to the air tank, reduces the pressure of the air and provides a steady flow, so the diver can breathe easily. The second stage, a piece of equipment no larger than your fist, is part of the mouthpiece. It contains one-way valves that direct the flow of air. A good look at the diagram of the mouthpiece shows you the ingenious way the second stage works.

The amount of air a diver uses depends on the depth and on the exertions of the diver. For example, when simply swimming near the surface, about 25 L/min are needed, so the air in a typical tank would last more than 2 h. At a depth of 20 m, the diver needs about 75 L/min, so if the dive is to take considerable time, two or even three air tanks might be carried along.

For safety, divers usually take along a compass, a depth gauge, and a pressure gauge attached to the air tank. By reading the pressure gauge, they can tell how much air is left in the tank and allow themselves plenty of time to return safely to the surface.

The Self-Contained Underwater Breathing Apparatus has enabled people to enter a world that once was hostile to them. Today, divers repair boat hulls, look for salvage, study sea creatures, and explore sunken ships. New materials and improved technology continue to make diving easier and more enjoyable for hobbyists and scientists alike.

Think About It

1. At what other times, besides when diving, do people need portable oxygen supplies?
2. Divers' lungs are squeezed by water pressure. What other body parts might also be affected?
3. What are some other uses for compressed air?

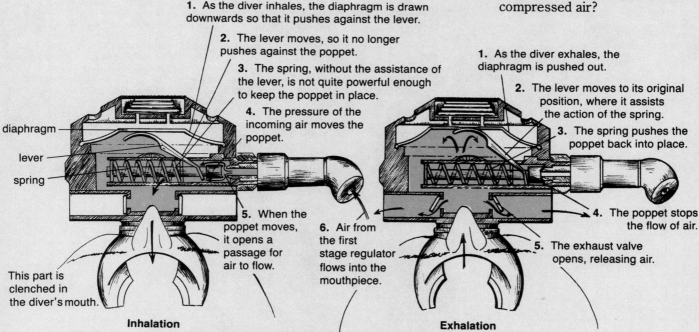

1. As the diver inhales, the diaphragm is drawn downwards so that it pushes against the lever.

2. The lever moves, so it no longer pushes against the poppet.

3. The spring, without the assistance of the lever, is not quite powerful enough to keep the poppet in place.

4. The pressure of the incoming air moves the poppet.

diaphragm
lever
spring

This part is clenched in the diver's mouth.

5. When the poppet moves, it opens a passage for air to flow.

6. Air from the first stage regulator flows into the mouthpiece.

Inhalation

1. As the diver exhales, the diaphragm is pushed out.

2. The lever moves to its original position, where it assists the action of the spring.

3. The spring pushes the poppet back into place.

4. The poppet stops the flow of air.

5. The exhaust valve opens, releasing air.

Exhalation

The diagrams show an exposed view of the inside of the mouthpiece, looking down from above the diver's head. When left at rest, the strong spring pushes against the poppet, preventing air from flowing through. Follow the arrows inside the hose and mouthpiece to see the opening and closing of valves, and how the spring, lever, poppet, and diaphragm are involved in the action as the diver inhales and exhales.

Fluids at Work

Fluids are used in braking systems in vehicles, in the lifting mechanisms of front-end loaders, and in various devices that involve liquids moving along in pipes. These are all **hydraulic systems**—mechanisms that work because of the movement of a liquid or the force of a liquid in a closed system. In this Topic, you'll investigate two of these: the hydraulic press, used in braking systems and lifting devices, and hydraulic systems for transporting liquids.

The Hydraulic Press

Because liquids are practically incompressible, a force applied to a confined liquid will be transmitted throughout the liquid. The pressure produced in this way is the same everywhere in the liquid and is exerted in all directions. These two characteristics of liquids —incompressibility and their ability to transfer pressure—are what make the hydraulic press work.

You can make a model hydraulic press using two modified syringes like the ones you used earlier in this Unit. The diagram shows the arrangement. Both syringes, and the tube joining them, contain water; there is *no air* in the tubing or in either syringe. If your thumb pushes down on the piston of one syringe, the pressure is transferred through the liquid and pushes upwards on the other piston. Like a hydraulic press, this apparatus transfers the force from one place to another. If you put an object on the platform of one syringe, you can raise it by pushing down on the other. In this arrangement, both pistons have the same surface area, so the same force is needed on either side.

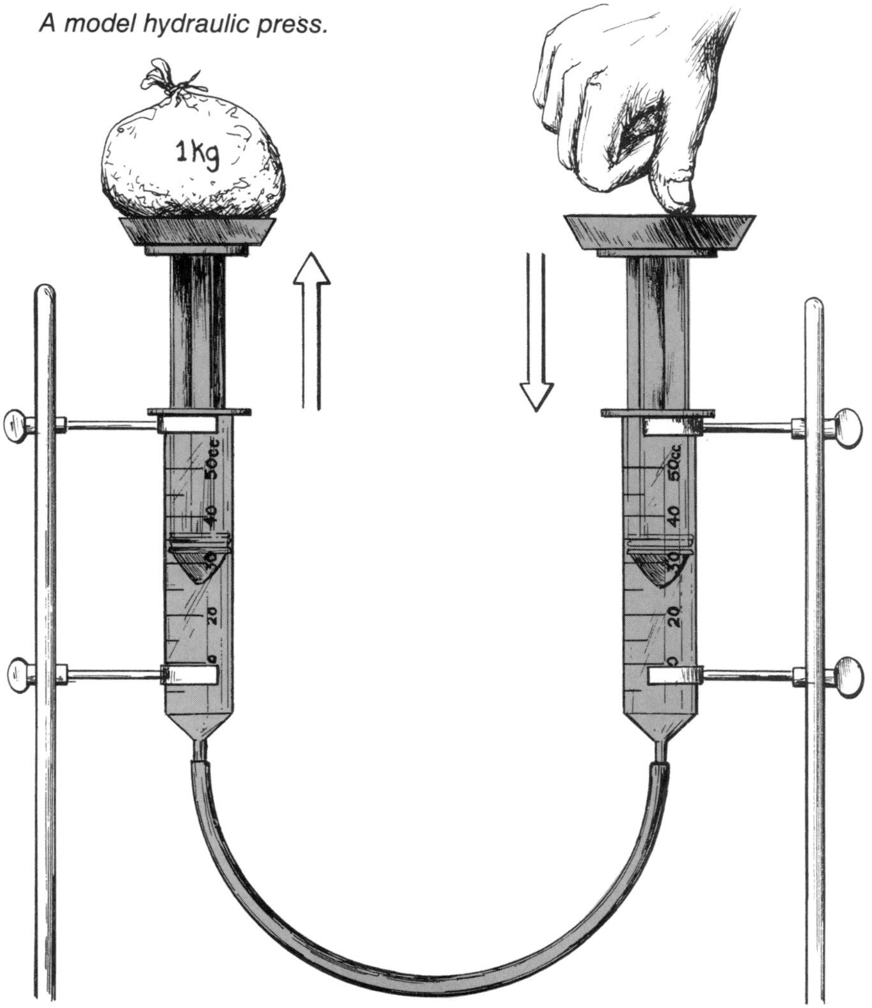

In the next Activity, you'll make a model hydraulic press like this one and feel what "standard force" is needed to raise a standard mass. Then, you'll replace one syringe with a smaller one, whose piston has a smaller surface area. How do you think the smaller surface area will affect the force needed to move the standard mass?

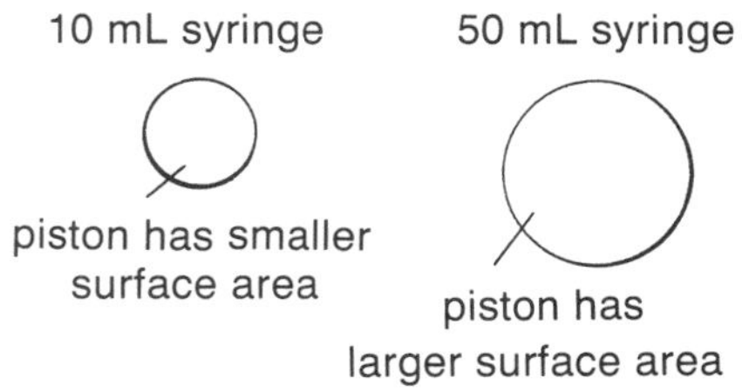

A model hydraulic press.

10 mL syringe

piston has smaller surface area

50 mL syringe

piston has larger surface area

The piston of the smaller syringe has a smaller surface area.

A Model Hydraulic Press

Problem

How can a hydraulic press transfer and multiply a force?

Materials

2 modified syringes (50 mL)
10 mL modified syringe
narrow rubber tubing
   approximately 50 cm in
   length
2 support stands
4 burette clamps
1 kg mass
250 mL beaker

Procedure

1. Use the rubber tubing to connect the two 50 mL syringes. Fill both syringes with water as shown in the illustration, and mount them on the support stands.
2. Ensure that the pistons in all the syringes move freely and do not stick. If they do stick, loosen them by alternately pressing one, then the other, until they slide freely.
3. Place the 1 kg mass on one platform.
4. While your partner steadies the support stand, push on the other platform with your thumb until you move the mass.
5. Remove the 1 kg mass and place it on the other platform. Push down on the empty platform and feel how much force is required to move it.
6. Make a note to help yourself remember how much force was required in this arrangement. Call this "standard force."

Setting Up a Model Hydraulic Press

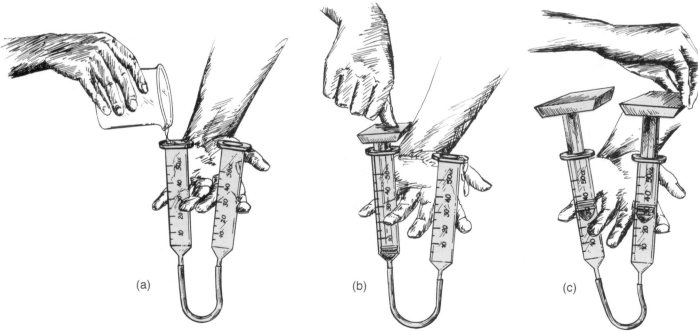

(a)

(b)

(c)

(a) Remove both pistons. Keeping the two syringes at the same level, pour water into one side until both syringes are completely full.

(b) Insert the piston into one syringe. Keeping the two syringes at the same level, push the piston all the way to the bottom, catching the overflow in the beaker.

(c) Insert the second piston. Push it half-way down. Both pistons should now be part-way down the barrel, and there should be no air in the system.

7. Disconnect the apparatus.
8. Use the rubber tubing to connect the 10 mL syringe to one of the 50 mL syringes. Fill them with water as before, but depress the piston on the 50 mL syringe only half-way before inserting the piston in the 10 mL syringe. Mount them on the support stand.
9. Repeat Steps 3 to 5, using the different-sized syringes. Feel how much force is required to move the mass on each side of the device. In your notebook, make a copy of Table 2-8, and fill it in, using "greater than standard" or "less than standard."

Analysis

1. (a) With how many newtons of force was the 1 kg mass pushing down on the piston? (b) Was this force the same for each trial? Explain.
2. In this investigation, what arrangement allowed you to raise the 1 kg mass using the least force?
3. Suppose you needed to raise a 1 kg mass using even less force. How could you modify your model hydraulic press?

Compare the force required when you push on the small syringe with the force required when you push on the larger one.

Table 2-8

SURFACE AREA OF PISTON		FORCE NEEDED
PUSHED ON BY THUMB	SUPPORTING 1 kg MASS	
same area	same area	standard force
smaller area	larger area	
larger area	smaller area	

Further Analysis

4. Suppose you filled the system with air instead of water. How would your observations differ? Give a reason for your answer.
5. Most hydraulic presses use a liquid other than water. Water has two drawbacks: it is not a good lubricant, and it can cause parts of the system to rust. Suggest a liquid that might work better than water. Give reasons for your answer.

Extension

6. (a) Measure as best you can the diameters of the ends of the large and small pistons. For each, calculate the radius (the radius is half the diameter).
   (b) Find the area of the end of each piston, using the following formula:
   $$A = \pi r^2 \quad (\pi = 3.14)$$
   (c) Calculate the pressure exerted on the water by each piston when it had the 1 kg mass on it:
   $$P = F/A$$
   (d) Which of the two pistons exerted more pressure on the system?
   (e) Write a general statement relating the surface area of a piston to the amount of pressure it causes in a hydraulic system.

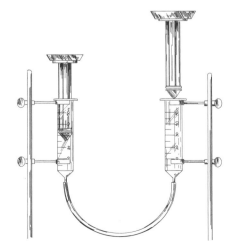

Uses of the Hydraulic Press

In a hydraulic press, the fluid is in a confined space and is used again and again to move a load or a machine part. Examine the diagram below to see how this device can be used to hoist a car. The effort force, supplied by an electric motor, pushes on the small movable piston. The downward movement of the piston increases the pressure of the liquid everywhere in the enclosed system. The increased pressure causes the large piston to move upwards.

How can a small effort force produce a much larger load force? You can explain this using the formula from page 91:

$$P = F/A$$

Rearranged, this formula becomes:

$$F = P \times A$$

That is, the force is equal to the pressure multiplied by the area. The pressure is the same on all the inner surfaces throughout the system. The formula tells you that at a surface where the area is small, the force exerted is small. At a surface with a larger area, the force exerted is larger. Thus the force is multiplied because of the different areas of the two pistons.

In the illustration, the arrow at the small piston represents the downward effort force. The equal spacing and equal size of the arrows inside the press indicate that the pressure is the same throughout the system. The area of the large piston is five times greater than the area of the small piston. Thus, the upward force is five times greater as well. The upward force that lifts the car is represented by the long arrow at the right.

This simplified diagram of a hydraulic lift shows that the force used to raise the car is five times as large as the force applied on the small piston. (Hydraulic car lifts actually multiply the force many times more than this.)

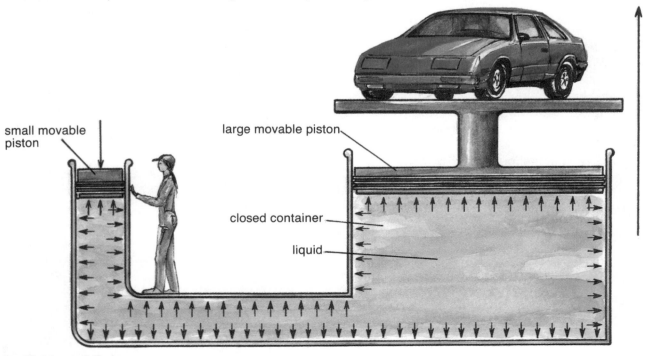

small movable piston

large movable piston

closed container

liquid

Heavy tractor-trailers use a different kind of braking system —air brakes. Find out how an air brake system works. Why are air brakes a safer system for a tractor-trailer?

Other machines that use hydraulic presses to lift or move loads include fork-lift trucks, bulldozers, hydraulic jacks, tipping mechanisms on garbage trucks, and retractable undercarriages on airplanes. In these and many other devices, the effort force for a hydraulic system is supplied by a pump. The piston of the pump is much smaller than the piston that operates the lifting mechanism. By moving back and forth many times, the pump's small piston can put enough pressure on the large piston to raise the load. In most hydraulic presses, the energy for pumping is supplied by a gasoline engine or by an electric motor.

One device that does not use a motor to power the pump is a hydraulic jack; this device is operated by the energy of human muscles. Hydraulic jacks are sometimes used in home repairs, to support a verandah or a basement ceiling while repairs are made. By pushing up and down on the handle of a hydraulic jack, almost anyone is able to raise a load of several tonnes.

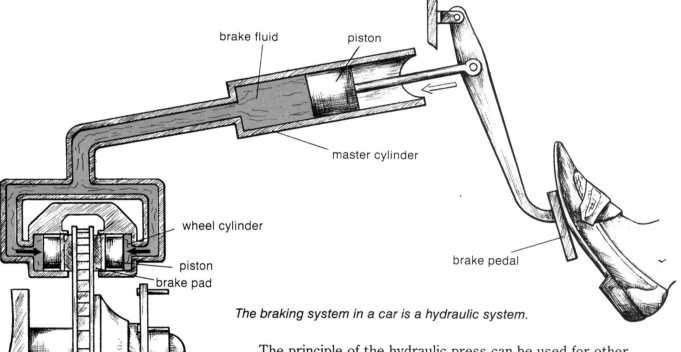

The braking system in a car is a hydraulic system.

The principle of the hydraulic press can be used for other purposes as well. For example, in an automobile braking system like the one shown, a small effort force on the brake pedal is transferred to the brake fluid in the master cylinder. This force places the brake fluid under pressure. Through hydraulic lines, the pressure is transferred to brake cylinders attached to each wheel. In the cylinders, the pistons are pushed outwards by the fluid pressure, so they push the brake pads against the discs attached to the wheels. The friction makes it difficult for the wheel to turn. The wheel turns more slowly, and the vehicle slows down.

Hydraulic Systems for the Movement of Fluids

In the type of hydraulic system that supplies water to homes, a force is applied to move the fluid itself. The system is confined, to maintain enough pressure to push the fluid along in pipes, but has taps, to allow the fluid to exit.

Water supply systems may be quite complex — even within a single building there are several subsystems. Many buildings have water heaters, with separate delivery systems for hot and cold water. Outlets may include taps, showers, toilets, washing machines, and dishwashers. If a home has its own well, a pump will be a necessary part of the system.

In towns and cities, the movement of fluids in underground systems usually goes unnoticed until repairs are needed.

In this drawing of a water supply system, try to trace the path of hot and cold water. Why do you think the main water supply for buildings is usually underground?

Valves

Valves are devices that regulate the flow of a fluid. In your home, taps similar to the one shown in diagram (a) are probably the most common type of valves. These allow you to adjust the rate of flow of water. Other kinds of valves that simply start or stop the flow of fluid, like those shown in (b) and (c), have many other applications. In a toilet tank, the valve (d) is designed to allow water to flow into the tank to a certain depth, then to stop the flow.

Some valves allow the fluid to flow in one direction only. For example, in the human heart (e), pressure from one direction causes the valves to open, while pressure from the opposite direction closes them. The artificial heart valve (f) serves the same function.

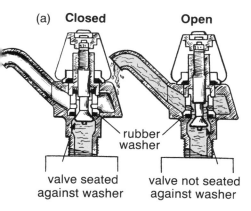

(a) **Closed**   **Open**

rubber washer

valve seated against washer   valve not seated against washer

(a) A common type of valve used in household taps (faucets). The valve is opened or closed by turning the handle.

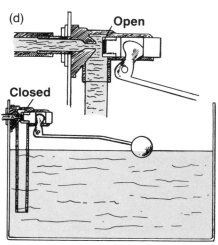

(d) **Open**

Closed

(d) A ballcock valve in a toilet tank allows water to flow while the water level is low. When the tank fills up, the floating ball causes the valve to close.

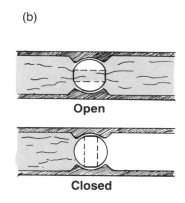

(b)

Open

Closed

(b) A ball valve in a garden hose. The ball has a hole through it. When the hole is aligned with the hose, water flows; when it is turned, it blocks the flow.

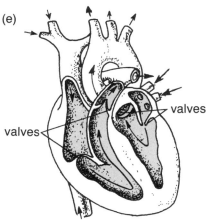

(e)

valves

valves

Arrows show the direction of blood flow through the heart.

(e) The valves in the human heart ensure that the blood flows in the correct direction.

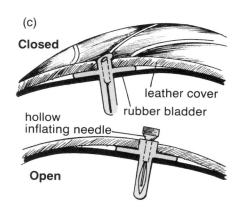

(c)

Closed

leather cover

rubber bladder

hollow inflating needle

Open

(c) This type of valve keeps air inside an inflatable ball. To deflate the ball, you insert a pin that pushes open one side of the valve. The pin is hollow, so air can pass through it.

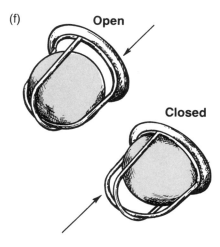

(f) **Open**

Closed

(f) A caged-ball valve can be used to replace a damaged heart valve.

Valves

Problem

How can you make a simple valve?

Materials

short length of tubing (e.g., rubber tubing, drinking straw, section of garden hose)

variety of easily obtainable materials, such as clamp style paper clip, laboratory pinch clamp, small marble, ball bearing, balloon, beakers

Procedure

1. Using any materials of your choosing, design and build a valve that will allow you to start and stop the flow of water through a piece of tubing. Test your valve, and improve it if you can.

2. Design and build a valve that allows water to flow through the tubing one way but not the other.

3. On paper, design a valve that would work if you used glass or metal tubing instead of tubing made from a flexible material.

Analysis

1. Compare your valves with the ones shown on page 109. In what ways are they similar, and how are they different?

A straw works because of air pressure.

Pumps for Hydraulic Systems

In an urban water supply system, the water reservoir is located high above the buildings it must serve. The downward pull of gravity on the water in the reservoir puts all the water in the system under pressure. This pressure pushes the water along the water mains into the buildings whenever a tap is opened. Because of gravity, no pump is needed in the delivery system.

But how did the water get up into the reservoir? It had to be pumped upwards against the pull of gravity. A **pump** is a device for moving a fluid. In cities, pumps move water into elevated reservoirs; in rural areas, pumps bring water up from deep underground wells. Pumps are also used to move oil and natural gas through pipelines. Automobiles have fuel pumps to deliver gasoline to the engine. Bicycle owners use pumps to push air into the tires. All of these pumps cause a force to be applied to a fluid and put the fluid under pressure.

A drinking straw is a means of moving a liquid from a glass into your mouth, so your mouth acts as a kind of pump. What force causes the liquid to move upwards? The arrows in the illustration show air pressure. As the drink sits on the table, air pressure pushes on the surface of the liquid. Because the pressure is the same everywhere, the liquid doesn't move. When you suck on the straw, you remove some of the air and lower the air pressure inside it. The air pressure outside, which acts in all directions, *pushes* the liquid upwards. Other kinds of pumps work in a similar way. Examine the workings of two kinds of pumps in the following Activity.

Activity 2-18

Pumps

1. Examine the illustration at the bottom of the page, showing stages in the operation of an old-fashioned water pump.
   (a) How many valves are there in the pump?
   (b) As the handle moves downwards in stage (b), in what direction does the water move? What causes this motion?
   (c) Describe the positions of the valves in stage (c).
   (d) What happens to the water in stage (c)?
   (e) Explain the positions of the valves in stage (d).
   (f) Why might it be necessary to pump several times before water comes out of the pump?

2. A force pump like the one shown below sends out water with more force than a lift pump; it also sends the water out continuously.
   (a) Explain the function of valve A and valve B.
   (b) What is the function of the air in the dome? (Hint: gases are compressible.)
   (c) Explain why the water can be delivered continuously.
   (d) What is the advantage to having a very narrow opening in the nozzle?

Extension

3. Pumps that rely only on air pressure cannot raise water more than about 10 m. Consult a reference book or a local manufacturer or sales outlet to find out what kind of pump is used to pump water from wells deeper than 10 m.

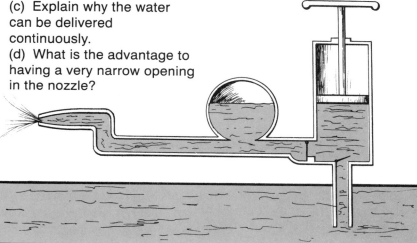

A force pump.

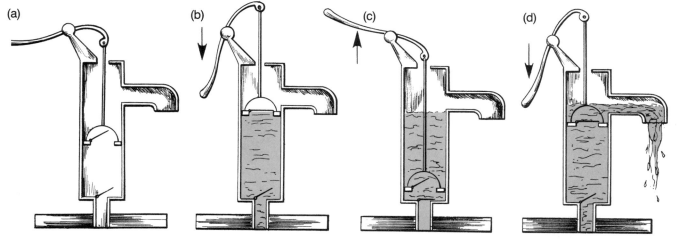

A lift pump.

Diagrams (b), (c), and (d) show stages in operating the lift pump. The arrows indicate the direction of motion of the handle.

Measuring Pressure in a Fluid System

In any fluid system, suitable pressure is necessary in order for it to function properly. For example, large ocean-going ships are driven by systems that contain superheated steam. To increase the speed of the vessel, steam pressure must be increased, but too much pressure could cause a serious accident. Such systems have pressure gauges permanently attached. In cars, abnormal oil pressure is a sign of trouble and can cause serious damage to the engine. Automobiles have internal pressure gauges that turn on a warning light if the oil pressure becomes too low.

In the human body, blood must be kept under pressure so that it can be pumped to all parts of the body. The highest pressure occurs just where the blood leaves the heart. In the hands and feet, the pressure is much lower. Within the blood vessels carrying blood away from the heart, the pressure increases and decreases with every heartbeat. Immediately after the heart contracts, a surge of blood causes high pressure. Then, before the next contraction, the pressure falls, only to increase again at the next heartbeat. Doctors measure both the maximum and minimum pressure using a special type of manometer called a *sphygmomanometer*. A normal blood pressure for young adults is about 120/70. The two numbers represent the highest pressure and the lowest pressure in the blood vessel where the measurement is taken. (These numbers give the pressure in an old-fashioned unit of pressure, millimetres of mercury, or *mm Hg*. These units are still used simply because sphygmomanometers have been calibrated that way. Blood pressure could just as easily be expressed in kilopascals: 100 mm Hg = 13.3 kPa.)

Probing

Find out what factors affect a person's blood pressure, and why a pressure that is too high or too low can cause problems. Find out as well why a sphygmomanometer is placed where it is on the arm and how exactly it can indicate blood pressure.

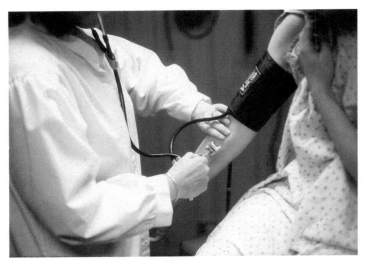

Using a sphygmomanometer to measure blood pressure.

Friction in Fluid Systems

Anytime a fluid flows along in a restricted space, its particles brush against each other and against the container. This friction slows the particles, and the fluid gradually loses its pressure. For example, water that travels through a very long hose may not have enough pressure to operate a sprinkler. City water systems must have extra pumping stations to restore pressure lost on long routes. Similarly, gas and oil pipelines have pumps at intervals along their length.

A compressor station on a gas pipeline—necessary to keep the gas flowing over long distances.

Friction is greater on rough surfaces than on smooth ones. Any irregularity in a pipe's inner wall will slow down the fluid. Rust on iron pipes, mineral deposits on copper pipes, and any changes of direction, all interfere with smooth flow.

Smooth linings and straight pipes help to minimize pressure losses in fluid systems. Where pipes must be bent, designers make the changes in direction as gradual as possible. Keeping the system clean is also important. City waste lines are usually flushed out once a year. Large natural gas pipelines are cleaned out even more frequently by a mobile computerized unit called a "pig." The pig is pushed along by the pressure of the flowing gas. It has brushes to clean the pipe, sensors to check conditions in the pipe, and a means of recording the information. This ensures that customers who depend on natural gas for cooking and heating receive a pure product at a suitable pressure.

Working on the Ocean Floor

Peering a few metres ahead in the beam of light, Verena Tunnicliffe sees what few humans have seen before—the fascinating and unusual animals that cluster around hot water vents on the ocean floor. To do her research, Verena squeezes into a small underwater vehicle, called a submersible, and takes a journey down two kilometres beneath the waves. In this deep-water world, under great pressure and in continual darkness, live animals such as giant tube worms, crabs, clams, and vent fish. As a marine zoologist, Verena wants to discover what adaptations these animals have. "How do they survive under the special conditions—the high pressure and continuous darkness—they face in this environment?" asks Verena.

Verena has always been fascinated by the ocean. Her university studies of geology and biology, and a previous job as a biologist, were good preparation for her present work. The dives in the submersible are part of her job as a teacher and researcher at the University of Victoria.

Planning for a dive takes a lot of time and a lot of care. "It can take up to two years of planning to get one or two dives," says Verena. "Using a research vessel and crew costs thousands of dollars a day, so everyone must know exactly what they will be doing." Several scientists share a research voyage that may last three to four weeks. They take turns going for ten hour dives in the submersible.

As the submersible descends, the sunlight rapidly fades. The researchers can see only what is in their beam of light, a small area about 6 m × 3 m. "It's like exploring the Rockies at night with a flashlight. We could pass within metres of something exciting and never know it." The researchers use more than just their eyes. Sensors on the outside of the vessel measure temperature, sample the water, and analyse its chemistry. Hydrophones pick up sounds. Other devices map peaks and valleys on the ocean floor. A stream of data is displayed on onboard computers. The team has a list of chores and experiments to carry out. They can manipulate rocks using remote-controlled claws, collect samples, and adjust the sensors.

Verena's work is an example of "pure research." The aim of her studies is simply to find out more about the world we live in. "There is nothing quite like experiencing the unknown and trying to understand it. It's a lot of fun, but science should be fun."

Verena Tunnicliffe 2000 m below sea level.

The water in this hot vent 1500 m deep off the British Columbia coast is at 285°C!

Checkpoint

1. Explain the meaning of the following terms: confined fluid, compressible, hydraulic, kilopascal, pressure, barometer, valve.

2. (a) In what way do gases and liquids behave differently when they are under pressure?
   (b) How are they similar?

3. (a) What is pressure?
   (b) In what units is pressure measured?

4. Which of the two fish tanks shown below contains a larger mass of water? Which one exerts the greater pressure on the tabletop? Explain your answer.

5. Refer to diagram (c) on page 92. If the 10 cubes were of sea water (density = 1.03 g/mL) rather than pure water (density = 1.00 g/mL), what would be the pressure at the bottom of all 10 cubes?

6. (a) Make a sketch of a mercury barometer.
   (b) How would a mercury barometer be affected if you took it to the top of a mountain? Explain your answer.
   (c) What change would occur if you took it to the bottom of a deep, open well?

7. (a) What is a manometer?
   (b) Describe how a manometer indicates pressure.
   (c) If the manometer shown on page 98 contained mercury instead of water, would it be more or less sensitive to changes in pressure? (Hint: mercury is much denser than water.)

8. The photograph shows a silo used to store corn. Explain why the horizontal bracing changes as the distance from the top of the silo changes.

9. To study conditions in the upper atmosphere, scientists often send up instruments by attaching them to balloons. If it was your job to fill the balloon with helium before it was sent aloft, how fully would you inflate it? Explain your answer.

A silo.

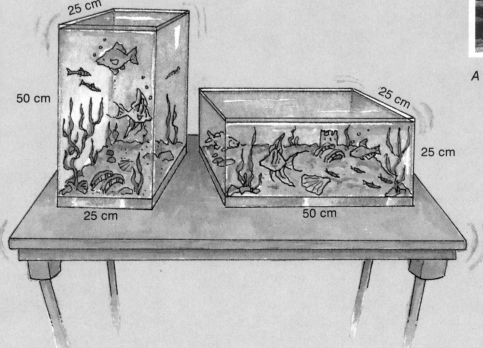

25 cm
50 cm
25 cm
25 cm
25 cm
50 cm

Focus

- Both liquids and gases are fluids.
- The viscosity of liquids can be determined by measuring the rate of flow.
- The shapes of vehicles are designed so that they can move easily through a fluid (air or water).
- Density is a characteristic property of substances.
- Hydrometers are used to determine density of liquids.
- A force puts a confined fluid under pressure.
- Gases are compressible, but liquids are nearly incompressible.
- Confined fluids exert pressure in all directions.
- Mercury barometers, aneroid barometers, and manometers are used to measure pressure.
- The hydraulic press, which makes use of properties of liquids, is widely used in many technological devices.
- A highly developed technology supports a variety of fluid distribution systems.

Backtrack

1. Ten words that are important in this Unit have had their letters scrambled. Rearrange the letters and use each in a sentence that tells something about fluids and pressure.

alaspickol	fowl
cityvoiss	seasg
criblesspoem	squidli
cubyanoy	tydesin
dulraichy	upsreers

2. Make a list of ten fluids that can be found in:
    (a) your science lab;
    (b) your kitchen.
3. (a) What is meant by "viscosity"?
    (b) How can you compare quantitatively the viscosity of two liquids? (Use a diagram in your explanation.)
    (c) Use the particle theory of matter to explain why liquids become less viscous at higher temperatures.
4. (a) Explain the meaning of "buoyant force."
    (b) Use the idea of buoyant force to explain why a cork floats in water, but a marble sinks.
5. Which of the following is more accurate: "Aluminum is lighter than lead," or "Aluminum is less dense than lead"? Explain your answer.
6. (a) What is density?
    (b) Why is it particularly easy to compare the density of other materials to the density of water?
7. Explain why a ship made of steel (density 8 g/cm³) can float on water (density 1 g/cm³).
8. Suppose two kinds of wood are available for you to use to construct a storage box for a motorboat. You find that a block of western red cedar (5 cm × 10 cm × 40 cm) has a mass of 740 g. A piece of birch (5 cm × 10 cm × 20 cm) has a mass of 660 g. You want to use the wood that is less dense; which one would you use? Show your reasoning.

9. A soaker hose has holes along its length. If you plugged half the holes, what effect would this have on the water coming out of the remaining holes? Explain your answer.

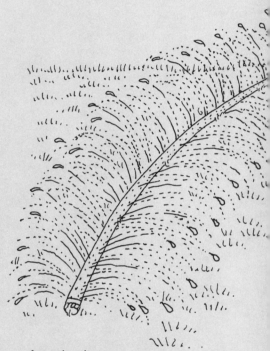

A soaker hose.

10. Describe how the air pressure changes inside a soccer ball when it is kicked.
11. (a) Explain why your body is sensitive to 30 kPa of water pressure at the bottom of a swimming pool, but doesn't notice 101 kPa of air pressure.
    (b) Describe how the water pressure pushes on your body.
    (c) What part of a diver's body might be hurt during a very deep dive? Why is this?

12. (a) Calculate the pressure exerted at the bottom of each of the two imaginary blocks of water shown in the diagram.
(b) What do you think the water pressure would be at the bottom of the 1 m length of pipe? Explain your answer.

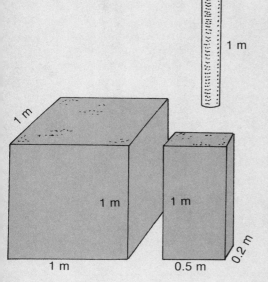

13. If you were designing the plumbing for a new house, what would you need to consider in order to make the water distribution system as efficient as possible?

Synthesizer

14. Skim milk is made by removing the fat from ordinary milk. A hydrometer that just floats in skim milk will sink in ordinary milk. What does this show about the density of milk fat? Explain your answer.

15. Small containers for carrying gasoline are often made of plastic. In what way must a tank for carrying liquid propane be built differently? (Hint: At normal temperature and pressure, propane is a gas.)

16. Would it be possible to drink through a straw on the moon, where there is no atmosphere? Explain.

17. The instructions on cake-mix packages include a special notice for some Alberta residents.
(a) Find out what these special instructions are.
(b) Decide if they apply to the community in which you live.
(c) Why are the special instructions necessary?

18. The illustration on page 108 shows a system for supplying fresh water to homes. Most homes have one or two other systems for moving fluids.
(a) What other fluid systems are common in homes?
(b) Describe how these systems are similar to or different from the water supply system.
(c) For one of the systems you have described, describe additional safety features, or different types of pipes, taps, or valves, that would be necessary.

19. (a) What is the advantage of using mercury rather than water in a barometer?
(b) In a mercury barometer, the column of liquid is 76 cm tall at normal atmospheric pressure. How tall would the column be in a barometer containing water?

20. The block of solid metal shown here measures 10 cm × 10 cm × 10 cm. The block has a mass of 2.7 kg.
(a) What is the density of the block in grams per cubic centimetre?
(b) Refer to Table 2-5 (page 80) and suggest the identity of the metal of which the block is made.

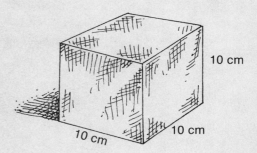

21. A block of iron measures 10 cm × 10 cm × 10 cm.
(a) What is the mass of the block? (Refer to Table 2-5 on page 80.)
(b) What is the weight of the block?
(c) How much pressure is the block exerting on the surface on which it rests?

Controlling Heat

Living organisms can exist only in moderate temperatures that are within a certain narrow range. If the Earth were much hotter or colder than it is, there would be no life on it. People have devised means of living comfortably by controlling heat in various ways. Skiers can enjoy themselves on the coldest days if they wear clothing padded with down feathers from geese or with special synthetic materials. A variety of heaters keeps homes comfortable. In greenhouses, plants thrive in the relatively warm temperatures under the glass.

Avoiding an excess of heat is just as important. For example, the overheating of a car engine may damage it. Designers of car radiators employ simple principles to construct efficient cooling systems. People use many ways to protect themselves from too much heat. On a sunny summer day, simply wearing a hat can help to keep a person from overheating. Where the heat is more intense, a special protective suit may be necessary.

What is heat? How does it transfer from one place to another, and how can you efficiently control it? How can it be stored? What happens to the heat when a hot object cools down? How can we harness the Sun's heat for immediate everyday use? In this Unit, you will find the answers to these questions.

Heat Conduction

You can safely stand near a stove with a red-hot element. You may feel some warmth, but you won't be uncomfortable. With care, you can stir a pot of boiling soup without hurting your hand. But— choose your utensil wisely! Which of the two spoons shown here would you choose for five minutes of continuous stirring?

Why does your hand get hot if you use a metal spoon? Why does the heat from the stove element burn you if you touch it, but not hurt you as you stand nearby? To answer these questions, you need to know how heat is transferred from one place to another. There are three distinctly different types of heat transfer; in this Topic, you will learn about one of them. First, try comparing the transfer of heat in different solids.

Comparing Heat Transfer in Various Solids

Problem

How do different solids compare in their ability to transfer heat?

Materials

safety glasses
glass rod and metal rod of approximately equal size
rods made of various metals (the metals available depend on the manufacturer; usually the metals are joined together at a common point)
2 support stands
2 clamps
wax
heat source: open flame
timing device

> **CAUTION: See Safety in the Classroom on page xii. Follow these rules when using an open flame.**

Procedure

1. Based on your experience with ordinary materials, predict how the following compare in their ability to transfer heat: glass, iron, aluminum, copper, brass, and wood. List these solids in order, from fastest rate of transfer to slowest rate of transfer. If possible, give reasons for your prediction.

2. To compare the transfer of heat by a non-metal (such as glass) to the transfer of heat by a metal (such as steel), use the method shown in the diagram. Place small beads of wax along equal-sized rods made of glass and metal. Support each rod horizontally at one end. Arrange the rods so that their other ends are heated. When all the wax beads on one rod have fully melted, remove the rods and record your observations.

3. To compare the transfer of heat by various metals, use the method shown in the second diagram. Place a bead of wax at the free end of each metal rod. Hold the apparatus so the point where the metals meet is directly above the open flame. Record how much time (in seconds) elapses before each bead fully melts.

4. Answer the questions on the next page.

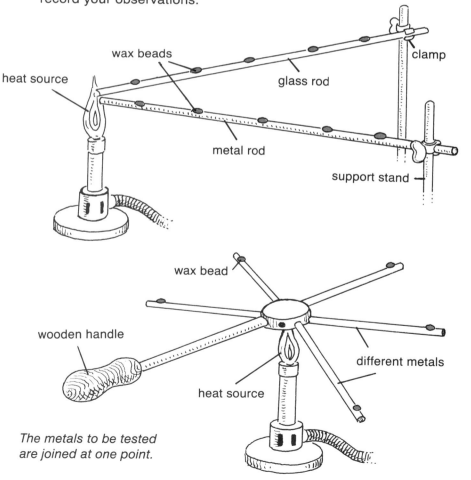

The metals to be tested are joined at one point.

Heat and Temperature

Analysis

1. (a) Did the wax melt more quickly on the metal or on the glass rod?
   (b) Explain what this result tells you about the transfer of heat through metal and glass.
2. List the metals tested according to their ability to transfer heat, from fastest to slowest.

Further Analysis

3. Compare your predictions from Step 1 with your results from Steps 2 and 3. Explain why you were or were not able to make correct predictions.
4. The ability of materials to allow heat transfer is an important consideration in certain products we use in our homes. Describe three products that involve the transfer of heat through a solid.
5. A person baking bread can choose between using a glass pan or a metal pan. Which pan would be more likely to cause the bread to have a firm brown crust on the bottom and sides? Explain.

Extension

6. Describe how you would use a set of several temperature probes connected to a computer to test how quickly various solids transfer heat. If possible, test the methods you have described.

"Heat" and "temperature" are common everyday words. Weather forecasters refer to "heat waves" and "tomorrow's temperature." You are probably familiar with several ways of measuring temperature. You may even refer to a thermometer every winter morning before venturing outdoors. But, to *explain* how heat and temperature work for us and against us, and to understand the difference between them, we need to use the words "heat" and "temperature" with their scientific meanings. In order to do this, you need to recall previous studies of matter and energy.

Of all the many kinds of energy, energy of motion is the most important in understanding heat. According to the particle theory of matter (page 62), particles move more quickly when they have more energy and move more slowly when they have less energy. Knowing this about energy and matter, we can explain the meaning of the words "cold" and "hot." Instead of saying something is "cold," we could say that its particles are moving relatively slowly. When it is "hot," its particles are moving more quickly. In any substance, some particles will be moving more quickly and some will be moving more slowly; that is, some have more energy and some have less. The average energy of the particles in the cold object is less than the average energy of the particles when it is hot.

What's the Difference?

- **Temperature** is a measure of the average energy of the particles that make up a substance.
- **Thermal energy** is the total energy of all the particles in a substance.
- **Heat** is the energy transferred because of a temperature difference. Heat transfers from a substance with particles of greater average energy to a substance with particles of lower average energy.

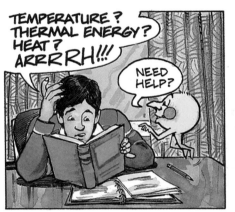

Heat Conduction in Solids and Liquids

Think about a metal spoon used to stir soup as it is heated on a stove. The temperature of the soup is higher than the temperature of the spoon or the temperature of your hand. So heat transfers from the soup to the spoon, from the bowl of the spoon to its handle, and then from the handle to your hand.

According to the particle theory of matter, the solid metal spoon is made of particles that constantly vibrate against each other. When energy from the soup is gained by the particles of metal in the spoon, the metal particles vibrate faster and collide more vigorously. This action continues like a chain reaction from particle to particle through the spoon. Thus, energy transfers from the warmer end of the spoon to the cooler end. The process of transferring energy through a material by direct collisions of the particles is called **conduction** of heat. A material that allows heat to transfer through it readily is called a **heat conductor**. As you have seen, metals tend to be good heat conductors.

So far you've been thinking about heat conduction in the spoon and other solid materials. But what about the soup? You are quite used to heating soup or other liquids in a pan on the stove. Because these liquids do get hot, you could infer that they conduct heat. But sometimes, particularly when you're very hungry, the soup seems to take forever to get hot. Are liquids such as soup good conductors? The particles in a liquid are not in fixed positions—how might this affect whether or not they conduct heat? Make a prediction about conduction in liquids, then do the next Activity.

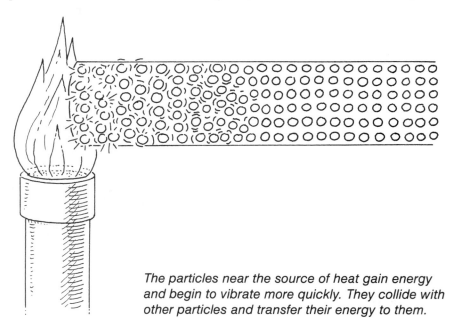

The particles near the source of heat gain energy and begin to vibrate more quickly. They collide with other particles and transfer their energy to them.

Heat Conduction in Liquids

Problem

Is water a good heat conductor?

Materials

safety glasses
large test tube with wax bead
samples of liquid (e.g., water or
    other non-flammable liquids)
support stand and clamp
source of heat: open flame
thermometer

CAUTION: See Safety in
the Classrom on page xii.
Follow these rules when
using an open flame. Heat
the test tube where there is
water; do not heat it above
the water.

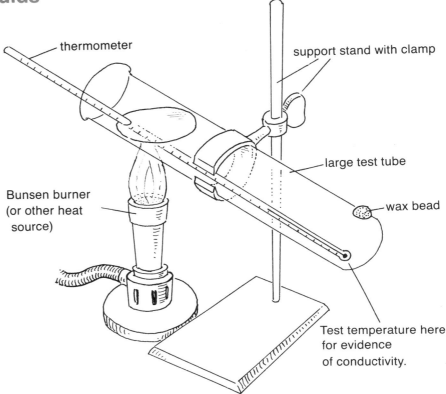

thermometer

support stand with clamp

large test tube

wax bead

Bunsen burner
(or other heat
source)

Test temperature here
for evidence
of conductivity.

Procedure

1. Make a note of your
   prediction about whether
   liquids are good or poor heat
   conductors.
2. Set up the apparatus as
   shown. Apply a flame to the
   top 3 or 4 cm of the water in
   the tube. Do not stir the
   water, and do not apply heat
   directly to the thermometer.
   Take temperature readings
   every minute for about
   10 min. Record your data in
   a table.
3. Repeat Step 2 using a
   different liquid, as suggested
   by your teacher.

Analysis

1. Did your results agree with
   your predictions?
2. Consider both the wax bead
   and the temperature of the
   water at the bottom of the
   test tube. What can you infer
   about the behaviour of the
   particles in the liquid?
3. In terms of heat conduction,
   are liquids more like solid
   metals or solid non-metals?
   Explain your answer.

Further Analysis

4. Suppose you carried out an
   identical experiment using a
   test tube made of metal
   instead of glass. How might
   your results be different?
   Explain your answer.
5. Predict whether gases would
   be good or poor heat
   conductors. Give a reason
   for your prediction.

Thermal Conductivity

"Therm" and "thermal" are used often in words and phrases that relate to heat, as in *therm*opane windows, *therm*ostats, *thermal* underwear, *therm*ometers, *therm*os bottles, and geo*thermal* energy. The ability of a substance to conduct heat is called its **thermal conductivity**. Good conductors, such as metals, have a high thermal conductivity. Poor conductors, such as non-metals, have a low thermal conductivity.

There are various ways to measure the thermal conductivities of substances. You used one method in Activity 3-1: You measured the time it took for heat applied at one end of a solid to reach the other end. Scientists have measured the thermal conductivities of a great variety of substances; they have found that liquids and non-metals have low thermal conductivities compared with metals. They have also discovered that gases have very low thermal conductivities — so low that they could not even be measured using the method you used in Activity 3-1.

By making very careful measurements, scientists have been able to rank the thermal conductivities of many substances. To prepare these rankings, the thermal conductivity of air was taken as a standard for making comparisons. It was assigned a conductivity of 1 (Table 3-1).

Table 3-1 *Thermal Conductivity of Materials Compared to Air**

MATERIAL	CONDUCTIVITY
air	1.0
down from birds	1.1
cork	1.8
human skin	8.7
glass	25
water	26
brick	30
concrete	73
ice	94
iron and steel	2 000
brass	4 600
aluminum	8 600
copper	16 000
silver	18 000

*At about 20°C

Applications of Heat Conduction

Pots and pans are usually made of metal, for efficient transfer of heat from the stove top to the food inside. Copper, one of the metals with the highest conductivity, is used in some of the best quality pots. The handles are usually made of non-metal, so you can safely hold a hot pot.

The tissues of the human body are relatively poor conductors of heat. Since this is so, heat from a heating pad or heat pack can be applied directly to the area of an injury. For stiff muscles, for example, heat causes increased blood flow and is sometimes recommended to speed up healing. In other injuries, the *removal* of heat from a small part of the body may be beneficial. For example, if you sprain your wrist, you could apply something cold, such as ice or a cold pack. This reduces the flow of blood to the area, causing less swelling.

Another useful application of heat conduction, one that has saved many lives, is in a safety lamp once used in mines. About 200 years ago, mines were lit by lamps with open flames. Many miners were killed by explosions when gases in the mine were ignited by the flames. Sir Humphry Davy (1778-1829), a chemist in England, invented a lamp with a wire mesh surrounding the flame. The metal wire conducted the flame's heat away from the flame, so that its temperature was too low to ignite the gases. Today, lamps with batteries are used—an even safer solution.

Wire mesh constantly conducts heat away from the flame.

The Davy lamp.

Conduction is also used to prevent overheating in car and truck engines. The combustion of gasoline in the motor produces a lot of heat. If this waste heat were not removed, the engine parts would expand so much that the engine would be damaged. In the cooling system there is a liquid called antifreeze, which circulates through the engine and then out to the radiator. Heat is conducted first from the engine to the liquid, then from the liquid to the radiator, which is made of a good conductor such as copper or aluminum. Since the radiator is a good conductor, heat is transferred from it to the air that rushes past as the car goes along.

An automobile radiator.

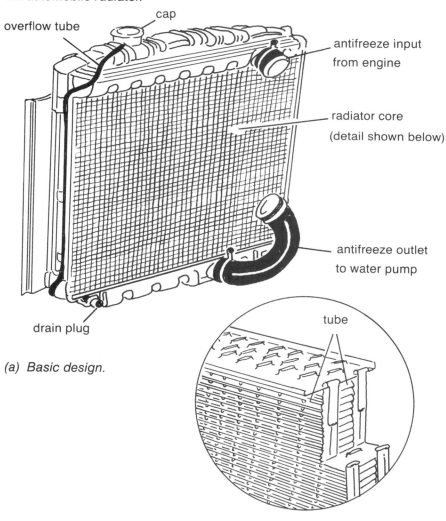

overflow tube

cap

antifreeze input from engine

radiator core (detail shown below)

antifreeze outlet to water pump

drain plug

(a) Basic design.

tube

(b) Detail of one type of radiator construction. As the hot antifreeze passes through the tubes, air rushes past the tubes and takes away heat.

Heat Conduction **127**

Convection of Heat

Liquids are poor conductors of heat, and gases are even poorer. But you know that if you heat a pot of water, all the water heats up, not just the water near the heat source. The heating system in your home warms the whole of your home, not just the patch of air around the heating outlet. There must be another type of heat transfer—one that acts where conduction doesn't, in gases and liquids. In this Topic, you'll see how heat transfer affects practical knowledge of swimming pools, sea breezes, and hang-gliders!

What does this hang-glider need to know about heat transfer?

In the following Activity, there are two alternative procedures suggested for trying to observe heat transfer in air. One involves observations only. The other involves taking measurements, tabulating data, and drawing a graph. The second alternative requires a large group of students.

Heat Transfer in Air

Problem

How can you observe heat transfer through air?

ALTERNATIVE A

Materials

safety glasses
box and chimney apparatus
candle
smoke paper

Procedure

1. Obtain a box similar to the one in the photograph. Place a candle in the holder under the right-hand chimney and light the candle. Predict what will happen when smoking paper is held above this chimney, then above the other chimney.

> CAUTION: See Safety in the Classroom on page xii. Follow these rules when using an open flame.

2. Use the apparatus to test your predictions.
3. Record your observations.

ALTERNATIVE B

Materials

cardboard box
5 thermometers (all of which read the same value at room temperature)
electric light bulb (source of heat)
graph paper
timing device

Alternative A.

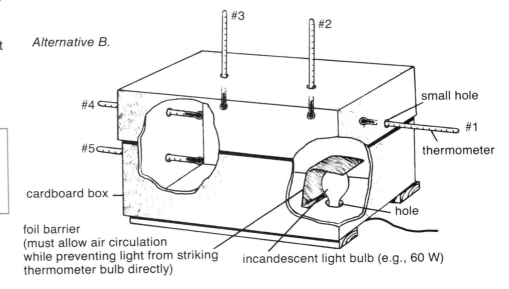

Alternative B.

#3 #2

#4

#5

small hole

#1

thermometer

cardboard box

foil barrier
(must allow air circulation while preventing light from striking thermometer bulb directly)

incandescent light bulb (e.g., 60 W)

hole

Procedure

1. Set up the apparatus as shown in the second diagram. Take the readings on all five thermometers.
2. When everyone reading the thermometers and recording the data is ready, turn on the light bulb and start the timing device.
3. After every minute, take a temperature reading on each thermometer.
4. Record everyone's observations in a data table.
5. Answer the questions on the next page.

Analysis

1. Although air is invisible, in this Activity you observed evidence that air moves in some kind of path.
   (a) Describe the path of the air movement.
   (b) How does this path relate to heat transfer in air?

2. Use the particle theory of matter to explain your observations in this Activity. Hint: Think about
   • the motion of particles;
   • the fact that substances expand when heated; and
   • the fact that less dense substances have more buoyancy than denser substances.

3. (a) If you performed Alternative B, plot all the data on a single temperature-time graph. Be sure to label the five different curves on the graph.
   (b) Describe what the graph indicates about the transfer of heat inside the box.

Further Analysis

4. If you were building a new home with electric heaters in each room, would you place the heaters near the floor or the ceiling? Explain.

5. On cold winter days, people with fireplaces may want to keep a fire burning for several hours. People who do this are advised to allow some fresh air in from outdoors. This is good advice, even if the air from outside is cold. Explain why. (Hint: Wood needs oxygen to burn. Humans also need oxygen.)

6. Do you think a path of moving particles could be set up in liquids? Could one be set up in solids? Why or why not?

Extension

7. (a) If you left the light on in the box, would the temperature inside continue to rise? Explain why or why not.
   (b) If you think the temperature would continue to rise, how high do you think it might go?
   (c) Compare your predictions with other students' predictions. If possible, explore these questions experimentally, and try to explain what you found out.

8. Use temperature probes and a computer to perform Alternative B of this Activity.

Convection Currents

You have seen evidence that air moves in a certain path when it is heated. This movement of air is an example of what is called a *convection current*. The convection current you observed in air appeared to rise upwards, away from the source of heat. We can infer that convection currents can be set up in other gases as well.

Can convection currents also be set up in liquids? If you were a swimming pool designer, you would need to know this. Suppose you have been asked to advise the Sweetwater Town Council about the pool in a new municipal recreation centre. The council members have decided that the water should be heated, but they want to know where the heater should be placed. Would you suggest that it be placed near the top, the middle, or the bottom of the water in the pool? After you do the following Activity, assess whether or not you gave good advice.

Heat Transfer in Water (DEMONSTRATION)

Problem

When heat is applied to one area of a container of water, how does it transfer to the rest of the water in the container? (This problem can be explored in various ways. Two alternatives are shown here, but you may be able to think of others.)

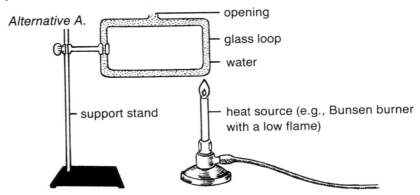

Alternative A.

opening
glass loop
water
support stand
heat source (e.g., Bunsen burner with a low flame)

Materials

safety glasses
cold water
source of heat: open flame
support stand and clamps
glass loop (Alternative A)
food colouring (Alternative A)
large beaker (Alternative B)
coloured chalk dust or ice cubes made with food colouring (Alternative B)

Procedure

> CAUTION: See Safety in the Classroom on page xii. Follow these rules when using an open flame.

ALTERNATIVE A

1. Set up the glass loop apparatus shown in the diagram above. Fill the loop with cold water and let the water stand for several minutes.
2. Add one or two drops of food colouring to the water at the top of the loop, which is open to the atmosphere. Begin gently heating the corner indicated, and record your observations.

ALTERNATIVE B

1. Set up the beaker so one end can be heated, as shown in the second diagram. Add cold water to the beaker. Then add either coloured ice cubes or coloured chalk dust. (You may have to "push" the chalk dust down to overcome the surface tension of the water.)
2. When the water and its contents have become still, turn on the source of heat and observe what happens for several minutes. Record your observations.

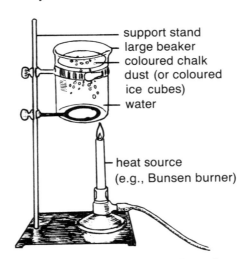

support stand
large beaker
coloured chalk dust (or coloured ice cubes)
water

heat source (e.g., Bunsen burner)

Alternative B. Be sure the flame is off to one side of the beaker.

Analysis

1. (a) Describe any evidence that there are convection currents in water.
   (b) Describe the direction of any convection currents you observed.
   (c) Describe any similarities or differences in the behaviour of air and water when they are heated.

Further Analysis

2. Do you think that convection currents could be set up in solids? Explain your answer in terms of the particle theory.
3. (a) If Sweetwater Town Council decides to place the heater directly in the pool, where in the pool would you install it?
   (b) The council has now decided to put the heater in a shed away from the pool. Design a circulating system to take advantage of convection currents. Draw a diagram showing the heater, a pump that forces the water to move, and the circulation system from the heater to the pool and back again.

Design a four-room, two-storey house that is to be heated by one electric heater. The air temperature throughout the house should be as uniform as possible. Build a model of the house, and use a low-power electric light bulb as the electric heater. Determine whether the design works well by checking the temperature in each part of your model. Describe how you would change the design if you were allowed to have one or two ceiling fans in the house.

Explaining Convection Currents

When a fluid is heated, the part near the source of heat gains energy. The added energy causes a flowing motion in a circulating path. As the fluid moves, it transfers the energy to other areas. The process of transferring heat by the circulating motion of a fluid is called **convection**. This circulating path is what you have observed as convection currents.

To explain why the currents are set up, you need to consider the particle theory of matter. Suppose, for example, you have a room with an electric heater but without a fan. The heater is located by a wall (see diagram). The particles of air near the heater gain energy, requiring more space in which to move and collide. They spread apart, with the result that the heated air becomes less dense than the cooler surrounding air. The warmer, less dense air rises and is replaced by the cooler, more dense air. The air thus begins to move in a circular path, in a convection current.

Convection currents in liquids are similar to convection currents in gases. Although the particles of liquids are much closer together than they are in a gas, they also gain energy, move faster, and require more space. As they spread apart, the liquid becomes less dense, and a convection current begins.

A convection current in a room with a heater along one wall.

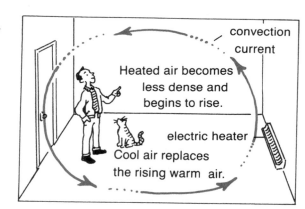

convection current

Heated air becomes less dense and begins to rise.

electric heater

Cool air replaces the rising warm air.

Convection and Conduction

A solid does not flow, so its particles cannot move in a convection current. Thus, we can state (at least for now) that heat transfers by convection in liquids and gases, and heat transfers by conduction in solids, particularly in metals. Both convection and conduction depend on the motion of particles to transfer heat. In convection, the particles move in a current and transfer the energy as they themselves move. In conduction, the particles vibrate back and forth, colliding with each other. These collisions transfer the energy from one particle to the next, but the particles themselves do not flow in a current.

Convection in Nature

Convection currents affect weather patterns near large bodies of water such as oceans or large lakes. These large convection currents are also known as "sea breezes." The diagram shows how a sea breeze arises on a hot, sunny day. The land warms up more quickly than the water. The air above the land warms up as well and becomes less dense than the cooler air above the lake. This cooler air forces the less dense air to rise, producing a convection current, blowing from the water towards the land.

As sunset approaches, the opposite occurs. The land cools down more rapidly than the water, so a "land breeze" flows from the land out over the lake. The water may become quite calm near the shores when a land breeze blows against the incoming waves.

Rising convection currents in the atmosphere are called *thermals*. Thermals generally occur on sunny days in the summer, or in areas such as deserts where the temperature is cold at night but rises quickly in the morning sun. Shortly after direct sunlight strikes the ground in the morning, the ground warms up and warms the air immediately above it. The warmer air expands and becomes less dense. This air is forced to rise by the surrounding cooler, more dense air. As that layer of cooler air warms up, it in turn is forced to rise. These up-drafts, or thermals, may continue until mid- to late afternoon, when the sun's rays become less direct.

A sea breeze.

warm air

cool air

sea breeze

water

land

In ancient days, warring nations would sometimes agree on a time and place to fight a battle. One such battle, the Battle of Salamis, occurred on the coast of Greece more than 2500 years ago, and sea breezes played a major role. The Greek leader, Themistocles, chose to fight the battle at a time when he knew the sea breezes would be greatest. At the agreed time, more than 1200 large Persian ships set out to attack a much smaller number of Greek ships. But the Persians had trouble manipulating their much larger ships in the wind and waves caused by the sea breezes. As a result, the Greeks were able to defeat the Persians and drive them away.

Birds make use of thermals to help them soar and glide. Large birds such as golden eagles may use thermals to help them lift off for flight. Some birds use thermals to conserve energy during migration. Like birds, hang-glider pilots also make use of thermals. By manoeuvering from one thermal to another, they can remain in flight for several hours.

Convection currents are also involved in the force of magnetism that surrounds the earth. Scientists have evidence that convection currents deep within the Earth cause its magnetism. The approximate structure is shown in the simplified diagram. The liquid outer core contains materials such as molten (liquid) iron. If these materials are not all at the same temperature, convection currents can be set up in them. The molten iron contains electrically charged particles that produce magnetism when they move. (This is similar to the way an electromagnet works; you will find out more about this in Unit Four.)

This diagram represents the internal structure of the Earth. Convection currents in its outer core cause the Earth's magnetic field. (Other convection currents in the mantle are responsible for volcanoes and earthquakes.)

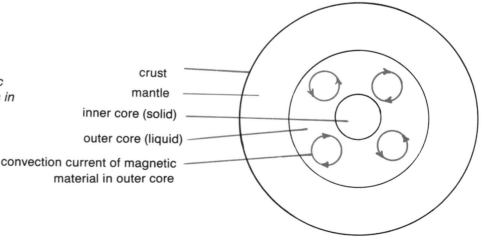

crust

mantle

inner core (solid)

outer core (liquid)

convection current of magnetic material in outer core

1. Explain why each of the following people would be concerned about heat transfer.
   (a) an architect
   (b) a manufacturer of winter boots
   (c) a chef in a restaurant
   (d) a runner on a cold day
2. Use the particle theory to describe the difference between:
   (a) solids and liquids;
   (b) temperature and heat;
   (c) conduction and convection.
3. Some cold water is poured into a cooking pot and the pot is placed on a hot stove. Use the particle theory of matter to explain:
   (a) how the heat is transferred to all the water in the pot;
   (b) why the temperature of the water increases as it is heated.
4. (a) What materials are commonly used to make automobile radiators?
   (b) Why are these materials effective for this purpose?
5. For each of the statements below, explain the principle being put into effect.
   (a) Cooking pots are often made with copper bottoms and stainless steel sides and lids.
   (b) Irons for clothes have plastic handles.
   (c) Inserting a metal skewer into a potato before it is baked will decrease the amount of time required to bake the potato. (Note: Do not try this with a microwave oven.)

6. If air were a good conductor, you would feel cold even on a day when the air temperature is 25°C. Why?
7. Use the particle theory of matter to explain why a dry sponge has a low thermal conductivity.
8. Explain the expression "hot air rises."
9. What are the advantages or disadvantages of placing the freezer compartment at the top of a refrigerator rather than at the bottom?
10. In Activity 3-2, you heated the water in the test tube near the top, to test for conduction. Would this experiment be a fair test for conduction if the water were heated near the bottom? Explain.
11. "Forced-air heating" refers to home heating systems that include both a source of hot air (such as a gas or oil furnace) and a fan to help the air circulate.
    (a) Why do homes with forced-air heating require return vents for cold air as well as hot air vents?

(b) In older homes with forced-air heating, the cold air return vents were located near the floor. In newer homes, the builders often place the return vents near the ceiling. Explain the difference.

Room in an Older Home

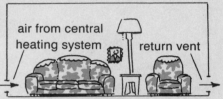

Room in a Newer Home

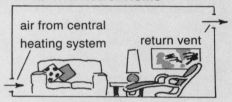

(c) Home heating consultants advise home-owners to use vent covers to direct hot air towards the middle of a room. However, they suggest that these covers be removed in the summer if a central air-conditioning system is turned on. Why?

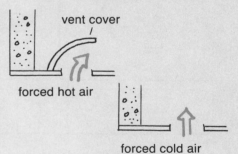

12. Zinc is found to conduct heat 200 times faster than glass. Where would zinc fit into Table 3-1 on page 126?

Radiation

Between the Earth and the Sun there is empty space — there are almost no particles. Yet through this space, this near-vacuum, we do receive heat. If we didn't, there could be no life on Earth. There must be a third way of transferring energy, one that requires no particles. This method is called *radiation;* **radiation** is the transfer of energy in a wave-like form.

Energy transferred by means of radiation is called **radiant energy**. **Heat waves**, or **heat rays**, are a form of radiant energy; they can also be called **infrared radiation**. All the forms of radiant energy — the entire *spectrum* — are shown in the diagram. Of all the forms, radio waves have the lowest energy, and X rays and gamma rays have the highest. All of them occur in nature and can be used in technology. We use radio waves commonly — they are everywhere in the air around us. Higher energy waves, such as ultraviolet light waves, are only used with extreme care. In this Topic, you'll use both infrared radiation (heat) and visible light. Activity 3-5 is an investigation of how well different materials absorb heat rays and visible light.

Radiant energy reaches the Earth from the Sun, through 149 million kilometres of a near vacuum.

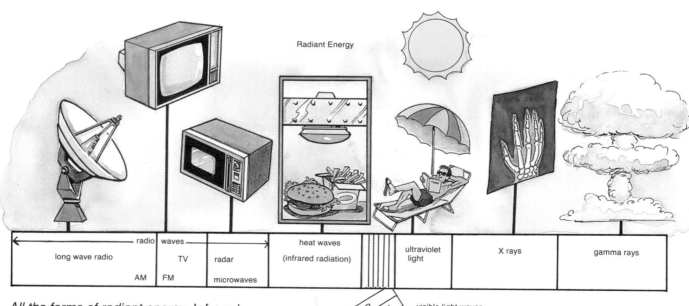

Radiant Energy

	radio waves			heat waves (infrared radiation)		ultraviolet light	X rays	gamma rays
long wave radio		TV	radar					
	AM	FM	microwaves					

visible light waves

RED
ORANGE
YELLOW
GREEN
BLUE
VIOLET

All the forms of radiant energy. Infrared means "below red" (below the red end of the visible spectrum). Ultraviolet means "beyond violet" (beyond the violet end of the visible spectrum).

The Effects of Radiant Energy

Problem

How are different types of materials affected by heat and light?

Materials

2 or more metal containers with contrasting types of surfaces (e.g., one dull and one shiny surface; or one dark and one white surface)
thermometer for each container
stirring rod for each container
200 W lamp or flood lamp
timing device

Procedure

1. Set up the apparatus as shown in the diagram. Add equal quantities of water to the containers. Be sure the lamp is an equal distance from each container.

2. In your notebook, make a table to record the temperature of your samples of water every minute for 15 min.

3. Predict what will happen to the temperature of each sample of water.

4. Record the initial temperature of each sample of water (they should be the same). Then, at the same time, start timing and turn on the lamp. While gently stirring with the stirring rod, measure and record the temperature every minute for 15 min.

5. On a single graph, plot the data recorded in the Activity. Assuming that the points for each sample form a straight or nearly straight line, draw a "best fit" line. (For a review of "best fit" graphs, see Skillbuilder Four, *Graphing*, on page 354.) Use a line of a different colour for each sample, and make a colour key for your graph.

Analysis

1. The lamp emits both visible light and infrared rays. What appears to happen to the light rays
   (a) on reaching a dark surface?
   (b) on reaching a bright shiny surface?

2. What factors(s) affected the amount of radiant energy absorbed?

3. When radiant energy is absorbed by a material, what happens to the average energy of the particles of the material?

4. How does the slope (or steepness) of each line on the graph relate to how quickly each type of material absorbs radiant energy?

5. What evidence indicates whether or not heat rays can travel through a gas?

Extension

6. If each sample of water in the two containers were at the same high temperature and were allowed to cool, which do you predict would cool faster? Why? Perform an investigation to discover if your prediction is correct.

7. Research and report on the following questions.
   • What type of radiant energy causes sunburns?
   • What harmful effects can result from overexposure to this type of radiant energy?
   • How can these harmful effects be reduced while you remain outdoors?

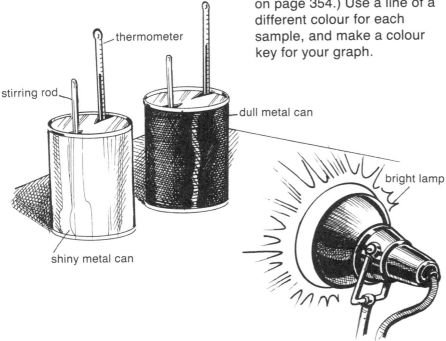

thermometer

stirring rod

dull metal can

shiny metal can

bright lamp

Properties of Radiant Energy

Scientists argued for hundreds of years about radiant energy—what it is and how it can travel in a vacuum. A little more than 100 years ago, a theory was developed that radiant energy is a form of wave with two important parts, an electric part and a magnetic part (that is, it is an *electromagnetic wave*). Once the two-part wave starts moving, the parts keep each other in motion, even in a vacuum, until the wave strikes some matter. Shortly after this theory was proposed, it was used successfully in a technological application—sending a message by radio waves.

The various forms of radiant energy have several properties in common. *First*, waves of radiant energy can travel in a vacuum, such as the vacuum of outer space. Radiant energy does not require particles to transfer heat, as conduction and convection do. *Second*, the waves all travel at an extremely high speed. In a vacuum, the speed of light and other radiant energy is 3×10^8 m/s (that's 300 million metres per second). The speed of light in air is about the same. (This means that light could travel all the way across Canada in about 0.02 s!) *Third*, radiant energy travels in a straight line. This is the reason we can't see around corners—the light rays don't bend. Scientists have discovered that these three properties apply to all kinds of radiant energy.

In this photograph the Sun is the source of light. From what direction is the Sun shining?

Because of the curvature of the Earth, radio signals cannot be sent directly across Canada. Instead, they are sent in a straight line to an orbiting satellite, and then sent in a straight line back to Earth.

Research to learn more about the operation of microwave ovens. Some questions to consider as you are researching are:

- Is the stirrer always on the top of the oven? What would happen if it were at the bottom?
- Is a turntable for the food useful, or is it just an unnecessary gadget?
- Why should a microwave oven be kept clean?
- What are "hot spots" and can you check for "hot spots" in the oven?

All kinds of radiant energy can interact with matter, but the different types of energy interact each in their own way. The energy can be *reflected* (bounced back), *absorbed* (taken in), or *transmitted* (passed through) by matter. For example, light *reflects* from a mirror, is *absorbed* by a black object, and is *transmitted* by a pane of glass. In Activity 3-5, you observed these three kinds of interaction, using heat rays from a lamp.

- Reflection occurs if the radiant energy striking an object cannot get past the particles at the surface. For example, infrared rays are reflected off a shiny smooth surface such as a metal pail. When most of the energy is reflected, there is little rise in the temperature of the object.
- Absorption occurs if the radiant energy penetrates part way into an object, where it can cause the particles to gain energy. When the radiant energy is absorbed, it increases the average energy of the particles and therefore increases the temperature of the object. For example, radiant energy from the Sun is absorbed by a dark, dull surface, such as asphalt pavement, causing the surface to warm up.
- Transmission occurs if the radiant energy can penetrate and pass through an object without being absorbed by the particles. No temperature increase is observed. For example, heat rays are transmitted through air.

All these properties of radiant energy are applied in microwave cooking. Microwaves are *reflected* off metals, so they reflect from the interior of the oven onto the food being cooked. Microwaves are *transmitted* through paper, plastic, and glass without heating them. So the food to be cooked is placed in containers made of one of these materials. But why do the containers remain relatively cool while the food cooks?

Microwaves are *absorbed* by water molecules, and most foods contain water. The waves penetrate almost 4 cm into the food, where they are absorbed. This causes the water molecules to vibrate billions of times each second. The energy of the vibrating water molecules is transferred to the food and the food heats up. If the food sample, such as a roast of meat, is thicker than about 8 cm, the outside will be cooked by radiant energy, and the inside will be cooked by heat conduction. Heat conduction may continue for a short time even after the food is removed from the oven.

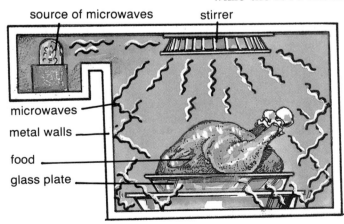

source of microwaves   stirrer

microwaves
metal walls
food
glass plate

A microwave oven. Microwaves are reflected in all directions off the rotating stirrer. (The wavy arrows are used to indicate the wave nature of the radiant energy.)

The amount of heat that radiates away from an animal's body depends partly on the total surface area of the body. A smaller surface area helps to retain heat; a larger surface area helps to radiate heat away. You can see in the illustration how three closely related species of foxes have adapted to their different environments. Small ears reduce the amount of body heat lost to the surroundings, so are an advantage in a cold climate. In a hot climate, large ears are an advantage.

arctic tundra

deciduous forest

desert

Absorbing and Emitting Radiant Energy

Dull, dark objects, which absorb radiant energy readily when they are cool, emit radiant energy readily when they are hot. You may have experienced this if you have ever walked on a black asphalt surface on a hot sunny day. The dark surface absorbs heat from the Sun. You receive both the heat from the Sun *and* the heat re-emitted by the asphalt.

By contrast, light-coloured, shiny objects or surfaces, which do not absorb radiant energy readily, also do not emit radiant energy readily. Bright, shiny tea kettles, for example, emit rays slowly, and thus prevent the water in the kettle from cooling too quickly. In general, a good absorber is a good emitter, and a poor absorber is a poor emitter.

Since black objects absorb heat more readily than light ones, wouldn't black fur help a polar bear absorb radiant energy? Wouldn't the bear be warmer if it had black fur? In fact, the polar bear *is* black—at least its skin is black. The individual hairs are transparent to visible light, but the fur appears white because it scatters much of the visible light that strikes it. (For the same reason, a single crystal of table salt appears clear, although a pile of salt appears white.) The fur has a feature that transfers another type of radiant energy, ultraviolet waves, to the skin, where it is absorbed. Each strand of hair has an outer coating that is transparent to ultraviolet radiation. The rays pass through the outer coating, bounce down the length of the rough shaft inside the hair and are absorbed by the black skin, providing warmth.

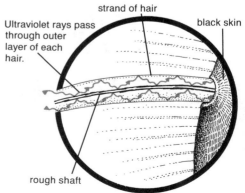

strand of hair

black skin

Ultraviolet rays pass through outer layer of each hair.

rough shaft

Controlling Heat Transfer

In modern society we obtain much of our heat from non-renewable resources such as oil and gas. For the sake of future generations, we shouldn't waste these resources. By controlling heat transfer—making it more efficient where we need it and preventing it where we don't—we can conserve our energy resources. In the illustration, you should be able to identify many situations where heat transfer is being encouraged or inhibited.

Assessing Alternatives

Conservation is just one reason why we should learn to control the transfer of heat. You can probably think of many more. In the next Activities, you'll explore *how* to control heat transfer. These Activities require creativity and a lot of planning on your part. In Activities 3-6 and 3-7, groups of students will try to keep the temperature of water or air as high as possible. Activity 3-8 involves a controlled scientific experiment; different individuals or groups will be responsible for testing different materials for their ability to prevent heat transfer. Read all three Activities before deciding which ones your class will do.

Activity 3-6

Keeping Water Hot

Problem

What are the best ways of preventing heat from transferring away from a sample of hot water? (Do not use commercial devices such as thermos bottles or ice chests.)

Procedure

1. As a group consider how you will try to prevent:
   - heat conduction,
   - heat convection,
   - heat radiation.
2. As you develop your plan, your group should consider:
   - size of the water sample to use,
   - cooling time to allow,
   - amount of time to allow between temperature measurements,
   - safety precautions required,
   - how you will present your results to the rest of the class,
   - what apparatus and materials will be provided at school, and what materials will come from home.
3. Carry out the experiment in one class period. As the cooling occurs, record the temperature and time data in a table.
4. Ilustrate your results on a graph.
5. Present your results to the class.

Analysis

1. Describe which features of your device were best for preventing:
   (a) conduction,
   (b) convection,
   (c) radiation.
2. Describe how you could improve your experimental design.
3. Based on the temperature-time graph of the data, what happens to the rate of cooling as the temperature gets closer to the temperature of the air in the room?
4. How can you apply what you learned in this Activity to the control of heat transfer in
   (a) appliances?
   (b) buildings?
5. Describe three ways you could apply what you have learned to the concept of energy conservation. (For example, how could you reduce the amount of energy needed to maintain a high temperature in a home hot-water tank?)

Extension

6. Perform this Activity in reverse by seeing how long your group can keep a chunk of ice from melting completely.

Activity 3-7

Hot Air in a Box

Problem

What are the best ways to prevent heat transfer from a cardboard box?

Procedure

1. Follow all the Procedure steps as in Activity 3-6. The diagram shows a suitable structure that uses an electric light bulb to heat air to an appropriate temperature (e.g., 40°C). Once the air has reached that temperature, the light can be turned off, and the cooling will begin.

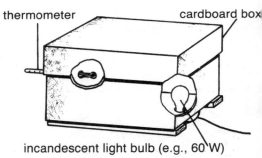

A box like this can be used in both Activities 3-7 and 3-8.

Analysis

Answer the five Analysis questions in Activity 3-6.

Testing the Ability of Materials to Prevent Heat Transfer

Problems

(a) What materials are best for preventing heat transfer?
(b) How does a material's thickness affect heat transfer?

Materials

cardboard boxes of the same size (e.g., the kind of box in which schools receive paper)
thermometers (one per box)
string, tape, and scissors
timing device
materials to be tested (e.g., cloth made of natural fibres such as wool, cotton, and linen; cloth made of synthetic materials such as rayon, nylon, and polyester; newsprint; aluminum foil; commercial insulating materials such as styrofoam and fibreglass)
incandescent light bulbs (one per box)

Procedure

1. As there are many conditions that can be tested, divide the class into groups. Decide which groups will test the different materials available.
2. Cut a hole in the bottom of the box just large enough for the light bulb to fit in. Poke a small hole in the side of the box for the thermometer. As shown in the diagram on page 142, have the thermometer as far away from the light bulb as possible.

3. Turn on the light and allow the air temperature to reach an appropriate value (35°C or 40°C).
4. Switch off the light and, at the same time, start timing.
5. In a table, record the temperature each minute until it has dropped to a predetermined value (such as 25°C). These values are the "control data" for the box used. Other data for the box will be compared to these.
6. Cover the outside of the box with one layer of the material to be tested. Make holes in the material so the light bulb and the thermometer can be in the same position as they were in the control. Repeat the measurement procedure described in Steps 3, 4, and 5. Use the same high and low temperatures. Record the temperature and time data in a table.
7. Choose a material for which you can use one layer, then two layers, then three and four. For each layer of material added, perform the measurements described in Steps 3, 4, and 5. Record the temperature and time data in a table.

Analysis

1. On a single graph, plot the temperature and time data from both Steps 5 and 6. In each case, the *manipulated variable* is the time, and the *responding variable* is the temperature of the air inside the box. There will be two lines on the graph, one for the control data (Step 5) and the other for the data with one layer of the material you were testing (Step 6). Label each line.
2. Compare your graph with the graphs of all the other groups who tested a single layer of a different material. What do you conclude about the relative ability of each material to prevent heat transfer?
3. On a single graph, plot the temperature and time data for the various thicknesses of the material you were testing in Step 7. What do you conclude about the thickness of a material and its ability to prevent heat transfer?
4. How could you apply what you have learned in this Activity to insulating homes?

"Wet suits" are worn by divers to help protect them in cold water. Research and report on the properties of the materials used in wet suits.

Heat Insulators

A **heat insulator** is a material that helps to prevent heat transfer. Some insulators prevent conduction, some prevent convection, some prevent radiation, and some are able to prevent two or even three methods of heat transfer. (A heat insulator can also be called **insulation**.)

A potholder is an effective insulator. It prevents conduction of heat from the hot dish to your fingers.

Insulating homes and other buildings is an important way to conserve valuable energy. Proper insulation helps keep heat inside during the winter and outside during the summer. As you did in Activity 3-8, scientists experiment to determine which materials make good insulators. In their more complex experiments, they measure the temperature both inside and outside the material being tested. They also measure the surface area and the exact thickness of each sample tested. Then they use temperature and time data to give each type of material a number called its **RSI value**. The "*R*" means resistance to heat conduction, or resistance value, and the "*SI*" means that the numbers used in the calculation are based on the metric system, *Système International*. The *RSI* values are usually stated for each centimetre of material (Table 3-2).

Installing insulation.

Table 3-2 *RSI Values*

MATERIAL	RSI/cm
polyurethane	0.38
polystyrene (blue)	0.35
polystyrene (white)	0.29
cellufibre (newsprint)	0.28
fibreglass	0.24
vermiculite	0.16
still air	0.15*
plywood	0.087
gypsum board	0.035
glass	0.017
brick	0.014
aluminum	0.000 05

* The *RSI* value of still air rises to about 0.18 when the air space is 1.5 cm thick. The *RSI* value remains at 0.18 as the thickness increases to about 10 cm. If the air space becomes larger than 10 cm, the *RSI* value drops because convection currents begin.

Waterfowl such as ducks, geese, and swans have soft feathers called *down* on their under-bodies. Down differs from other feathers in that it is lighter and much fluffier. A small mass can therefore fill a large volume of space. Every gram of the best quality down has about 70 000 fluffy filaments that overlap and interlock to form a protective layer of dead-air space.

Information about the *RSI* value of various building materials is important to people working in the construction industry. They need to know which materials are good insulators. And, what is just as important, they need to know which materials are poor insulators. They might be able to substitute alternative material for a poor insulator. They might use the poor insulator in combination with a good insulator. Or, they might find ways to use it that will improve its *RSI* value.

As an example of this third possibility, consider the insulating ability or resistance value of a window with a single pane of glass, 3 mm (0.3 cm) thick. The *RSI* value of this glass is found in the following way.

$$\frac{RSI \text{ value}}{\text{cm}} \times \text{thickness in cm}$$

$$\frac{RSI\,0.017}{\text{cm}} \times 0.3\text{ cm} = RSI\,0.0051$$

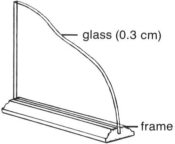

Single-glazed Window

glass (0.3 cm)

frame

Now consider the total resistance value of a double-glazed window, one with two panes of glass separated by a 1.0 cm air space. (In other words, air is trapped between the panes of glass.) The *RSI* value of the glass and air combined is calculated in the following way.

Photograph of a single-glazed and a double-glazed window.

		RSI
Glass (0.3 cm)	=	0.0051
Air (1 cm)	=	0.15
Glass (0.3 cm)	=	0.0051
Total:		0.1602

The total *RSI* value is about 0.16.

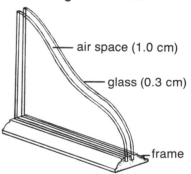

Double-glazed Window

air space (1.0 cm)

glass (0.3 cm)

frame

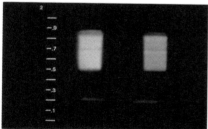

Thermograph of the same two windows taken at the same time. Light blue indicates the greatest heat loss, dark blue shows less heat loss, and black shows very little heat loss.

You can see that the double-glazed window provides more than 30 times the insulating ability of a single-glazed window. Although glass itself is a poor insulator, a properly constructed window is a good insulator. These calculations apply only if the air space is still. A still air space is commonly called a ''dead-air'' space. If the air space were larger, convection currents could transfer heat readily. Dead-air spaces provide insulation not only in double-glazed (and triple-glazed) windows, but also in clothing such as scarves, gloves, and thermal underwear, and in down-filled objects such as sleeping bags, bed covers, and ski jackets.

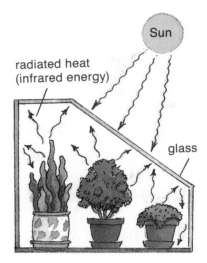

radiated heat (infrared energy)

Sun

glass

The greenhouse effect.

The Greenhouse Effect

Heat insulators control heat transfer mainly by preventing conduction and convection of heat. The transfer of radiant energy can also be partially controlled. In the **greenhouse effect**, there is a warming of the temperature of the air because of the trapping of radiant heat. Radiant energy from the Sun is composed mainly of waves that are easily transmitted through glass. Plants and other objects in the greenhouse absorb the radiant energy, and this causes an increase in their temperature. At the higher temperature, the plants and other objects themselves emit radiant energy as infrared rays. This type of radiant energy has less energy (see illustration at left) and tends to be reflected from the glass rather than passing through it. The energy cannot escape; it becomes trapped inside the greenhouse, keeping it relatively warm even in winter.

The greenhouse effect occurs in the Earth's atmosphere as well. A portion of the light energy that strikes the Earth is absorbed and re-emitted as infrared energy. There is no ceiling of glass to trap this energy, but there are carbon dioxide molecules in the atmosphere. These molecules absorb the energy and re-radiate it, so that much comes back to Earth.

Carbon dioxide is a natural part of Earth's atmosphere. However, its concentration is increasing, at least partly because of human activities such as the burning of fossil fuels. Increasing carbon dioxide concentration means increasing absorption and re-radiation of radiant energy. This is expected to increase the average temperature of the Earth. Even a minor increase could have a serious effect on the climate everywhere on our planet.

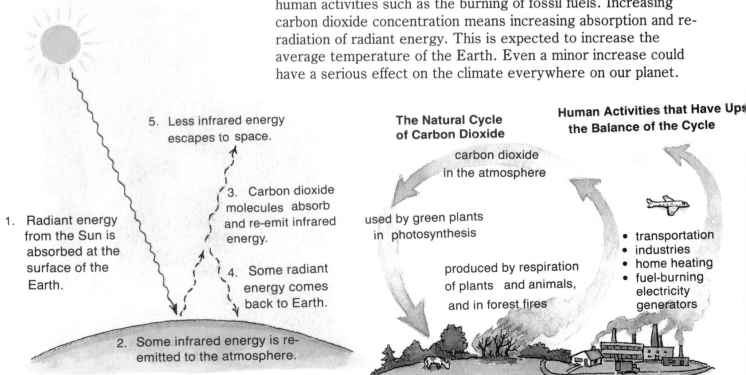

5. Less infrared energy escapes to space.

3. Carbon dioxide molecules absorb and re-emit infrared energy.

1. Radiant energy from the Sun is absorbed at the surface of the Earth.

4. Some radiant energy comes back to Earth.

2. Some infrared energy is re-emitted to the atmosphere.

The Natural Cycle of Carbon Dioxide

Human Activities that Have Upset the Balance of the Cycle

carbon dioxide in the atmosphere

used by green plants in photosynthesis

produced by respiration of plants and animals, and in forest fires

- transportation
- industries
- home heating
- fuel-burning electricity generators

(a) The greenhouse effect in Earth's atmosphere.

(b) Carbon dioxide build-up in the atmosphere.

1. List three properties of radiant energy.
2. Describe the type of surface that tends to:
   (a) absorb infrared energy well;
   (b) reflect infrared energy well;
   (c) emit infrared energy well.
3. It is said that Benjamin Franklin, a famous American scientist, placed white cloths and black cloths of equal size near each other on snow one sunny day. Describe what you think happened to the snow beneath each piece of cloth.
4. In the spring and fall, why is frost less likely to occur on a cloudy night than on a clear night? (Hint: Consider the interaction of infrared radiation and water molecules.)
5. If you were to use aluminum foil to keep a baked potato hot, should you have the dull side or the shiny side inside, next to the potato? Explain.
6. Suppose you plan to buy a "heat lamp" to install in a bathroom. In one area of the store, you discover infrared lamps and ultraviolet lamps.
   (a) Which type of lamp is the "heat lamp" you are there to buy?
   (b) What is the purpose of the other lamp?
   (c) What are the dangers associated with each type of radiation?
7. (a) What is meant by the term "energy conservation"?
   (b) Why is it important?
   (c) How do insulators help in energy conservation?
8. Explain why down is an excellent insulator.

9. A thermos bottle can keep cold liquids cold and warm liquids warm. Use the information in the diagram to describe how a thermos minimizes conduction, convection, and radiation.

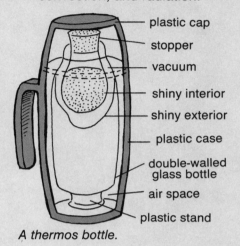

plastic cap
stopper
vacuum
shiny interior
shiny exterior
plastic case
double-walled glass bottle
air space
plastic stand

A thermos bottle.

10. The photograph shows an astronaut in space, where temperatures are extremely cold. Describe ways in which the astronaut is protected from the cold.
11. Use the data in Table 3-2 on page 144 to determine the *RSI* value of:
    (a) 2.0 cm of plywood,
    (b) 5.0 cm of fibreglass,
    (c) 8.0 cm of brick.
12. Calculate the total *RSI* value of a triple-glazed window. Each air space is 1.0 cm wide and each pane of glass is 0.3 cm thick.
13. Calculate the total *RSI* value of a ceiling consisting of 1.0 cm of gypsum board and 18 cm of fibreglass insulation.

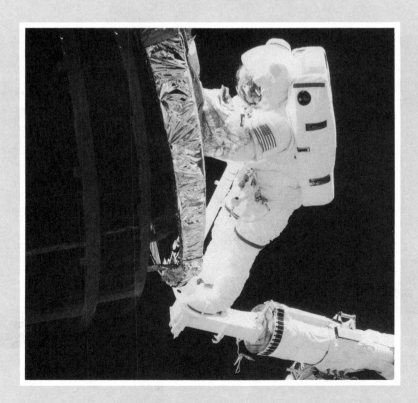

The R-2000 Home

In Canada's cold climate, we must have heat for our homes. But when we burn fuels, we add carbon dioxide to the atmosphere As well, when we burn oil or gas, we are using up a non-renewable resource (see page 157). We should therefore try to conserve energy so less fuel is burned.

With this in mind, the federal government has a program to encourage the use of improved insulation in homes and other buildings. The program is designed to ensure that, by the year 2000, all buildings being constructed have the best insulation possible. The program is called R-2000: R refers to the resistance of heat transfer, and 2000 represents the target year.

The R-2000 program is the biggest of its kind in the world. The program began when Canada's first super energy-efficient house was completed in 1977. This house is located in Regina, Saskatchewan, where the winters tend to be long and cold. In the first year of operation, the entire space-heating bill for the house was only $50.

Following the success of the Saskatchewan house, several other energy-efficient houses were designed and built in other provinces. The energy consumed per home was about 70% lower than the energy consumed by similar homes without the R-2000 technology. The extra cost needed to improve the homes was soon paid for by the savings in energy costs. By the end of the 1980s, the R-2000 program was fully in place in all provinces.

There are six main features of an R-2000 home. These features promote the efficient use of energy and the availability of clean, fresh air in the home.

1. *Insulation and Wall Construction*
R-2000 homes have two to three times more insulation in the basement, roof, and exterior walls than conventional homes do.

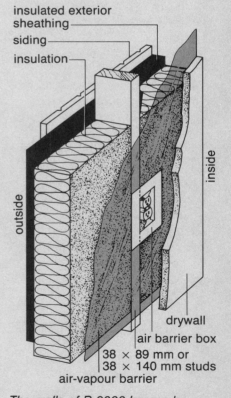

The walls of R-2000 homes have excellent insulating properties. Even electric outlets are well insulated.

The Saskatchewan Conservation House was the first super energy-efficient house built in Canada.

2. *Reduced Air Leakage*

R-2000 homes have a sealed air-vapour barrier to reduce air leakage and the build-up of moisture. The barrier is made of a plastic (such as polyethylene) and is placed on the inside of the wall. This prevents drafts, and in winter it helps prevent moisture from inside the house condensing on the outside wall.

3. *Ventilation and Heat Recovery*

R-2000 homes are designed to provide healthy air indoors every day, all day long. Each house is equipped with a continuous ventilation system to bring fresh air from outdoors. Some homes also have a method of transferring heat from the incoming fresh air. The moisture content of the air is controlled as well.

4. *Windows and Doors*

Windows are either double-glazed or triple-glazed. Doors are made from good insulating materials, or a double-door system is used. All windows and doors have tight weather seals.

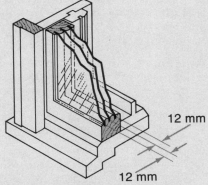

12 mm

12 mm

Triple-glazed windows provide excellent insulation; there are tight seals along the edges, and dead-air spaces between the panes of glass.

5. *Direction of Construction*

Whenever possible, an R-2000 home faces towards the Sun (east-west, with southern exposure) and uses a special design to take advantage of solar energy (see page 158).

6. *Appliances*

Each R-2000 home has modern appliances and lighting designed to consume as little energy as possible. Water heaters have extra insulation to prevent heat loss.

Think About It

What are some features of R-2000 homes that are being used in building construction in your area?

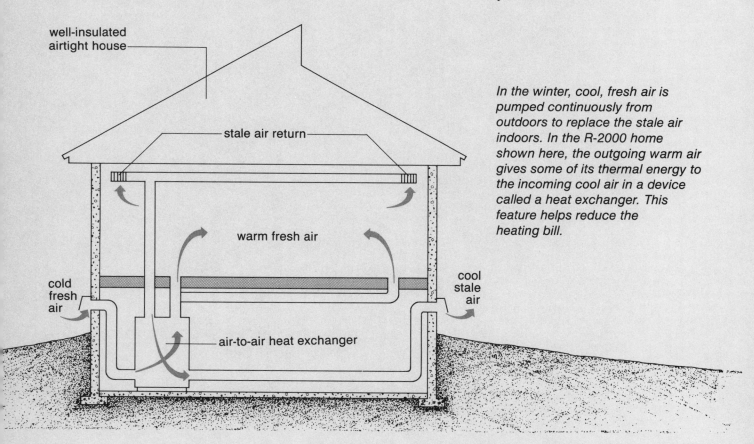

well-insulated airtight house

stale air return

warm fresh air

cold fresh air

cool stale air

air-to-air heat exchanger

In the winter, cool, fresh air is pumped continuously from outdoors to replace the stale air indoors. In the R-2000 home shown here, the outgoing warm air gives some of its thermal energy to the incoming cool air in a device called a heat exchanger. This feature helps reduce the heating bill.

Heat Capacity

Below the map of Canada on this page, the graph shows the average January temperature in several cities. You might think that Edmonton is so cold because it is the farthest north, but why are Regina, Winnipeg, and Quebec City so cold? What makes Victoria, Toronto, and Halifax so much warmer in winter?

One thing these warmer cities have in common is that they lie near large bodies of water. As the weather gets colder in the fall and winter, the water takes a longer time to cool down than the land does. Or to put it another way, water has the capacity to retain thermal energy for a longer time. The large bodies of water keep the nearby cities relatively warm.

Numerous experiments have shown that water has a high capacity to hold heat. *Any substance that has a high capacity to hold heat also requires a large amount of heat to raise its temperature.* How do other substances compare with water in this regard? You can explore the answer to this question by heating various liquids.

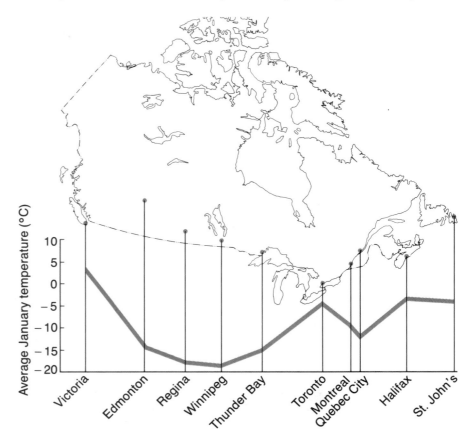

Heating Liquids

Problem

How do various liquids compare in the amount of heat needed for identical increases in temperature?

Materials

safety glasses
various liquids (e.g., water, vegetable oil, peanut oil, other non-flammable liquids)
hot plate (or other source of constant heat)
Pyrex glass beakers (one for each liquid)
thermometer
stirring rod
timing device
balance
tongs or gloves

CAUTION: Avoid spilling any liquid on the heater. Use tongs or gloves to hold a hot container.

Procedure

1. Turn on the hot plate so that it can heat up and reach a constant temperature while you are carrying out the next step.
2. In your notebook, draw a table to record data. Use two columns for each liquid, one to record time every 30 s, and the other to record temperature, starting at 30°C. (That is, you will consider time to be 0 s when the temperature is 30°C.)

3. Use a balance to measure 100 g of the liquid to be tested.
4. Place the beaker and liquid on the heater. As you stir with the stirring rod, observe the temperature of the liquid. As soon as the temperature reaches 30°C, start taking readings of the temperature every 30 s. Continue until the temperature reaches 70°C, then carefully remove the beaker from the heat source.
5. Repeat Steps 3 and 4 for each of the other liquids.
6. On a single graph, plot the temperature-time line for each of the liquids, using a different coloured line for each liquid. Label each line or make a key for your graph.

Analysis

1. Which liquid took the longest time to heat to 70°C?
2. List the liquids in order of the time it took each to heat from 30°C to 70°C.

Further Analysis

3. Which liquid required the largest amount of heat to increase its temperature from 30°C to 70°C? (That is, which liquid had the highest capacity to hold heat?)
4. List the liquids used in this Activity in order from greatest to least capacity to hold heat.
5. In what ways was this Activity a controlled experiment?

The Special Case of Water

When you bite into a hot apple pie, which is more likely to burn your tongue—the crust or the filling? When the pie is removed from the oven, the crust and the filling are at the same temperature, but the filling remains hot for a longer time. Even the bottom crust cools off more quickly than the filling. The filling contains more water, so it cools more slowly—this is one example of the effect of high heat capacity.

Because of its high capacity to hold heat, water is an ideal substance to use in a number of practical situations. For example, a small amount of water can absorb a fairly large amount of energy without undergoing a large temperature change. Because of this property, water is often used in the cooling systems of engines and other devices that produce heat from friction. Heat must be removed to prevent damage. Water's high heat capacity allows it to remove a large amount of heat. Because only a small amount of water is required, the cooling system does not need to be excessively bulky.

Probing

Predict your results in the following experiment. Suppose you use a single constant heat source (such as a hot plate that has been on for several minutes) to heat different masses of water. Each sample—100g, 150g, and 200g—is in an identical beaker and is originally at 20°C. You heat each sample until it reaches 60°C and record how much time you need for each. Which sample will reach 60°C most quickly? *How much* more time will be required for the slowest than for the fastest? If you have your teacher's permission, try this and make a graph of your results.

Specific Heat Capacities

The data you obtained in Activity 3-9 gave you a general comparison of the heat capacity of various substances. The **heat capacity** of a substance is either the amount of heat required to raise its temperature by 1.0°C, or the amount of heat released when it cools by 1.0°C. In other words: Heat capacity is the amount of heat transferred when the temperature of a substance changes by 1.0°C. Heat is measured in joules (J). (In the *Système International*, all forms of energy are stated in joules.)

In order to make comparisons of the heat capacities of various substances, an equal mass of each substance must be used. You did this in Activity 3-9, using 100 g of each type of liquid tested. The standard mass commonly used to compare heat capacities of substances is a mass of 1.0 kg. The **specific heat capacity** of a substance is the amount of heat transferred when the temperature of 1.0 kg of the substance changes by 1.0°C. (The temperature change can be either an increase or a decrease.) In mathematical form, this is written:

$$\text{specific heat capacity} = \frac{\text{energy}}{\text{mass} \cdot \text{temperature change}}$$

Table 3-3 *Specific Heat Capacities of Common Substances at 25°C*

SUBSTANCE	SPECIFIC HEAT CAPACITY [J/(kg·°C)]
hydrogen gas	14 400
helium gas	5 300
water	4 200
concrete	3 000
ethanol	2 500
ethylene glycol	2 200
ice (at 0°C)	2 100
steam (at 100°C)	2 100
vegetable oil	2 000
air	995
aluminum	920
glass	840
sand	790
iron	450
copper	390
brass	380
silver	240
lead	130

When a pie is removed from the oven, it loses heat and cools down; the kitchen gains heat and warms up.

The experimentally determined value for the specific heat capacity of water is 4200 J/(kg·°C). This means that it takes 4200 J of energy to increase the temperature of 1.0 kg of water by 1.0°C. It also means that if the temperature of 1.0 kg of water drops by 1.0°C, it has released 4200 J of energy.

If you know the amount of energy used, the mass of the substance, and the change in temperature, you can calculate the specific heat capacity of any substance. For example, suppose in an experiment like Activity 3-9 your heater gives 8000 J of energy to warm the vegetable oil. You used a mass of 0.1 kg (100 g) and changed the temperature by 40°C (from 30°C to 70°C). So, to calculate the specific heat capacity:

$$\text{specific heat capacity of vegetable oil} = \frac{8000\,J}{0.1\,kg \cdot 40°C}$$

$$= 2000\,J/(kg \cdot °C)$$

You probably found that water has a higher specific heat capacity than the other liquids you tested. In fact, water has a higher specific heat capacity than most substances. The specific heat capacities of several common substances are listed in Table 3-3.

Heat Transfer

The equation you used to calculate specific heat capacity can be rearranged to express energy in terms of the other quantities.

energy = mass · specific heat capacity · temperature change

This equation is a short, mathematical way of saying that the amount of energy required to heat a sample of a substance depends on the mass of the sample, the type of substance, and the temperature change. The amount of energy released when a substance cools depends on the same three factors.

You know from earlier studies that scientists have concluded that *energy is never created or destroyed* in any physical or chemical change. This is such an accepted idea that it is stated as a law—the **Law of Conservation of Energy**. In countless investigations using both simple and sophisticated apparatus, scientists have found that heat never just "disappears"; it is transferred from one substance to another. Of course, in experiments in science classrooms, this idea is difficult to prove without error. When samples of hot and cold water are mixed, for example, most of the energy of the hot sample is transferred to the cold water, but some is transferred to the thermometer, the stirring rod, the container, and the air surrounding the container.

Scientists state this as the **Principle of Heat Transfer**.

In a perfectly insulated system, when two substances at different temperatures are mixed, the heat released by the hotter substance equals the heat gained by the colder substance.

In mathematical terms:

$$\text{heat released} = \text{heat gained}$$

This equation is useful in calculating the specific heat capacity of substances. Metals, for example, would be difficult to use in an experiment like Activity 3-9, but you can determine their specific heat capacity as described in the following problem.

SAMPLE PROBLEM

Suppose you put a piece of metal with a mass of 500 g into boiling water for several minutes. Then you use tongs to transfer the metal very quickly into 200 g of water at 20°C. The final temperature of both water and metal is 30°C.

(a) How many joules of energy were absorbed by the water as it was heated?

(b) How many joules of energy were released by the metal as it cooled?

(c) What is the specific heat of the metal?

The solutions to these questions are on the next page.

When a piece of hot metal is put into cool water, the heat released by the metal is gained by the water.

CALCULATION

(a) mass of water = 200 g = 0.2 kg
specific heat capacity of water = 4200 J/(kg·°C)
change in temperature = 30°C − 20°C = 10°C
energy = mass · specific heat capacity · temperature change

$$= 0.2 \text{ kg} \times \frac{4200 \text{ J}}{\text{kg} \cdot °C} \times 10°C$$

$$= 8400 \text{ J}$$

Energy absorbed by the water = 8400 J

(b) According to the Principle of Heat Transfer,

heat released = heat gained

therefore, energy released by the hot metal = 8400 J

(c) mass of metal = 500 g = 0.5 kg
The boiling point of water is 100°C, so the initial temperature of the metal was 100°C.
change in temperature = 100°C − 30°C = 70°C

$$\text{specific heat capacity} = \frac{\text{energy}}{\text{mass} \cdot \text{temperature change}}$$

$$= \frac{8400 \text{ J}}{0.5 \text{ kg} \cdot 70°C}$$

$$= \frac{8400 \text{ J}}{35 \text{ kg} \cdot °C}$$

$$= 240 \frac{\text{J}}{\text{kg} \cdot °C}$$

The specific heat capacity of the metal is 240 J/(kg·°C). Comparing this value with the values in Table 3-3 on page 153, you can see that the metal is likely to be silver.

Using Symbols in Equations

In mathematics, symbols are often used to simplify calculations. For example:

c represents specific heat capacity in J/(kg·°C).
E represents energy in J.
m represents mass in kg.
ΔT (pronounced "delta T") represents the amount of change in temperature, in °C. (The Greek letter Δ is used to indicate "change in," and T represents temperature in °C.)

The equation for specific heat capacity above becomes:

$$c = \frac{E}{m\Delta T}$$

Rearranged to solve for energy, it becomes:

$$E = mc\Delta T$$

If the subscript $_M$ represents "metal," the energy released by a hot metal is:

$$E_M = m_M c_M \Delta T_M$$

Similarly, the energy gained by the water is:

$$E_W = m_W c_W \Delta T_W$$

Using the Principle of Heat Transfer, the combined equation is:

$$m_M c_M \Delta T_M = m_W c_W \Delta T_W$$

If you know any five of the six quantities, you can use the equation to find the unknown one.

In the next Activity, you can apply these concepts to calculate the specific heat capacity of some metals. The calculations are similar to those carried out by scientists when they test a new material. For example, a researcher might be trying to develop a new metal alloy (a mixture of metals) with a low specific heat capacity to be used as shielding on a spacecraft. The researcher would apply the principle of heat transfer to determine the specific heat capacity of the new alloy.

The specific heat capacity is an important property of metals used in spacecraft.

Specific Heat Capacity

Use your knowledge of specific heat capacity to solve the following questions.

1. Suppose you have 2.5 kg of mercury at 22°C. You find that it requires 3200 J of energy to raise its temperature to 32°C. Calculate the specific heat capacity of mercury.

2. To raise the temperature of 400 g of gold from 50°C to 75°C, 1.3 kJ of energy is needed. What is the specific heat capacity of gold?

3. To cool 400 g of a solid substance having an initial temperature of 85°C, you put it into 200 g of water at 15°C.

(a) If, immediately afterwards, the temperature of the mixture is 35°C, determine the specific heat capacity of the unknown substance.

(b) Use Table 3-3 on page 153 to suggest the identity of this substance.

Solar Heating

Until half a million years ago, the only energy resources people used were food and sunlight. Once they learned to make fires, they were also able to produce heat from the chemical energy stored in wood. For only a few hundred years, humans have been obtaining heat from fossil fuels such as coal, oil, and gas. These last three are in limited supply on Earth. They are *non-renewable* energy resources, and they are being used up. In the future, our society will have to depend more on *renewable* energy resources, such as **hydroelectricity** (the electricity generated from the force of moving water) and **solar energy** (the radiant energy that reaches Earth from the Sun). In this Topic, you will design and build a model home heated by solar energy.

Solar heating technology involves conduction, convection, radiation, and heat capacity. In your model solar home you will be able to make use of everything you have learned about controlling heat. The information in this Topic about different types of solar heating will help you with your final design.

Green plants use solar energy to carry out photosynthesis. But animals, as well as plants, have found ways to use radiant energy from the Sun. Termites, for example, build their mounds so that the broad side faces east. Because of this, a large area of the mound receives the direct rays of the Sun early in the morning; the mound heats up quickly after a cold night. The narrow sides of the mound face north and south. This is an advantage in the extreme heat at noon, as then only a small area of the mound receives direct rays from the Sun.

Passive Solar Heating

Solar energy can be used to heat buildings and swimming pools directly. One way to do this is *passive solar heating*. The word "passive" means that the system simply lets the solar energy in and prevents much heat from getting out. Passive solar heating is popular because it is not expensive and it is easy to maintain. In fact, most houses have always used some form of passive heating, for example, by allowing sunlight to shine in windows.

The diagram shows the basic features of a home with passive solar heating. The large windows face towards the south to take advantage of solar energy in the winter. These large windows have wide overhangs to prevent the Sun's rays from entering in the summer, when the Sun is much higher in the sky. Deciduous trees are located on the south side of the home. These trees lose their leaves in winter, allowing the Sun's rays to enter the house, but provide shade in the summer. Another feature is the use of some material, such as rock or cinder block, that absorbs radiant energy during the day and emits it later. (Remember this feature. It will make a big difference to the model solar home you design later.) Still other features include window shutters that are closed at night to help prevent heat loss, as well as the features discussed in the special section on page 148 about the R-2000 home.

Probing

Design and build a solar cooker to be used on a warm, sunny day. Test it by trying to cook a hot dog or a marshmallow.

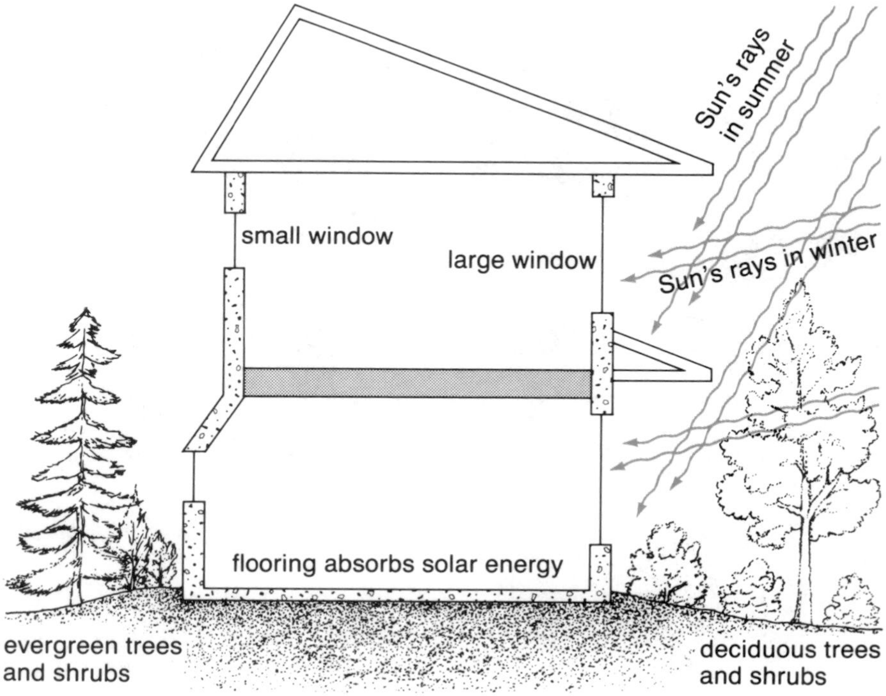

small window

large window

Sun's rays in summer

Sun's rays in winter

flooring absorbs solar energy

evergreen trees and shrubs

deciduous trees and shrubs

A home with passive solar heating.

Active Solar Heating

A second way to heat buildings directly is called *active solar heating*. The word "active" means that the heating system is designed so that as much solar energy as possible is absorbed by a material, usually a liquid, which is then distributed throughout the building. The diagram shows a typical active solar heating system, with a solar collector on the roof. Notice the pumps in the diagram. These pumps need electricity to operate. An active solar heating system usually requires another source of energy besides the Sun.

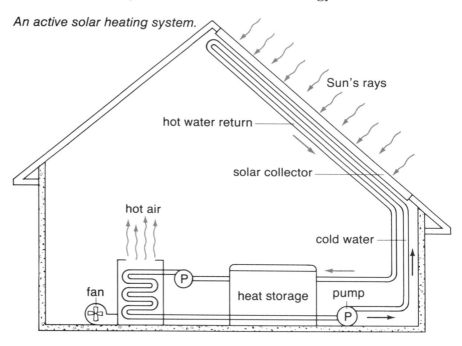

An active solar heating system.

Sun's rays

hot water return

solar collector

hot air

cold water

fan

P

heat storage

pump

P

There are several types of solar collectors in active solar heating systems. The system shown in the diagram uses a flat collector. This is placed on a slanted roof facing south, to absorb as much solar energy as possible. The energy passes through a clear plastic or glass covering and is absorbed by a dark-coloured collector plate as shown in the diagram. The insulation under the collector plate helps prevent loss of heat. In this example, the solar energy heats a liquid, which then circulates in a convection current with the help of pumps.

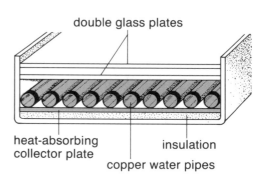

double glass plates

heat-absorbing collector plate

insulation

copper water pipes

A flat solar collector.

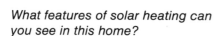

What features of solar heating can you see in this home?

Another type of solar collector uses a highly reflective curved surface. When the solar energy strikes the collector, it reflects towards a central point. In some collectors of this type, there is an automatic tracking system that keeps it moving all day, so that it always faces the Sun.

What liquid do you think is suitable for active solar heating systems? Water is cheap and readily available, and water is an excellent liquid for storing thermal energy because it has a high specific heat capacity. However, water freezes when the temperature drops below 0°C, so antifreeze (a liquid called ethylene glycol) is added to provide a useful liquid mixture. The mixture has a lower specific heat capacity than water, but it is still capable of storing a great deal of thermal energy.

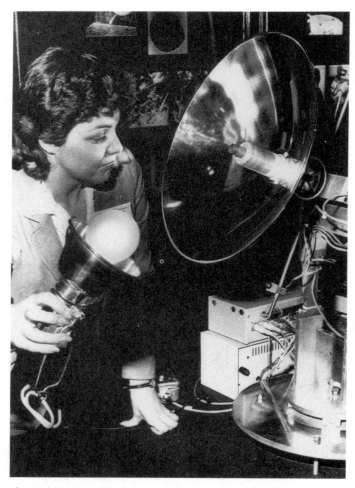

A working model of a curved solar collector, which automatically tracks a bright light source. This model, built by Canadian university students, was displayed at an international environment conference in Hamilton, Ontario.

Building a Model Solar Home

Problem
Design and build a model home that uses solar energy.

Procedure
1. While developing a plan, your group should consider these questions.
   - What size should the model home be?
   - What types or thicknesses of materials should be used?
   - How should the homes be evaluated?
   - Should the home have movable parts, for example, windows, or shutters that close, or walls that could be moved?
   - Should heating abilities be evaluated separately from insulating abilities? (This is important. A home that has a temperature increase of only 4°C will have a slow rate of cooling, so it might appear to have good insulation.)
   - What alternative testing procedure can be used if there are several days without sunshine when you want to test the homes?
   - Should the model use passive solar heating only, active solar heating only, or both?
   - What other questions should you consider?

2. Discuss with the other members of your group how you are going to build the model home. The diagram may give you a few useful ideas. Decide on the tasks for each member of the group.

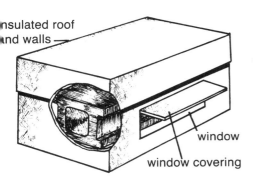

Starting a model home that uses passive solar heating.

3. Design and build the model home. To test it, you could allow sunlight to strike the home directly for a predetermined amount of time and find the maximum temperature inside. Then you could remove the home from sunlight and continue to take temperature readings.

Analysis

1. Discuss the advantages and disadvantages of the various model homes built and tested.
2. What special considerations would be important in a real solar-heated home in your own region? Assume no other sources of heat are allowed.

Extension

3. Describe and evaluate solar heating in your area.
   (a) Find out how many hours of sunshine there are on an average day in summer and winter.
   (b) Describe how climate affects the type of solar heating suitable for buildings in your area.
4. Describe how solar heating could be used in two or more of the following places: apartment buildings, schools, hospitals, barns, industrial buildings, and swimming pools.

5. There is much research and development being done in solar technology. Use a suitable reference to find out about any one particular project. You should consider the following questions.
   (a) Who is financing the research and development?
   (b) Where is the project located?
   (c) What are the advantages of this location?
   (d) How is the solar energy being stored or used?

At this solar-powered generating plant in southern California, enough electricity is produced during daylight hours to supply about 1500 homes. Sunlight, reflected by 1818 racks of mirrors, heats water in a boiler atop a 70 m tall tower.

Working with Heat

Alan Nursall is an atmospheric scientist at a science centre (Science North in Sudbury, Ontario). His job involves teaching, entertaining, and acting as a scientific authority. Besides supplying weather forecasts to the local TV station, he develops exhibits and presents special programs to groups of visitors. As a meteorologist, he is interested in the atmosphere, its motion, its composition, its effects on living and non-living things, its history, and its future. Meteorologists do much more than produce weather forecasts; they might study hurricanes, chase tornadoes, build computer models of the atmosphere, monitor acid rain, or study ice in Antarctica for clues to climate in the past.

How did Alan become interested in the career he chose? "In my first year at university I had no strong sense of where I wanted to head, other than something in the sciences. I took a few courses in geography in my second year, and they introduced me to climatology. I was intrigued by the physics of the atmosphere and how all the pieces of this incredibly complex puzzle, the clouds, wind, rain, sky colour, and so on, all fit together so neatly. If air rises somewhere, then it has to sink somewhere else. As it rises, it cools; as it sinks, it heats up again. It was this fascination with how the atmosphere worked that sent me into climatology. I concentrated on this first, but focused on meteorology in my Master of Science degree. That allowed me to pick up the atmospheric physics that I had missed earlier."

What advice does Alan give to students considering various career choices? "Try to incorporate into your career those activities you truly enjoy doing. You might as well choose something you'll look forward to each day. If you're interested in a career in science, you'll never regret taking math. A good understanding of algebra, calculus and geometry are important to a scientific career, especially in the computer age."

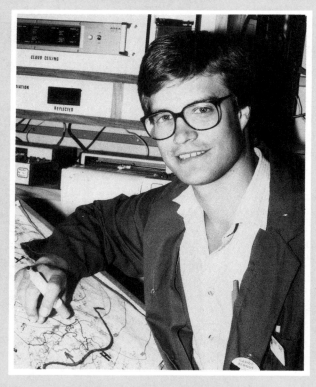

1. A baked potato (with unbroken skin) and some meat of equal mass are cooked in the same oven. When they are removed, which food cools more slowly? Explain why.

2. The graph shows the results of an investigation in which 1.0 kg of three different liquids, A, B, and C, are heated. Which liquid has the highest capacity to hold heat? Explain.

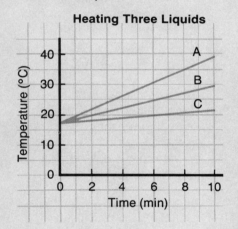

Heating Three Liquids

3. Suppose you are at a beach on a clear, hot day.
   (a) Which becomes warm faster during the day, the sand or the water?
   (b) Which cools down more rapidly at night?
   (c) Explain your answers by referring to Table 3-3 on page 153.

   Answers to the question on page 157:
   Non-renewable: coal, oil, natural gas. Renewable: food, water power (hydroelectricity), Sun (solar energy), wood, wind.

4. For each example below, calculate the amount of energy needed to raise its temperature from 20°C to 80°C (see Table 3-3 on page 153).
   (a) 3.5 kg of helium gas
   (b) 600 g of iron

5. Which type of substance would likely make a better heat conductor, one whose specific heat capacity was low or high? Explain.

6. Apply the Principle of Heat Transfer to determine the mass of the substance in each case below. Show your calculations.
   (a) When 140 g of water at 95°C are mixed with an unknown mass of water at 25°C, the final temperature of the mixture is 45°C.
   (b) An unknown mass of hot ethylene glycol at 83°C is mixed with 210 g of the same liquid at 20°C. The final temperature of the mixture is 48°C.
   (c) An engineer wants to determine the mass of water needed to cool each kilogram (1.0 kg) of hot iron to a temperature of 60°C. The initial temperature of the iron is 1250°C, and the initial temperature of the water is 10°C. (Refer to Table 3-3 on page 153 for specific heat capacities.)

7. List several advantages of passive solar heating in homes.

8. How can the greenhouse effect (described on page 146) be applied in the use of passive solar heating?

9. How are convection currents set up in the liquid in an active solar heating system?

10. Examine the photograph showing panels of solar collectors. Describe features that allow them to use radiant energy from the Sun. Mention features you can see here, as well as others that could be used with these panels.

Focus

- The particle theory of matter is used to explain many aspects of heat.
- Temperature, heat, and thermal energy are distinct concepts, each having its own definition in scientific terms.
- Heat can be transferred by conduction, convection, and radiation.
- Conduction of heat occurs when the particles in one part of a solid (or liquid) collide with the particles near them, passing the energy along.
- Transfer of heat by convection occurs when the particles of a liquid or gas move in a path called a convection current, carrying energy with them.
- Energy that can be transferred without particles is called radiant energy.
- Controlling the transfer of heat is important for comfort, efficiency, and conservation of energy.
- The specific heat capacity of a substance is the amount of heat transferred when the temperature of 1.0 kg of the substance changes by 1.0°C.
- The Principle of Heat Transfer is an example of the Law of Conservation of Energy.
- Solar energy provides heat in passive solar heating systems and in active solar heating systems.

Backtrack

1. The pieces of a set of dominoes are arranged in a row in a standing position. When the first domino is tipped over, it hits the next one, and so on until the entire row is knocked over. For which method of heat transfer might this be used as a simple scientific model? Explain.

2. Name the most efficient method of heat transfer through each of the following:
   (a) a metal,
   (b) a vacuum,
   (c) a liquid.
3. State which method of heat transfer
   (a) does not require particles;
   (b) works because particles interact with their neighbours;
   (c) travels at the speed of light;
   (d) works when particles circulate in a path.

4. Describe the motion of particles in
   (a) conduction.
   (b) convection.
5. Explain why wearing tight jeans on a cold day does not provide good insulation.
6. A manufacturer claims that a certain insulating material is useful "to keep the cold out." Is this expression scientifically accurate? Explain.
7. A down jacket temporarily loses some of its insulating ability when it becomes wet.
   (a) Explain why this occurs.
   (b) Explain why puffing up the down after it has dried will restore its insulating ability.
8. You enter a car that has been parked in the sunlight on a cold winter day. The air in the car is much warmer than the air outside the car. Explain why.
9. What does the word "specific" refer to in specific heat capacity?
10. Determine the specific heat capacity of each of the following, and identify what substance each could be made of. (Refer to Table 3-3 on page 153.)
    (a) It takes 1920 J of energy to increase the temperature of a 100 g spoon from 20°C to 100°C.
    (b) An empty container of mass 400 g releases 50 400 J of energy as it cools from 170°C to 20°C.

11. Calculate the heat released or gained in each case below. (Specific heat capacities are given in Table 3-3 on page 153.)
(a) A sample of 5.0 kg of ethanol is warmed from 15°C to 25°C.
(b) A 3.0 kg copper plate is cooled from 100°C to 60°C.
(c) A 500 g chunk of iron is cooled from 80°C to 0°C.

12. Describe the main differences between active solar heating and passive solar heating.

Synthesizer

13. The graph shows the results of an experiment in which 1.0 kg samples of three different liquids, A, B, and C, were heated with an immersion heater.
(a) Which liquid required the greatest heat input for a certain temperature change?
(b) Which liquid has the highest specific heat capacity?
(c) Use Table 3-3 on page 153 to suggest the identity of each liquid.

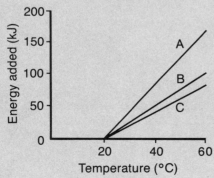

Heating 1.0 kg Samples of Three Liquids

14. People generally associate snow with cold temperatures. However, snow can act as an excellent insulator.
(a) Explain why snow is a good insulator.
(b) Describe how the insulating properties of snow are useful for plants, animals, and people in Canada's northern regions.

15. How does thermal underwear help prevent the loss of body heat?

16. Describe how a cold-weather survival suit, shown in the photograph, prevents the loss of heat.

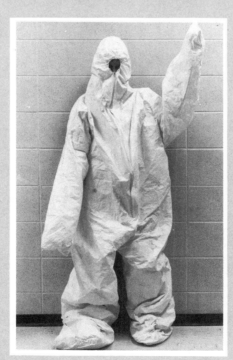

A cold-weather survival suit.

17. Use concepts learned in this Unit to explain why manufacturers of televisions, radios, and other appliances print on them a caution stating that an air space should be left around each appliance when in operation.

18. What methods have been used in your own home to help prevent the loss of heat in winter? What could be done to improve the insulating properties of your home?

19. (a) How do outdoor swimming pools use a lot of energy?
(b) Describe features you would include if you were designing a cover for an outdoor swimming pool.

20. Why should our society be concerned about conserving energy?

21. In your notebook, make a table with three columns, with headings "Metal," "Conduction Time," and "Thermal Conductivity." List iron, aluminum, and copper in the first column. Fill in the second column with your data from Activity 3-1. In the third column, copy the appropriate data from Table 3-1 on page 126.
(a) Compare the conduction times and the thermal conductivities of the three metals.
(b) Explain how this comparison indicates whether or not your results agree with the scientists' results.

Using Electricity

Suppose this morning you had a glass of milk, an egg, and some slices of toast for breakfast. As you got ready for school, you listened to some music on a portable radio. Later in the day, you used a hand-held calculator to do your math homework, then watched a show on television.

These small events are examples of only a few of the ways you use electricity on any day. An electric refrigerator cooled your milk, and an electric toaster browned your bread. Batteries powered the portable radio and calculator. The broadcast of your favourite show came from an electrically operated television.

In the technology of today's world, few things are as vital as electricity. As you can see, you depend on electricity for a great many activities. If you have ever experienced a power failure in your neighbourhood, you know just how much you rely on it!

In this Unit, you will investigate how electrical energy is produced, put to use, measured, and safely controlled. You will accomplish practical tasks by building electrical devices and experimenting with them.

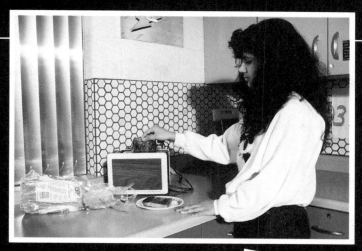

Electricity — Static and Current

If you have ever received a shock by touching something after walking on a rug, or had freshly laundered clothing cling to your body, you have had a first-hand experience of electricity. Long before scientists understood electricity or put it to use, they were aware of this form of electricity.

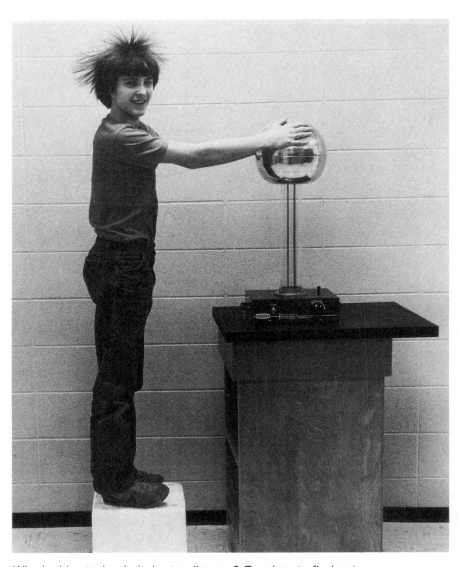

Why is this student's hair standing up? Read on to find out.

In 600 B.C., a Greek philosopher named Thales wrote about a hard, yellowish material called amber. People found this material in dead trees. He described how amber, when rubbed with fur, would attract hair, feathers, and other small objects. Since the Greek word for amber is *elektron*, modern words having to do with electricity stem from Thales' discovery.

Electric Forces and Static Electricity

The forces caused by the build-up of electricity are called **electric forces**. These forces are responsible for lightning, nature's most spectacular show of electricity.

As you study the electric forces that cause electricity, you will notice a certain word being used repeatedly. This word is *charge*. An object acquires static electricity when it has an excess of **electric charge**. In other words, **static electricity** is the build-up of electric charge on an object.

An object having an electric charge can be discharged in ways you have likely experienced. For example, when you walk across a rug on a dry day, you gain an electric charge. Later, when you touch something and feel a shock, you are being discharged. Thus, an **electric discharge** is the removal of an electric charge from an object. Large electric discharges occur during thunderstorms. The movement of drops of water in clouds causes the build-up of electric charges. The sudden discharges appear as flashes of lightning.

How can an object acquire an electric charge? Is there a single type of charge, or are there several? How do electric charges on different objects affect one another? You will investigate these questions in Activity 4-1.

Lightning is caused by electric forces.

Electric Charges

Problem

What are the properties of electric charges?

Materials

support stand
right-angle clamp and rod
string and straightened paper clip
2 strips of vinyl
2 strips of clear cellulose acetate
wool cloth (or fur)
cotton cloth (or rayon)
plastic comb and pen

Procedure

1. Prepare a table similar to Table 4-1. As you read each step, be sure to predict each result and to record this prediction in the table. (There are three possible results: attraction, repulsion, or no effect.)
2. Suspend the vinyl strip as shown. Hold the suspended strip near the top, and rub the whole strip with wool. Rub the second vinyl strip with wool, and hold this strip close to one end of the suspended strip. Do this a few times. Decrease the distance between the strips and under *Comments* note what happens when you do so.

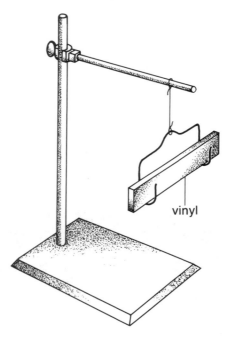

vinyl

3. For this step, assume that the charge acquired by the vinyl is called *negative* and the charge acquired by the wool is called *positive*. Rub both ends of the suspended vinyl with wool. Now hold the part of the wool that you just used for rubbing close to one end of the vinyl. Under *Comments*, note what happens.
4. Recharge the suspended vinyl with wool. Now rub an acetate strip with cotton, and hold the strip close to the suspended, negatively charged vinyl. Note the effect.
5. Suspend an acetate strip from the stand, and rub both ends with a cotton cloth. Rub a second acetate strip with cotton and hold this strip close to the suspended one. Note the effect.

Table 4-1

PROCEDURE STEP	OBJECT SUSPENDED	OBJECT IN HAND	PREDICTED RESULTS (ATTRACTION, REPULSION, OR NO EFFECT)	COMMENTS
2	vinyl	vinyl		
3	vinyl	wool		
4	vinyl	acetate		
5	acetate	acetate		
6(a)	vinyl	comb		
6(b)	vinyl	pen		
6(c)	vinyl	?		

6. Suspend the vinyl strip once again from the stand. Charge the strip negatively, so that it can be used to test other charges. Determine the size (small, medium, or large) and type (positive or negative) of charge on the following:
(a) a comb rubbed through your hair;
(b) a plastic pen rubbed on a piece of clothing;
(c) various other objects rubbed on a material of your choice.

Analysis

1. What evidence from this Activity suggests that
(a) two like charges repel each other?
(b) two opposite charges attract each other?

Further Analysis

2. State the type of charge that results on each of the following:
(a) vinyl rubbed with wool;
(b) wool used to rub vinyl;
(c) acetate rubbed with cotton;
(d) cotton used to rub acetate.

A Theory of Electricity

Your observations from Activity 4-1 can be combined with the atomic theory (see *Inside the Atom*, pp. 16-18) to develop a theory of electricity.

You have observed that there are two types of charge. An American, Benjamin Franklin (1706-1790), introduced the idea of calling them *negative* and *positive*. He named the type of charge on amber rubbed with fur "negative," and we still use that term today. It is the same type of charge as that acquired by vinyl when it is rubbed with fur or wool. The opposite type of charge, which is on the wool, is called "positive." An object with no charge is said to be *neutral*.

When two different types of matter are brought into close contact, one may lose electrons to the other, as shown in the diagram below.

You can see the effect of this loss of electrons when you comb your hair. If the material your comb is made of has a stronger attraction for electrons than does your hair, your comb will take electrons from your hair. Your comb thus becomes negatively charged (it has excess electrons), and your hair becomes positively charged (with a shortage of electrons). As shown below, vinyl and wool behave in a similar way.

Before going any further, you should know the three "laws" of electric charges:

1. *Opposite charges attract each other.*
2. *Similar (like) charges repel each other.*
3. *Charged objects attract neutral objects.*

(a) Before rubbing, both objects are neutral.

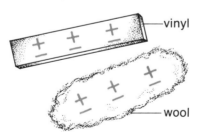

(b) Rubbing causes electrons to transfer from the wool to the vinyl.

(c) The negatively charged vinyl and positively charged wool attract each other.

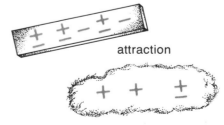

attraction

Static electricity has numerous industrial applications. *Electrostatic spray painting* is one of them. The object to be painted has been given a positive charge. As the paint leaves the nozzle, it becomes charged. The resulting attraction not only prevents the wasting of paint but also helps bond the paint and create a smoother surface.

Electrostatic spray painting.

Current Electricity

In Activity 4-1, the electric effects you observed were due to static charges; that is to say, the electricity produced did not move. Once electrons transferred from one object to another, the charges thus produced were "at rest."

Now consider the form of electricity that operates a portable radio and calculator. **Conduction** of electricity through the wires that are in contact with the batteries inside those two devices is responsible for their operation. The electrons do not remain stationary after a charge has been produced, but move from one place to another. **Current electricity** is electricity in which the charged particles move.

Energy from another source is needed to produce energy in the form of an electric current. As you will see later in this Unit, there are numerous devices that convert other forms of energy into electrical energy. They are most commonly chemical batteries or generators, but can also be light-activated solar cells, heat-sensitive thermocouples, and even sound-operated crystal microphones.

In carrying out the Activities that use these devices and others, remember that safety is crucial. Provided that you follow the instructions, all Activities in this unit are electrically safe. *Never experiment freely with appliances and devices at home.* Electric currents in the household can be extremely dangerous, even fatal.

Detecting Current

You can find out if there is current in a wire by using a simple "current detector." This device also allows you to compare the strengths of different electric currents.

Making a Current Detector

Problem

How can you make a device that will detect and compare currents?

Materials

compass
2 m of insulated copper wire (approximately 26 gauge)
piece of emery cloth (5 cm x 5 cm)
tape
D-size cell in holder
small lamp with 3 V bulb
2 connecting clips

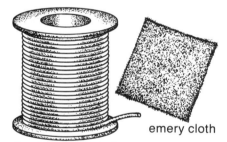

2 m fine insulated copper wire

compass

tape

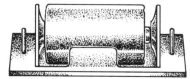

1.5 V cell in holder

Procedure

1. Using the emery cloth, remove about 2 cm of insulation from each end of the wire. Make a connecting wire by attaching a connecting clip to each end.

2. Place a section of the wire *over* the needle of the compass as shown below, so that both the wire and needle are aligned in a north-south direction. Tape the wire in place.

3. *Briefly* connect each end of the wire to a terminal of the cell until the needle comes to rest. Observe the position of the needle and then disconnect the wire from the cell.

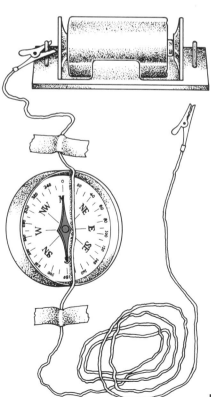

4. Sketch or note how the wire and needle looked
   (a) before you connected the cell.
   (b) after you connected the cell.

5. Repeat Steps 3 and 4, reversing the connections of the ends of the wire to the cell.

6. Place the wire *under* the compass as shown below, but keep it aimed in a north-south direction. Repeat Steps 3 to 5, then continue with Step 7 on the next page.

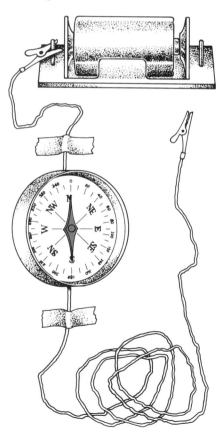

Electricity — Static and Current **173**

7. Wrap a single turn of wire around the compass as shown below, lining up the coil of wire in a north-south direction.

8. *Briefly* connect the cell to the wire.

9. Sketch or note how much the needle is deflected from its original position.

10. Reverse the connections of the end of the wire to the cell and repeat Steps 8 and 9.

11. Wrapping first two, then five, then ten turns of wire around the compass, repeat Steps 8 and 9.

12. Cut the wire, bare the two new ends, and connect them as shown to a light bulb and a battery.

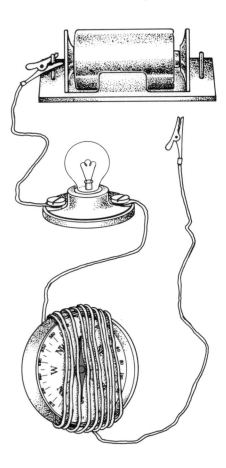

Analysis

1. What happened to the direction of the compass needle deflection when the wire was moved from over the compass to under it?

2. Did the deflection of the compass needle increase, decrease, or stay the same when

   (a) more turns of wire were wrapped around the compass?

   (b) a light bulb was inserted?

Further Analysis

3. What evidence suggests that current in a wire produces a magnetic effect around the wire?

4. How can this magnetic effect be increased?

Extension

5. Make a permanent version of your current detector.

6. How can you modify it to make it

   (a) more sensitive, so you can detect weaker currents?

   (b) less sensitive, so you can detect stronger currents?

Measuring Current

A *galvanometer* is normally used to measure weak electric currents. An *ammeter* is generally used to measure stronger currents. The electric current that is measured consists of moving electrons. The greater the number of electrons that move past a point in one second, the greater is the current. The unit by which current is measured is called the **ampere**. Its SI metric symbol is A.

Examine a galvanometer or an ammeter closely. How are they similar to the "current detector" you built in Activity 4-2? How are they different?

A commercial galvanometer.

The galvanometer gets its name from an Italian scientist, Luigi Galvani. Galvani made the first recorded discovery of electric current. He attached a brass hook to the spinal cord of a dead frog laid on an iron plate. When he pressed the other end of the hook to the iron plate, the frog's legs "jumped."

Luigi Galvani (1737–1798) experimenting with electricity.

Cells and Batteries

Where does electricity come from? In Canada, most current is produced by hydroelectric generators. The energy from great quantities of water flowing downhill is used to produce electricity. As well, electricity can be produced from coal- or oil-fired generators. Although generators are the most common source of electricity, there are others. For example, two different kinds of metal strips stuck into a fruit or vegetable can provide enough electricity to operate a clock like the one shown here. Similar materials are used to operate the many gadgets we use each day — from flashlights to wristwatches. You will investigate this common source of electricity through Activities that follow.

An Electric Lemon

Problem

What causes a current, and what determines its strength?

Materials

fresh lemon
sampling of other fruits and
    vegetables
2 large paper clips
2 copper wires (15 cm each)
galvanometer
2 connecting wires
250 mL beaker

Procedure

1. Straighten out a paper clip.
2. Insert both the paper clip and the copper wire to a depth of about 3 cm into the lemon, keeping them about 1 cm apart.

3. Connect a galvanometer as shown. Observe and record how many divisions the needle moves on the scale.
4. Making the following substitutions, repeat Steps 2 and 3. Use:
    (a) a second copper wire instead of the paper clip;
    (b) a second paper clip instead of the copper wire;
    (c) different fruits or vegetables instead of the lemon; and
    (d) tap water in a beaker instead of the lemon.

Analysis

1. According to the reading on the galvanometer, which combination of materials produced
    (a) the strongest current?
    (b) the weakest current?

Further Analysis

2. What materials are essential to produce an electric current?
3. Would this apparatus work using a raisin? Why or why not?

Extension

4. Use two lemons in the Procedure above, and connect them in different ways. Compare your galvanometer readings to those obtained earlier.

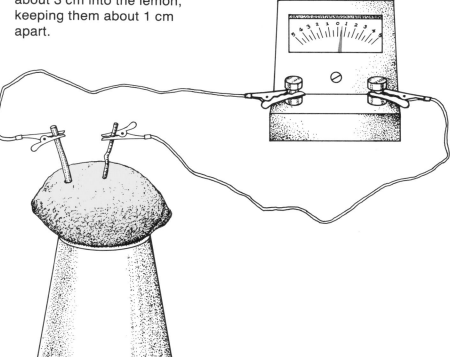

Voltage of Cells and Batteries

The lemon that you hooked up in Activity 4-3 was acting as an *electric cell*—a chemical device for producing electric current. In other words, a **cell** is a device that converts chemical energy into electrical energy.

The "batteries" you buy for a portable radio or hand-held calculator are, strictly speaking, not batteries but cells. Two or more cells joined together to produce an electric current constitute a **battery**. If you have to use more than one cell in the portable radio or calculator, then you have created a battery.

Examine the markings on any commercially manufactured cell, such as the D-size cell you used in Activity 4-2. You will notice that the cell is described as a 1.5 V cell. A **volt** (**V**) is a unit in which you can measure the energy of charges delivered by a cell. The more *voltage* a cell has, the more energy is supplied to the electrons that leave the cell to produce an electric current. The voltage in a battery will equal the sum of the voltages in the cells that make up the battery. For example, a typical flashlight runs on a set of two D-size 1.5 V cells. The voltage of the battery within the flashlight is therefore 3 V.

To measure the voltage of a battery, you use a *voltmeter*. It looks very much like a galvanometer, but it is designed to measure voltage rather than current. When you connect a voltmeter to a commercial cell, the positive terminal of the cell is connected to the positive post on the voltmeter; the negative terminal of the cell is connected to the negative post. Assuming the cell is fully charged, the reading on the voltmeter should be the same as the voltage indicated on the label of the cell.

In the next Activity, you will design and build your own cell. In doing so, you will find out what factors affect its voltage.

Inside this 9 V battery are six 1.5 V cells joined together.

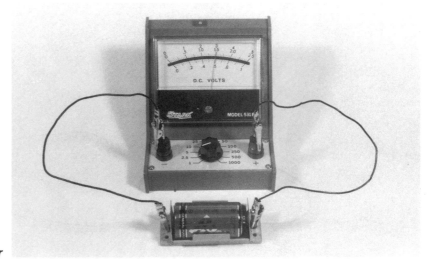

A voltmeter

Building a Better Cell

PART A

Problem
Which of the suggested liquids is best for making an experimental cell?

Materials
voltmeter
2 connecting wires
electrode holders
carbon strip
zinc strip
empty beaker (250 mL)
beakers containing these liquids: tap water; salt solution; sugar solution; bleach (diluted); vinegar; soap solution; dilute ammonia; dilute ammonium chloride solution

Procedure
1. Using electrode holders, insert a zinc strip and a carbon strip into the empty beaker, keeping them apart as shown.
2. Pour the beakerful of tap water into the beaker containing the zinc strip and carbon strip.
3. Using the connecting wires, connect the positive (red) post of the voltmeter to the carbon strip, and the negative (black) post of the voltmeter to the zinc strip.
4. Record the voltage, if any, of your cell.
5. Disconnect the strips from the voltmeter. Rinse and dry them.
6. Empty the tap water into its original beaker. Rinse and dry the beaker used to hold the strips.
7. Return the strips to the empty beaker and set up again as shown.
8. Substituting the other liquids for the tap water, repeat Steps 2 to 7. For every trial, be sure to fill the beaker to the same level and to record the type of liquid and the corresponding voltmeter reading.

CAUTION: Handle all chemicals with care.

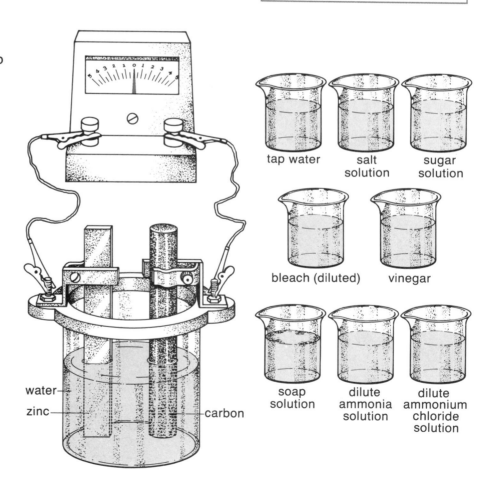

water
zinc
carbon

tap water   salt solution   sugar solution

bleach (diluted)   vinegar

soap solution   dilute ammonia solution   dilute ammonium chloride solution

PART B

Problem

What metal or combination of metals available in a school laboratory is best for making an experimental cell?

Materials

voltmeter
2 connecting wires
empty beaker (250 mL)
2 strips each of carbon, zinc, copper, nickel, aluminum, iron, etc.
beaker of appropriate liquid (see Step 1 below)

Procedure

1. In the beaker containing the liquid that gave you the highest voltmeter reading in Part A, set up the zinc and carbon strips as shown. Note the voltmeter reading.

2. Disconnect the strips from the voltmeter and from the electrode holders. Rinse and dry them.
3. Substituting other combinations of metal strips for the zinc and carbon strips, repeat Steps 1 and 2. For every trial, be sure to record the metal(s) used and the corresponding voltmeter reading. *Note that for some combinations, you may have to reverse the connections to the voltmeter in order to get a reading.* The sequence in which you test the metals should ensure that one strip of the pair is tested consecutively with each of the others.

Analysis

1. Which three liquids in Part A gave the highest voltmeter readings?
2. Which three pairs of metals in Part B gave the highest voltmeter readings?

Further Analysis

3. What seems to have the most effect on the voltage of the cell: the liquid used, or the metal or combination of metals used?
4. To ensure the tests were fair, what conditions had to be kept constant in
   (a) Part A?
   (b) Part B?

Extension

5. Working with other students, see what happens to the voltage when you connect your cells together. How is the voltage affected when you connect them in different ways?
6. Conduct a consumer test to see which brand of commercial battery lasts the longest in a particular battery-operated toy. Think carefully about how to keep the comparison fair. What conditions must you control?

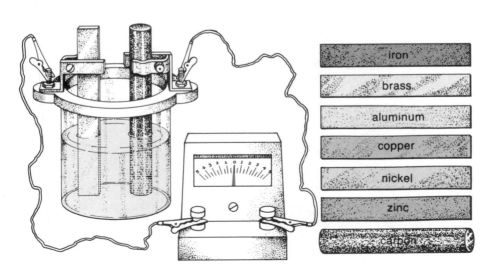

Types of Cells and Batteries

All cells have two features in common. One is the presence in the cell of two different metals. The other is the separation of the metals by a solution that can carry an electric current. These metals are called the **electrodes**, and the solution is called the **electrolyte.** Whatever their appearance or purpose, all batteries contain electrodes and an electrolyte.

A very common type of commercial cell uses a cylindrical zinc electrode with a carbon rod electrode down the centre. Between the electrodes is a watery paste containing the electrolyte, ammonium chloride. A chemical reaction in the cell causes a surplus of electrons to build up on the zinc electrode. At the same time, the carbon electrode loses electrons. When a conducting wire is connected between the two electrodes, electrons move along the wire from the zinc to the carbon, producing an electric current. After a time, the zinc becomes used up, and the cell must be safely discarded.

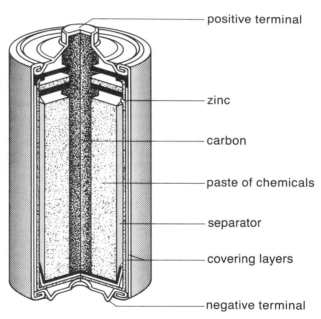

positive terminal

zinc

carbon

paste of chemicals

separator

covering layers

negative terminal

Cross section of a zinc-carbon cell. This type of cell has been widely used for more than a century.

Some cells can be recharged. Those with nickel and cadmium electrodes are widely used in devices such as camera flashes, portable radios, radio-controlled model cars and airplanes, and other devices where it would be too costly to use non-rechargeable batteries. (The electrolyte in nickel cadmium cells is very caustic; *never* attempt to cut open this type of cell.)

What design features would be
needed in a battery used in
(a) a spacecraft?
(b) a heart pacemaker?
(c) an automobile powered
entirely by batteries?
(d) a camera for taking outdoor
pictures in the Arctic?

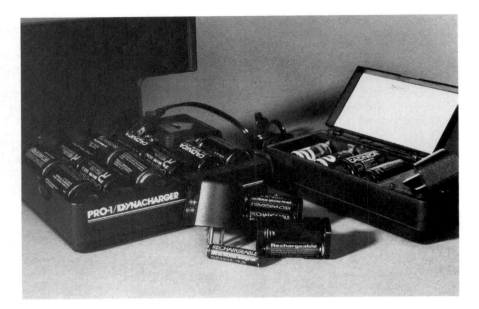

*Using the chargers shown, these nickel cadmium cells can be recharged
many times.*

Alkaline batteries have electrodes of zinc and manganese
dioxide. They are used in devices such as battery-powered toys
that require heavy currents over longer periods of time. Mercury
batteries have electrodes of zinc and mercury. They are often used
in cameras and watches.

A car battery uses lead and lead oxide as electrodes. This
battery produces a current to start the engine, but once the engine
is running, a generator is used to pass current back through the
battery. The chemical reaction that produced the battery current is
thus reversed and the battery is restored to full charge.

In the next Activity, you'll experiment with different
combinations of cells, bulbs, and wires in order to vary the
brightness of a light bulb.

Did You Know?

A research team at the
University of British Columbia
headed by Professor R. R.
Haering (born in 1934) has
developed new cells that store
several times as much energy as
the cells of the same mass in an
ordinary car battery. One type of
cell is called the MoLi cell (after
molybdenum and lithium, two of
the elements in these cells). The
MoLi cell is rechargeable.
During testing, it outlasted
existing cells used to supply
power to devices that draw large
currents.

Building a Flashlight

PART A

Problem
How can you make a simple flashlight?

Materials
2 D-size cells
holder for 1 cell
holder for 2 cells
small lamp
1.5 V bulb
2 V bulb
3 V bulb
connecting wires

Procedure
1. Connect the small lamp to a single cell as shown.

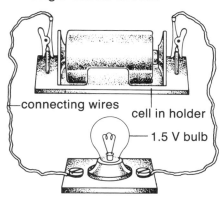

connecting wires
cell in holder
1.5 V bulb

2. Substituting first a 2 V bulb and then a 3 V bulb for the 1.5 V bulb, repeat Step 1. Observe any difference in bulb brightness.
3. Using whichever of the materials you require, experiment to create a cell and bulb arrangement that will make a 3 V bulb glow brightly.

PART B

Problem
How can you vary the brightness of your flashlight?

Materials
2 D-size cells in holder
small lamp with 3 V bulb
connecting wires with clips
very fine nichrome wire (50 cm)

Procedure
1. Connect the lamp and cells as shown.

> **CAUTION: The nichrome wire may become too hot to touch while it is connected.**

2. Slide one clip back and forth along the nichrome wire. Observe what happens to the bulb as the clips are moved closer together and further apart.

Analysis
1. What was the effect on the brightness of the 3 V bulb when you
   (a) added more cells?
   (b) moved a clip on the nichrome wire?

Further Analysis
2. Which type of wire allows more current to pass, copper or nichrome?
3. If you want to increase the current passing through nichrome wire, should you increase or decrease the length of the wire?
4. Which combination of cells, bulbs, and wires would you use to make the "best" flashlight? Explain your answer.

Set-up for Part B.

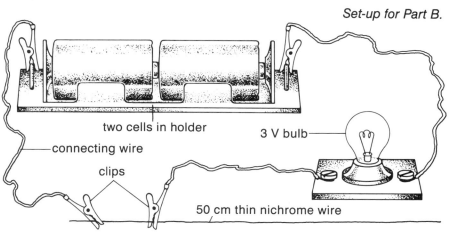

two cells in holder
connecting wire
clips
3 V bulb
50 cm thin nichrome wire

Resistance

As you observed in Activity 4-5, the amount of current in a wire is affected by the type of wire used and, in the case of the nichrome wire, by the length of the wire.

For current to exist in a wire, the electrons must move past the atoms within the wire. They have a difficult trip, frequently colliding with the atoms. A **resistor** is a device made of material that offers resistance to the flow of charges. In this way, it controls the amount of current. The greater the resistance of a material, the greater the amount of energy the electrons give up as they pass through it.

Thanks to resistors, a variety of devices supplies us with light and heat for daily use. For example, an ordinary light bulb operates because of the resistance of the wires in the bulb. Look at the diagram below of the inside of a bulb. The thick outside wires have low resistance and gain little energy from electrons. The very thin, coiled tungsten wire strung across the top has a high resistance to the flow of charges. Thus it gains much energy from the electrons, heats up, and gives off light. If the thin wire were replaced with straight copper wire, the bulb would not heat up and produce light. The heating elements in electric stoves, toasters, and electric kettles are also resistors. Television sets and radios may contain many resistors. Some are made of high-resistance wire coiled tightly on a form, but most are made of short lengths of compressed carbon. *Variable resistors*, or *rheostats*, can be used to control lights or motors, or to control the volume of sound from a radio or stereo.

Both the tungsten wire in a light bulb and the heating element in a toaster have high resistance to the flow of charges. As a result, they produce light and heat.

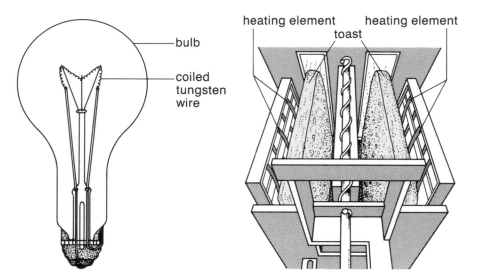

bulb

coiled tungsten wire

heating element   heating element
toast

Checkpoint

1. After taking clothes out of a clothes dryer, you may have experienced a situation like that in the cartoon. Use what you know about static electricity to explain why such situations occur.

2. Suppose you hear someone making the following statements. How would you change each statement to make it scientifically accurate?

   (a) "The galvanometer shows that there's a current of 1.5 V here."

   (b) "This light bulb will glow more brightly if I add a length of nichrome wire to hook it up to the battery."

   (c) "The second time, when I switched the connections of my current detector to the cell, the compass needle swung the other way; that tells me there was no current the second time."

   (d) "There was a blackout in my neighbourhood last night. We lit candles, then listened to a transistor radio to hear the news about it. It was strange not to use any electricity for that whole hour."

3. (a) What are the essential components of a chemical cell?

   (b) Which of the cells shown at the bottom of the page would produce no current? Why?

4. (a) What quantity does each of these devices measure?
   - ammeter
   - voltmeter
   - galvanometer

   (b) In what unit is each of these quantities measured?

5. Explain in what sense a battery is a system with its own subsystems.

6. When the copper wire in a "flashlight" set-up (similar to that in Activity 4-5) is replaced by another wire made from "metal X," the bulb glows less brightly. What can you say about the resistance of metal X compared with that of copper?

7. You want to design a living-room lamp whose brightness can be varied.

   (a) What components might be included in your design for the lamp?

   (b) What problems might your design cause?

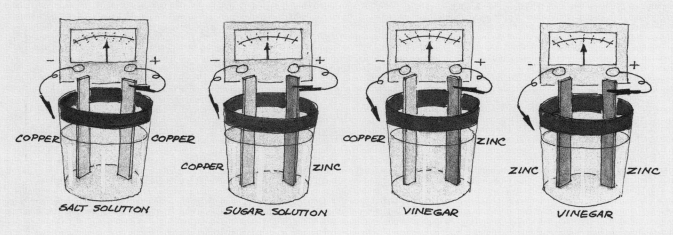

COPPER   COPPER   COPPER   ZINC   COPPER   ZINC   ZINC   ZINC

SALT SOLUTION   SUGAR SOLUTION   VINEGAR   VINEGAR

Generators, Magnets, and Motors

Although many electrical devices use batteries, most of the appliances in your home run on energy that comes from huge **generators.** Every time you turn on a light switch, plug in a toaster, or switch on your television set, you are most likely using electrical energy produced by a generator that is hundreds of kilometres away. Whereas a battery converts *chemical* energy into electrical energy, a generator converts *mechanical* energy into electrical energy.

In Activity 4-2, you saw a relationship between electricity and magnetism. Having wrapped wire carrying electric current around a magnetic compass, you observed that the compass needle was deflected from its normal north-south position. This observation was first made in 1819 by a Danish scientist, Hans Christian Oersted. He made the important discovery that a current produces a magnetic effect.

In hydroelectric generators, a magnet and coil apparatus is moved by the force of falling water.

This generator at Keephills, Alberta, produces electricity from the chemical energy of coal. Coal is burned, and the thermal energy is used to heat water, changing it to steam. The pressure of the steam turns large turbines. As the turbine shafts rotate, they cause the rotating part of the generator to spin, producing electricity.

Several years later, a British scientist, Michael Faraday, wondered if he could get the opposite effect to that observed by Oersted. Could he use a magnet to produce an electric current? In 1831, after many trials, he found he could do so. This discovery led to the development of modern generators. In Activity 4-6, you will experiment with magnets and wire to discover the best design for a simple generator. Then, in Activity 4-7, you will make use of a generator to light a bulb.

Activity 4-6

Generating Electricity

Problem

How can a magnet and a coil of wire be used to produce a current?

Materials

emery cloth
10 m of insulated copper wire (approximately 26 gauge)
tape
galvanometer
cardboard tube from paper towel roll
fresh magnet

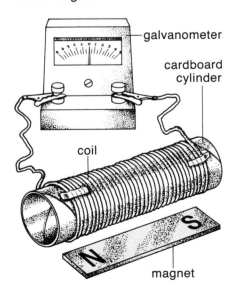

Procedure

1. Using emery cloth, remove about 2 cm of the insulation from each end of the wire.
2. Wrap at least 50 turns of the wire around the length of the cardboard tube.
3. Tape the coil in place so the wire will not unravel. Do not cut off any unused wire.
4. Connect the two ends of the wire to the terminals of the galvanometer.
5. Bring the north pole of the magnet close to the outside of the coil. Move it slowly, then rapidly, recording the galvanometer readings.
6. Using the south pole of the magnet instead of the north pole, repeat Step 5.
7. Experiment with moving the magnet in other ways around or inside the coil. Try moving the coil instead of the magnet. Record the galvanometer readings.
8. Investigate the effects of adding more turns to the wire coil. In stages, add 20 to 50 more turns and repeat Steps 5 to 7.

Analysis

1. Make a table to show what kind of galvanometer reading (positive or negative, strong or weak) you obtained for each positioning of magnet and coil. Use two columns, one to indicate the position of the magnet and coil, and the other to indicate the reading. Represent a positive reading as strong (+ +) or weak (+), and a negative reading as strong (− −) or weak (−).

Further Analysis

2. How can a current be obtained from the coil using a bar magnet?
3. What must you do to get a continuous supply of current?
4. List at least three ways to increase the amount of current you get from the coil.

Using a Generator

Problem

How can a generator be used to light a bulb?

Materials

small hand-held generator
galvanometer
3 V bulb in lamp
2 connecting wires

Procedure

1. Connect the generator to a light bulb.
2. Turn the crank of the generator first in one direction, then in the other. Vary the speed of rotation and observe the effects on the brightness of the bulb.
3. Replace the light bulb with a galvanometer as shown in the photograph, and repeat Step 2. Record the galvanometer readings.

Analysis

1. Using the generator, how can you make the bulb brighter or dimmer?
2. What happens to the brightness of the bulb when you change the direction in which you turn the crank?
3. What happens to the galvanometer needle when you change the direction in which you turn the crank?

Further Analysis

3. What factors determine the amount of electricity produced by a generator?

Extension

4. Design another method of turning the generator, using the energy of wind or running water. Use your design to light a bulb.

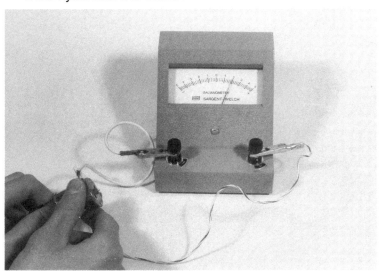

Types of Magnets

To generate electricity in Activity 4-6, you used a **permanent magnet.** This type of magnet is often made of a hard steel alloy. If treated with care, it will stay magnetized for a long time. As you built your own current detector in Activity 4-2, you made an **electromagnet**. This type of magnet is made from a coil of wire attached to a source of current. When an electric current is present, the wire has magnetic effects. As you will shortly see, an electric motor is made to work through the interaction of these two types of magnets.

Making an Electromagnet

Problem

How can you make and use a strong electromagnet?

Materials

compass
2 m of insulated copper wire
    (approximately 26 gauge)
piece of emery cloth
    (5 cm x 5 cm)
2 D-size cells in holders
2 connecting wires
pencil
iron spike
50 or more small finishing nails

Procedure

1. Using the emery cloth, remove about 2 cm of insulation from each end of the wire. Wrap 20 turns of wire around one end of the pencil.

2. Place this 2 cm from the north end of the compass needle, perpendicular to the needle.
3. Use the connecting wires to briefly connect the ends of the wire to a single cell.
4. Observe and record how many degrees, if any, the needle is deflected from its normal north-south direction.
5. Replace the pencil with an iron spike wound with 20 turns of wire. Repeat Steps 2 to 4.
6. Double the number of turns of wire on the iron spike, and repeat Steps 2 to 4.
7. Using the iron spike with 40 turns of wire, and two cells instead of one, repeat Steps 2 to 4.

8. Use the magnet you have created to lift some small nails. Consider how you can improve the design of the magnet to increase its lifting strength.

Analysis

1. In which Procedure step did you get
   (a) the greatest needle deflection?
   (b) the smallest needle deflection?
2. How can you increase the strength of an electromagnet? (Recall that the stronger the magnet is, the greater the deflection of the compass needle.)

Further Analysis

3. What advantage does an electromagnet have over a permanent magnet?

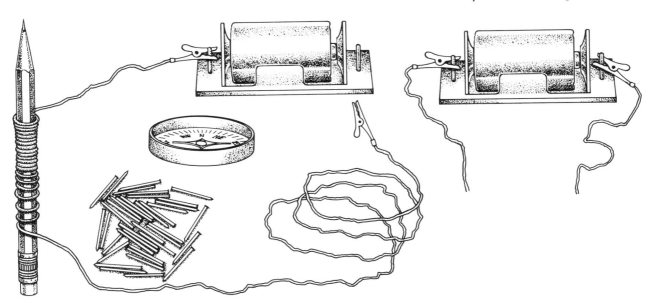

How a Simple Electric Motor Works

Find out how the beam of electrons in a television tube is made to trace out the picture you see. What type of force is used to shape the picture?

Electric motors work because of the interaction of a permanent magnet and an electromagnet. You will find these motors in many devices—in electric can openers, cassette players, food processors, electric hand mixers, and electric drills.

Like permanent magnets, electromagnets have north and south poles. The north and south poles of the electromagnet, however, are reversed if the direction of current flow through the wire coil is reversed. An electric motor uses this property to make an electromagnet move.

Examine the illustration of a St. Louis motor. This is a special motor designed to show how electric motors work.

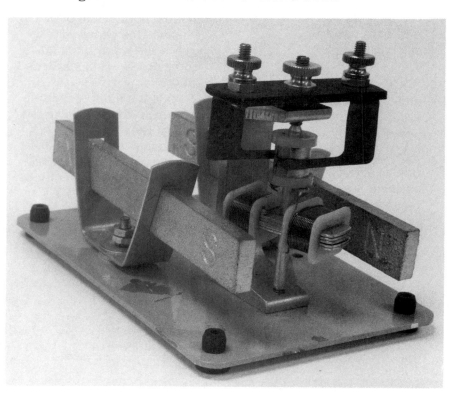

A St. Louis motor.

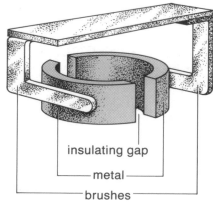

insulating gap
metal
brushes

A close-up view of the commutator and brushes. The brushes are in contact with the metal pieces of the commutator.

Observe the position of the two permanent magnets, which are fixed in place, and the electromagnet, or *armature* , which is free to turn about a pivot. The electromagnet's wire coil is connected to the bottom of each of the two metal half-rings of the *split ring commutator*, which turns along with the armature. Separating those two metal pieces are two insulating gaps. When in operation, the terminals above the brushes are connected to a battery power source.

When the motor is connected to a power source and the brushes are touching the metal parts of the commutator, there is an electric current. It extends from the connecting wires through one brush, through one side of the commutator, through the coils of the armature, back up the other side of the commutator, and through the second brush. However, if the brushes are in contact with the insulating gaps of the commutator, rather than the metal parts, no current exists in the coil.

Observe the illustrations below and note how the armature is made to move. The steps (a), (b), and (c) will repeat themselves over and over as long as the motor is connected to a power source. The spinning armature of a motor can be attached by means of a gear mechanism to many devices—for example, a drill bit, the disc drive of a computer, or the blades of a blender. In this way, a motor converts electrical energy into mechanical energy. You'll see an example of this conversion for yourself in Activity 4-9.

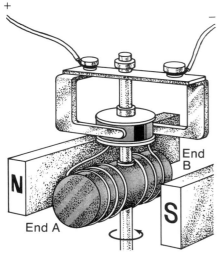

(a) When there is current in the coils of the armature, end A of the armature becomes a north pole, and end B a south pole. Since like poles repel each other, the permanent magnets repel the armature, causing it to move in the direction shown by the arrow.

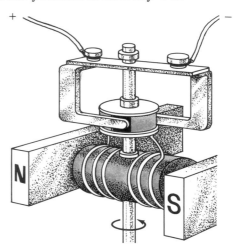

(b) As the armature turns, the insulating gaps pass by the brushes, so for an instant there is no current in the armature. Yet the armature continues to turn.

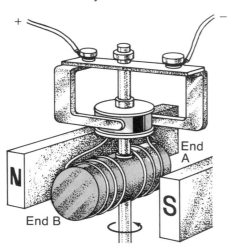

(c) The armature has now turned through 180°. The brushes are once again touching the metal portions of the commutator, so once again there is current in the armature. However, the current now flows through the coils in the opposite direction, so that end A is now the south pole and end B is the north pole. As in the first step, the repulsion of like poles causes the armature to turn.

Interacting Magnets
(DEMONSTRATION)

Problem

What effect does a permanent magnet have on an electromagnet?

Materials

large alnico magnet
support stand
right-angle clamp
insulating rod
80 cm bare copper wire
2 connecting wires
6 V battery

Procedure

1. Set up the magnetic swing as shown. Use the right-angle clamp to attach an insulating rod to the support stand. Fashion a "swing" out of copper wire and loop it over the insulating rod. Ensure that the swing can move freely.
2. Position the alnico magnet on the table so that the bottom of the copper wire "swing" sits between its poles, with the north pole on top.
3. Observe what happens to the swing when
   (a) the two ends of it are briefly connected to the battery;
   (b) the connections to the battery are reversed;
   (c) the alnico magnet is turned over, so the south pole is on top.

Analysis

1. (a) What happened to the wire swing when there was a current in it?
   (b) What change took place when
   • the direction of current flow was reversed?
   • the alnico magnet was turned over?

Further Analysis

2. What two factors affect the direction in which the magnetic swing is pushed by the permanent magnet?
3. What happens to an electromagnet when the direction of the current through its coils is changed?

Extension

4. Try using this effect to make your own electric motor. You will need a pair of *strong* bar magnets, insulated wire, and various other common materials.

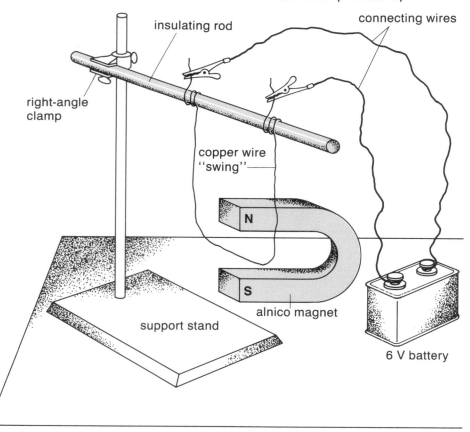

insulating rod

connecting wires

right-angle clamp

copper wire "swing"

N

S

alnico magnet

support stand

6 V battery

Electricity and Other Forms of Energy

Like all forms of energy, electrical energy can be produced *from* other kinds of energy, or it can be converted *into* other kinds of energy. In this Unit, you have so far seen three ways in which electricity is involved in such changes.
- Batteries convert chemical energy into electrical energy.
- Generators convert mechanical energy into electrical energy.
- Electric motors convert electrical energy into mechanical energy.

In this Topic, you'll examine other energy conversions involving electricity.

Electricity and Heat

The conversion of electrical energy into thermal energy takes place in many household devices—for example, toasters, electric kettles, and electric coffee makers. On the other hand, a device such as the electric thermometer in an oven makes use of the conversion of thermal energy into electrical energy.

How can you get heat from electricity? You already know that when electrons travel through a wire, they collide with the atoms of the wire. The greater the resistance in the wire, the greater the number of collisions, and the greater the amount of heat.

In Activity 4-10, you'll investigate this phenomenon, using high-resistance nichrome wire. Then, in Activity 4-11, you'll observe the conversion of thermal energy into electrical energy by means of a device known as a **thermocouple**.

The electric kettle operates by converting electrical energy to thermal energy.

Heating Water

Problem

How can you heat water by means of an electric current?

Materials

60 cm nichrome wire (28 gauge or thinner)
pencil
glass tubing
large test tube
water
support stand
utility clamp
thermometer
connecting wires with clips
battery of 4 D-size cells in holder

Procedure

1. Prepare a coil by winding ten turns of fine nichrome wire around a pencil, leaving about 15 cm of straight wire at each end of the coil.
2. To keep the straight ends of wire from touching, slip a length of glass tubing over one of them.
3. Insert the coil into the test tube as shown, add just enough water to cover it, and clamp the test tube to a ring stand.
4. Place a thermometer in the test tube and record the temperature of the water.
5. Using two connecting wires, connect the coil to the battery.
6. Record the temperature of the water every 30 s for 5 min.
7. Disconnect the battery.

CAUTION: The nichrome wire may become too hot to touch while it is connected.

Analysis

1. By how many degrees did the temperature of the water rise in 5 min?
2. (a) What effect did the current have on the nichrome wire?
   (b) From what evidence do you infer this effect?

Further Analysis

3. Prepare a graph to show the relationship between the length of time the coil was attached to the battery and the amount of change in the temperature of the water.
4. How could you modify the Activity to make the water heat up faster?
5. What property of nichrome wire permits the changes you observed?
6. Name a household appliance that makes use of the property of nichrome observed in this Activity.

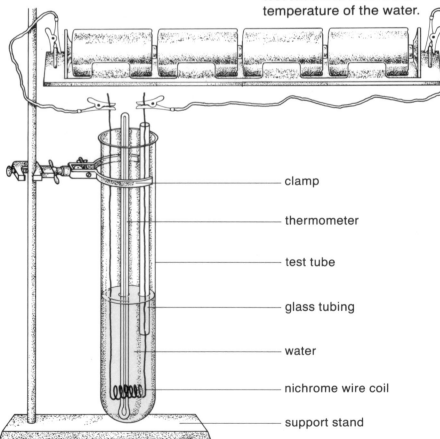

clamp

thermometer

test tube

glass tubing

water

nichrome wire coil

support stand

The Thermocouple

Problem

How can electrical energy be obtained from thermal energy?

Materials

20 cm bare thick copper wire
15 cm iron coat-hanger wire
2 connecting wires
galvanometer
heat source
pliers
beaker of ice-cold water

Procedure

1. Coil about 5 cm of the copper wire tightly around the coat-hanger wire as shown.
2. Attach the wire leads of the galvanometer to the copper-iron wire couple as shown.
3. Observe and record the galvanometer reading, if any.
4. Hold the coupled wires with a pair of pliers and, using the heat source, heat the junction of the wires; repeat Step 3.
5. Cool the junction in a beaker of ice-cold water; repeat Step 3.

Analysis

1. What factor(s) affected the current from the thermocouple?
2. Why do you think the apparatus you have just used is called a thermocouple?

Did You Know?

In 1821, a German scientist, Thomas Johann Seebeck, discovered the principle by which the thermocouple works. He joined two different metals so there were two junctions. He kept one junction at a steady temperature by placing it in an ice-water bath and heated the other junction in a flame. Seebeck observed that the higher the temperature at one junction relative to the other, the greater the current.

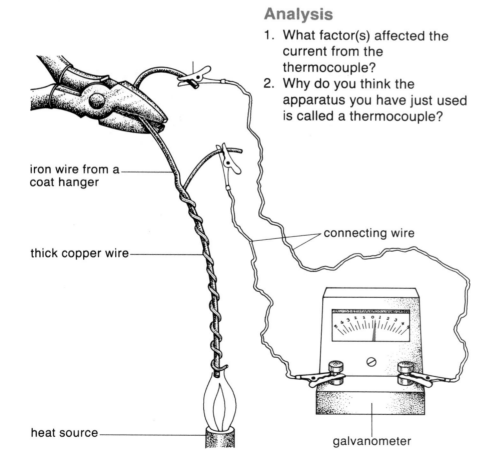

iron wire from a coat hanger

thick copper wire

connecting wire

galvanometer

heat source

Electricity from Solar Energy

The solar cell in a solar calculator turns the energy of light into electricity. In this next Activity, using a battery-powered light source to represent the Sun, you'll find out how much electricity is produced by a single solar cell.

A solar calculator.

Spacecraft, such as Ranger A *shown here, carry thousands of solar cells.*

The Solar Cell

Problem

How can electrical energy be obtained from solar energy?

Materials

solar cell
small battery-powered light
 source
metre stick
galvanometer
connecting wires

Procedure

1. In a lighted room, connect the two wires from the solar cell to a galvanometer. Note the reading on the galvanometer, then cover the solar cell with your hand so that no light reaches it and observe any change. Record both readings in your notebook.
2. Darken the room, turn on the light source, and move it towards the cell until you obtain a full-scale (maximum) reading on the galvanometer. Measure the distance between the light source and the cell. Record both the reading and the distance in your notebook.

3. Double the distance between the light source and the cell, and record the galvanometer reading.
4. Triple the distance between the light source and the cell, and record the galvanometer reading.
5. Adjust the distance between the light source and the solar cell until you obtain a reading on the galvanometer that is half the full-scale reading. Place a second, similar source beside the existing one, and record the galvanometer reading.

Analysis

1. What factors determine how much current the solar cell produces?

2. List some advantages and disadvantages of using this type of cell to produce electricity.

Further Analysis

3. One of the first uses for solar cells was in space probes. Why do you think solar cells are especially practical in space?

Extension

4. What happens to the current if you use different colours of filter (or coloured cellophane) over the cell?

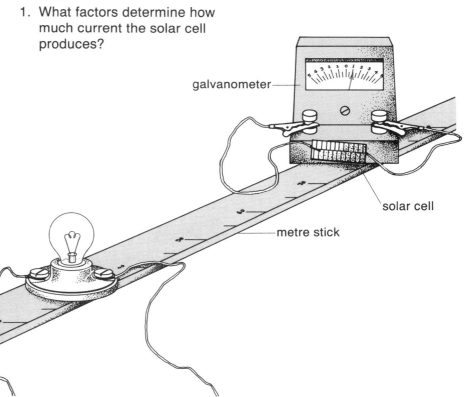

galvanometer

solar cell

metre stick

Electricity and Sound

Some crystals such as Rochelle salt and quartz will produce a current when they are squeezed or struck! Since these crystals respond to the slightest change in pressure, the changing pressure of sound waves causes these crystals to vibrate along with the sound. With each vibration, the crystal produces an electric current. The strength of the current changes according to the strength of the sound waves.

Crystal microphones use this relationship between current and sound waves, called the *piezoelectric effect*. When someone whistles into a crystal microphone, the sound is changed into an electrical signal, which can be shown as a wave pattern on a screen. If you have a tape recorder, it probably has a crystal microphone that uses the piezoelectric effect. Other devices, such as loudspeakers, telephones, and radios, also use electrical signals to produce sound.

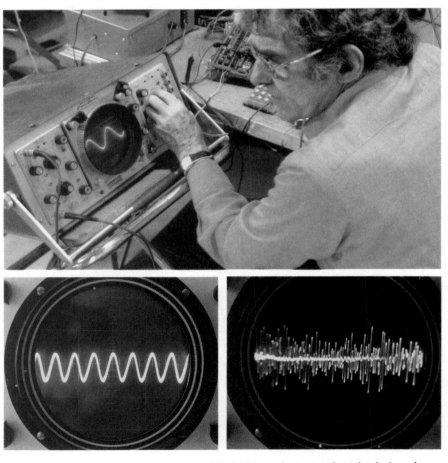

A crystal microphone changes a whistled tone into an electrical signal.

1. Use the definitions to find the correct words for the word puzzle. When you have completed the puzzle, the highlighted letters will spell out a word that has to do with a device that converts another form of energy into electrical energy.

    (a) a type of magnet in which the polarity can be reversed

    (b) a solution that conducts electricity

    (c) a device used to measure small electric currents

    (d) the unit in which electric current is measured

    (e) a device that converts electrical energy into mechanical energy

    (f) a type of magnet in which the polarity cannot be reversed (two words)

    (g) the rotating device in a St. Louis motor

    (h) a device that can convert thermal energy into electrical energy

    (i) a device that can be used to control the amount of current

    (a) ■■■■■■■■■□■■■
    (b) ■■□■■■■■■■□■■
    (c) ■■■■□■■■■■
    (d) ■■■□■■■
    (e) ■■■■□
    (f) ■■■■□■■■■■■■■■■
    (g) ■■■■□■■■
    (h) ■■■■□■■■■■■
    (i) □■■■■■■■

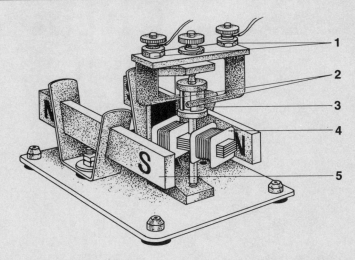

2. Someone has tried to make an electromagnet by wrapping 20 turns of copper wire around a drinking straw and attaching it to a 1.5 V D-size cell. The person is disappointed because it will not pick up anything. Suggest three ways to improve the strength of the electromagnet.

3. (a) Examine the drawing of a St. Louis motor above and identify the five parts that are labelled. In your notebook, list the names of the parts.

    (b) For each part you listed in (a), explain in one sentence why it is an essential part of the motor.

4. You turn on a St. Louis motor, but nothing happens. What parts of the motor would you check in trying to solve the problem?

5. (a) In a hand-cranked generator like the one shown below, what is the source of the electrical energy?

    (b) How could you increase or decrease the brightness of the light bulb?

    (c) What must occur within a generator, in order for a current to be produced?

    (d) How is a generator similar to a motor, and how is it different?

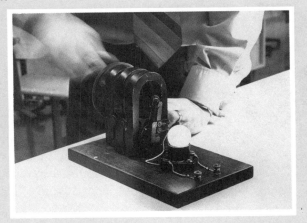

Circuits and Switches

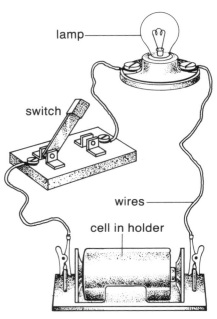

A basic electric circuit. Note the source (battery), conductor (wires), control (switch), and load (lamp).

What happens if you connect a wire from a single battery terminal to a small light bulb? Nothing happens—until you also connect a wire from the other terminal.

In order for electrons to flow and do work, there must be a complete pathway through which they can move. The electrons pass from the battery into the bulb and then back to the battery. Such a pathway is called an **electric circuit**.

When there is a flow of electrons throughout the circuit, the circuit is said to be *closed*. When this flow is interrupted and the electrons cannot move throughout the circuit, the circuit is said to be *open*. If a light bulb is connected to both terminals of a battery, a closed circuit exists. To make it an open circuit, you could remove one of the connections. However, there is another, more practical way to open and close a circuit—a **switch**. In household circuits, switches enable you to control which devices are on or off.

An electrical circuit can be regarded as a system. Within this system, there may be four subsystems: a **source**, a **conductor**, a **control**, and a **load**. The source changes some other form of energy into electrical energy. The conductor allows electrons to travel in the path. The control (usually a switch) is a device for starting and stopping the electron flow. The load changes the electrical energy into some other form of energy.

Series and Parallel Circuits

In some strings of decorative lights, you may remove one bulb and discover that the whole string of lights goes out. In this type of circuit, known as a **series circuit**, the electrons flow through each device along a single connecting path. The removal of one bulb interrupts the path of electricity through the circuit.

In most strings of lights, however, you can remove one bulb and the rest remain lit. In this type of circuit, known as a **parallel circuit**, the electrons can flow through two or more alternative paths. If one bulb is removed, the circuit still remains closed. In the next three Activities, you'll build a variety of circuits and determine where to connect switches.

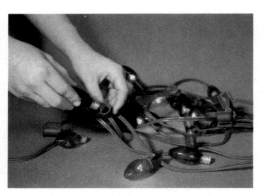

Are these lights connected in series or in parallel?

Lamps in a Series Circuit

Problem

How does the number of lamps in a series circuit affect their brightness?

Materials

battery of 2 D-size cells in a holder
3 lamps with 3 V bulbs
push-button switch
4 connecting wires

Procedure

1. Connect the circuit as shown in illustration (a). Push the button and observe how bright the single lamp is.
2. Connect a second lamp along the same conducting path as the first, as shown in (b). Push the button switch and observe how bright each lamp is.
3. Connect a third lamp along the same conducting path as the other two. Again, push the button and observe how bright each lamp is.

4. Predict what will happen if you unscrew one of the lamps in the series circuit. Try it.

Analysis

1. As you added lamps to the circuit, did their brightness increase or decrease?
2. What effect does the addition of more lamps have on the current in the circuit?
3. Does the addition of lamps to a series circuit increase or decrease the resistance offered by the circuit?

Further Analysis

4. (a) Do you think the appliances in your house are connected in a series circuit?
(b) Why or why not?

Extension

5. Try to make all three lamps as bright as the single lamp was, while still connecting them in a series circuit. (You will have to adjust at least one of the other three subsystems in the circuit.)
6. Repeat the Procedure, this time including a commercial ammeter in the circuit. Measure the current as you complete each step.

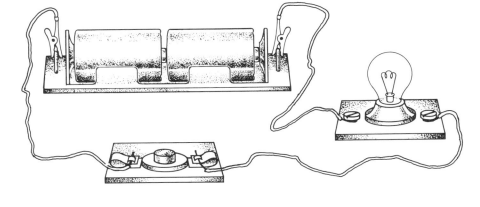

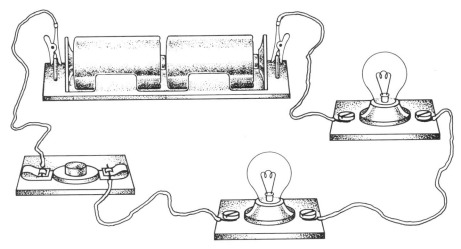

Lamps in a Parallel Circuit

Problem

How does the number of lamps in a parallel circuit affect their brightness?

Materials

battery of 2 D-size cells in a holder
3 lamps with 3 V bulbs
push-button switch
8 connecting wires

Procedure

1. Set up a circuit using two cells, one lamp, and a push-button switch.
2. Connect a second lamp in parallel as shown. Push the button switch and observe how bright each lamp is.

3. Predict what will happen to the brightness of the lamps when a third lamp is added in parallel circuit with the first two. Test your prediction.
4. Predict what will happen when you unscrew any one of the lamps. Test your prediction.

Analysis

1. As you added lamps to the circuit, what change occurred in the brightness of lamps already in the circuit?

Further Analysis

2. The brightness of a lamp depends on the amount of current supplied to it. What must be happening to the total current drawn from the battery when you use
   (a) two lamps instead of one?
   (b) three lamps instead of one?
3. (a) Would a battery last as long with two lamps connected in parallel, as it would with just one lamp?
   (b) Why or why not?
4. Is there an advantage to stringing lamps in parallel rather than in series?
5. (a) Are the circuits in your home wired in parallel or in series?
   (b) How do you know?

Extension

6. Add an electric motor in a parallel circuit with the three lamps. Compare how the motor runs in the circuit with how it runs when connected to a battery alone. What difference do you notice?
7. Find out how to measure the current in different parts of a parallel circuit. Repeat the Procedure, this time using an ammeter to measure the current in each branch of the circuit, and the total current leaving the battery.

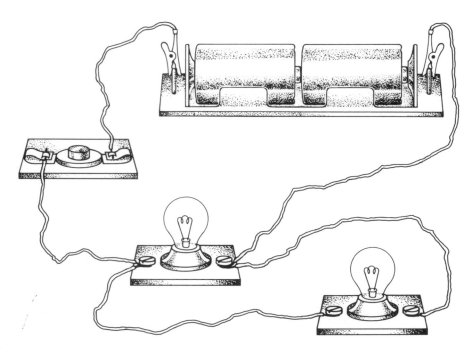

Positioning Switches

PART A

Problem

What are the best places to put switches in order to control the devices on a parallel circuit?

Materials

battery of 2 D-size cells in holder
2 lamps with 3 V bulbs
4 push-button switches
6 connecting wires
small toy motor

Procedure

1. Set up the parallel circuit as shown, including two lamps and a motor.
2. Plan where to place switches in the circuit so that you could turn on or off any one of the three devices.
3. Connect the switches and test your plan.
4. Predict where you could place a switch that would turn off all the devices at once. Test your prediction.

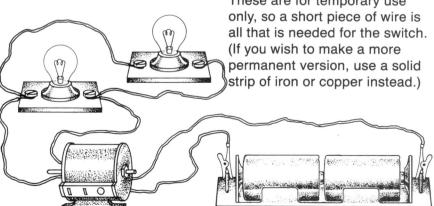

PART B

Problem

How can a single light be turned on from either of two different switches?

Materials

battery of 2 D-size cells in holder
lamp with a 3 V bulb
2 blocks of wood (about 10 cm square each)
6 screws
screwdriver
5 connecting wires
2 lengths of copper wire (10 cm each)

Procedure

1. Make two "two-way switches" from the wood and screws as shown.

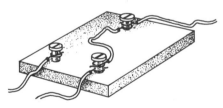

These are for temporary use only, so a short piece of wire is all that is needed for the switch. (If you wish to make a more permanent version, use a solid strip of iron or copper instead.)

2. Set up the circuit as shown in diagram (a). Record whether the lamp is on or off.
3. Connect the switches in the positions shown in diagrams (b) and (c). Record whether the lamp is on or off each time.
4. Experiment with different switch combinations until you understand how the light can be turned on or off with either of the two switches.

(a)

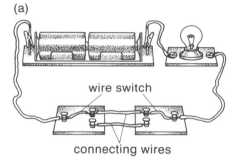

wire switch

connecting wires

(b)

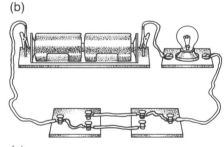

(c)

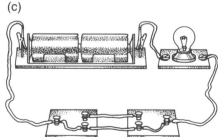

Three possible arrangements in a hall switch circuit are shown in (a), (b), and (c).

PART C

Problem

How can a single circuit be built for use with both a front doorbell and a back doorbell?

Materials

battery of 2 D-size cells in a holder
doorbell (or lamp)
2 push-button or knife switches
5 connecting wires

Procedure

1. Connect the battery in series with a doorbell (or lamp) and a switch.
2. Connect the second switch in parallel with the first switch.
3. Test your design to see if the doorbell will ring (or the lamp will light) if
   (a) each switch is pushed at a different time;
   (b) both switches are pushed at the same time.

PART D

Problem

How can a switch be used as a burglar alarm?

Materials

battery of 2 D-size cells in a holder
doorbell (or lamp)
small block of wood
2 nails or screws
aluminum pie plate
scissors
4 connecting wires

Procedure

1. Cut a strip from an aluminum pie plate to make a "switch" like the one shown. This switch will complete a circuit if it is pushed down to touch the top of the nail.
2. Connect the battery, doorbell (or lamp), and switch in series. Test your switch.
3. Design a burglar alarm using this switch and circuit. The alarm should be triggered when a door opens.
4. Modify your design so as to connect the alarm bell to three different doors in the house.

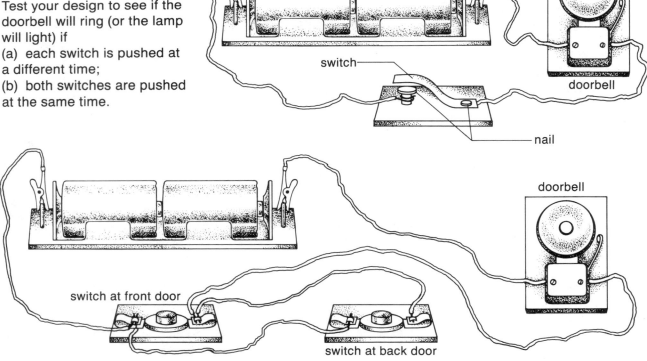

switch

nail

doorbell

doorbell

switch at front door

switch at back door

Circuit Diagrams

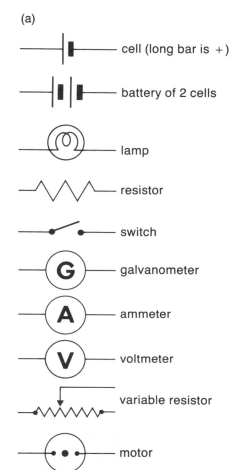

cell (long bar is +)

battery of 2 cells

lamp

resistor

switch

G galvanometer

A ammeter

V voltmeter

variable resistor

motor

Standard symbols are generally used to represent each part of a circuit when the plan for the circuit is drawn. These standard symbols are shown in diagram (a). Shown in (b), (c), and (d) are diagrams for circuits similar to some of those you have constructed in the Unit.

The circuit in (b) is three lamps in series with a switch and a battery of two cells. The switch turns all three lamps on or off simultaneously. The battery provides current.

The circuit in (c) is a motor in series with a switch, an ammeter, a variable resistor, and a battery of two cells. The switch turns the motor on or off. The ammeter indicates how much current exists in the circuit. The variable resistor controls the speed of the motor. The battery provides current.

The circuit in (d) is two lamps and a motor in parallel with three switches and a battery of two cells. One switch controls one of the lamps, the second switch controls the other lamp, and the third switch controls the motor. The battery provides current.

(b)

(c)   (d)

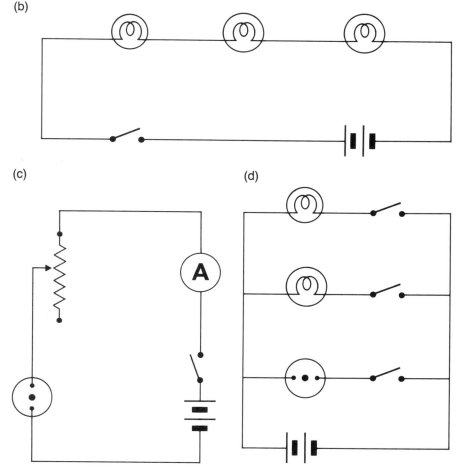

Drawing Circuits

Go back to Activity 4-15 on page 203. Choose the illustrations for any two Parts of the Activity and redraw them using standard circuit symbols. (For Parts C and D, make your own symbol to represent the doorbell.)

Electrical Inventions

Think for a moment about how many electrical devices you have used in the past few days. Someone had to invent each of these devices.

To be successful, an invention generally has to serve some useful purpose. As well, it must improve on devices that were previously used for that purpose. For example, modern laser disc players produce sound superior to that from phonograph or cassette players.

In the following Activity, you'll devise a useful electrical invention. It may be either serious or humorous, but it must be safe.

A laser disc player is a successful electrical invention.

You Be the Inventor!

Problem

Can you invent something useful and safe that runs on electricity?

Procedure

1. Examine the illustrations of the two inventions and read the descriptions of how they operate.
2. Design an invention, such as the one shown below, that uses a single circuit to perform a single task. Make a circuit diagram using the standard symbols, then construct your invention.
3. Design an invention that will perform two functions at once, or an invention that will perform one function that then leads to another. Make a circuit diagram using the standard symbols, then construct your invention. Use your imagination, or try one of these:

   - a model electric motor;
   - a model traffic light;
   - a model pinball machine, in which the balls rolling over switches causes lights to flash;
   - a lamp, motor, or doorbell that is turned on by an electromagnetic switch"relay";
   - a model scoreboard for team sports;
   - a model burglar alarm in which lights flash, bells ring, and a trapdoor opens to capture the burglar.

Analysis

1. Were there any trouble spots in your invention? Explain why they caused a problem.

Further Analysis

2. How could you improve your design?

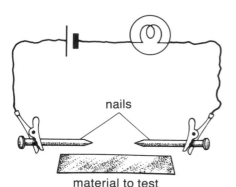

nails

material to test

This invention can test a material to see if it conducts electricity. The material is placed between the two nails, touching each of them. The better a conductor the material is, the more brightly the lamp will shine. If the material is a very poor conductor or cannot conduct electricity at all, the lamp will not glow.

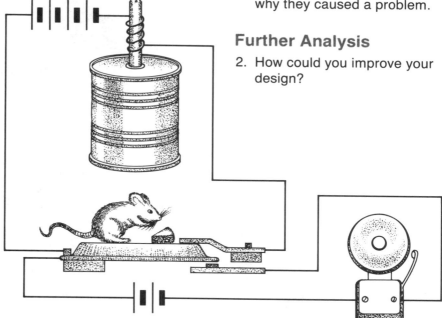

This invention performs two functions. First, an electromagnetic mousetrap captures a mouse, and second, an alarm sounds to alert the inventor of the mouse's capture. The trap is set by starting a current in the electromagnet that holds up the juice tin. When a mouse runs onto the aluminum pie-plate "switch," the circuit is broken. This turns off the electromagnet. The can, no longer held up, falls over the mouse, trapping it. At the same time, current to the bell is turned on, causing the bell to ring and thus signalling that the mouse has been caught.

Science and Technology in Society

Superconductors and the Future

Something is about to happen up at the front of the room. A styrofoam cup sits upside down on a table. A small black disc rests on the bottom of the cup. Liquid is poured out of a beaker over the disc, engulfing it in a cold fog. Plastic forceps then hold a tiny chunk of dark material just above the disc. The forceps are gently removed and—the unexpected happens. The tiny chunk floats in the air like a levitating person in a magic show. The crowd in the room gasps and claps.

The "magician," Harcharan Pardhan, explains, "You've just seen a superconductor at work." The questions start to fly as the audience tries to unravel the strange behaviour of the superconductor. Harcharan, not a magician at all, but a physics instructor at Red Deer College in Red Deer, Alberta, is ready to give them the answers.

For a few moments, the dark object hangs suspended in the air above the flat disc.

What's the black disc? "It's a specially made ceramic that, when cooled to extremely low temperatures, becomes a superconductor." How did you cool it? "I poured liquid nitrogen with a temperature of −196°C onto it." What makes the tiny chunk of material levitate? "The chunk of material is a magnet, so, like all magnets, it has a magnetic field of force around it. Placing this magnet near the superconductor causes a super-current within the superconductor. This super-current generates its own magnetic field. Then the two opposing fields, one from the chunk of magnet and the other in the superconductor, cause them to repel each other, just as do the opposite poles of two magnets. The magnetic force of the superconducting ceramic disc repels the magnet enough to keep it up in the air.

(a) "Here's another way to see the same effect. I've suspended the superconductor between the poles of this powerful horseshoe magnet."

(b) "Watch to see what happens after I dip the superconductor in liquid nitrogen for a few minutes."

(c) Suddenly the superconductor moves away from the magnet.

Watching the interaction of a superconductor and a magnet fascinates all those who see it. Scientists who are experimenting with superconductors have big dreams for their use. For example, they predict that superconductors could revolutionize our transportation. By winding superconductors into coils, they could produce powerful, efficient magnets. These could be used to levitate trains much as the electromagnetic Japanese ''bullet'' trains now do, increasing speed even more.

Maglev (magnetic levitation) trains use powerful electromagnets to levitate them above the ground. With much less friction than ordinary cars or trains experience, they have reached speeds of more than 500 km/h.

In addition to producing huge magnetic forces, superconductors do not lose any electrical energy as electric current passes through them. That's how they got their name: they are superior conductors of electricity.

Today, every time you flick a switch or turn on an appliance, the wires through which the electricity passes offer resistance. As much as one-fifth of the electric energy can be lost as heat. By cooling some materials to a very low temperature, they have no resistance to electric current, and they become superconductors. Because there is no resistance, there also is no transformation of electric energy to heat which can be lost to the surrounding environment, or can cause overheating. Thus, superconductors would be big energy savers. For example, use of superconducting materials to make cables for transmission of electrical energy from power stations to towns and cities would prevent the loss of energy as heat from these power lines.

Problems to Overcome

Superconductors are not new. In 1911 it was discovered that mercury would become a superconductor if cooled to the temperature of liquid helium ($-269°C$). Because of the expense of keeping temperatures this low, applications of superconductivity were very limited. Over the years, more and more materials have been found to be superconductors at very low temperatures. Then, recently came a breakthrough. In the late 1980s, ceramic materials were prepared that would superconduct at the temperature of liquid nitrogen ($-196°C$). This still is very cold, of course, but a lot warmer than liquid helium. Liquid nitrogen is less expensive to prepare and store so more applications are possible. What is the aim now? To try to find materials that will superconduct at room temperature, or close to it. Then the possible applications will be enormous.

Think About It

1. What invention has revolutionized the way we live today as much as superconductors may do in the future? Describe how you think this invention has changed our lives.
2. Collect articles on the current status of superconductor research. What are some other uses people are hoping to put superconductors to?

Electricity and Safety in the Home

Electrical devices in the home can be grouped into four categories, based on the main purpose for which the electrical energy is used.

- *Light-producing* devices include lamps and flashlights.
- *Heat-producing* devices include kettles, coffee percolators, stoves, hair dryers and curlers, clothes dryers, and automobile block heaters.
- *Mechanical* devices (those that convert electricity to mechanical energy) generally contain motors. They include electric can openers, furnace and air-conditioner fans, refrigerator fans, food processors, vacuum cleaners, drills, electric saws, floor polishers, and electric clocks.
- *Audio-visual* devices use electricity to produce sound and/or a visual image. Generally used for entertainment or education, they include radios, cassette recorders, disc players, phonograph record players, television sets, and computers.

Modern appliances offer us convenience and variety, but they do consume energy.

The Measurement and Cost of Electrical Energy

The next time you use a toaster or radio, try to find the label on the appliance that tells you how much electrical energy it uses in one second. Recall from your previous studies that a unit of energy is called a **joule** (J). If a device uses energy at the rate of one joule per second, then its power rating is one **watt** (W). In other words, one watt equals one joule per second. This equation could be expressed as:

$$1\,W = \frac{1\,J}{1\,s}$$

This equation could be expressed another way:

$$1\,W \cdot s = 1\,J$$

Energy can be thus measured in either joules or in **watt seconds** (W·s). This unit is, however, too small to be practical for measuring electrical energy consumed in a household. Instead, a larger energy unit, called the **kilowatt hour** (kW·h), is used. One kilowatt equals 1000 W, and one hour equals 3600 s. Therefore,

$$1\,kW \cdot h = 1000\,W \cdot 3600\,s$$
$$1000\,W \times 3600\,s = 3\,600\,000\,W \cdot s$$

Suppose your local electric company charges you seven cents for each kilowatt hour of electricity. Suppose also that in your home there is a television set that consumes at a rate of 200 watts and that the television set is on for 150 hours in an average month. How much would the monthly cost of power for the television be? First, change the watts to kilowatts.

$$200\,W = 0.2\,kW$$

Next, multiply the number of kilowatts by the number of hours the television runs.

$$0.2\,kW \times 150\,h = 30\,kW \cdot h$$

Last, multiply the number of kilowatt hours by the cost per kilowatt hour.

$$30\,kW \cdot h \times \frac{\$0.07}{kW \cdot h} = \$2.10$$

In Activity 4-18, estimate the power used by various appliances in your home, and calculate the cost of the power.

NOW, HOW MUCH DOES IT COST ME?

Surveying the Cost of Electrical Energy

1. Make a list of several electric appliances in your home.
2. On a single day, record which appliances are used and for how long they are used.
3. Using Table 4-2, or the power rating indicated on the appliances, estimate the total number of kilowatt hours consumed by the appliances in a 24 h period.
4. Find out the cost per kilowatt hour of electricity in your home. Use your result from Question 3 to calculate the cost of running appliances in your household over a period of 24 h.

5. Think of some benefits to cutting back your consumption of kilowatt hours.
6. What could you do to reduce the number of kilowatt hours consumed by
   (a) your household?
   (b) your school?

Table 4-2 *Power Rating of Typical Household Appliances*

ELECTRIC APPLIANCE	POWER RATING
kitchen stove	12 000 W
clothes dryer	4 600 W
kettle	1 500 W
toaster	1 000 W
coffee maker	600 W
vacuum cleaner	500 W
colour TV	200 W
desk lamp	60 W
clock	4 W

Most smaller, relatively low-power household electrical devices, such as desk lamps, have the common two-prong plug. Other devices, such as home computers, have three-prong plugs. Very high-power appliances, such as electric stoves, have four-prong plugs. Find out why some appliances must use three- or four-prong outlets instead of a two-prong outlet.

A Safe Supply of Electricity

To measure the actual amount of electricity you use, your house or apartment building is equipped with a meter. If electricity is delivered through overhead wires, the meter is likely to be on the outside of the building. If the electricity comes through underground cables, the meter may be in the basement. If there are several households in the building, there may be a separate meter for each one. The meter consists of a motor with a counting mechanism. The more current that is being used, the faster the counting mechanism turns, indicating the amount of energy that is being used. Your household is billed by the electric company on the basis of a reading taken from the meter.

There are several safety features in any properly wired household. At the location of the meter, there are three wires, two hot wires and one neutral wire. After passing through the meter, the three wires go to the service entrance panel, or fusebox. There, a master switch can be used to turn off all the power in the household. Once the electricity goes into the household circuits, devices such as fuses and circuit breakers ensure that the electricity has a safe passage.

Neutral and Hot Wires

One of the three wires that enters your home is connected to the ground; that is, it is **grounded** and is electrically neutral. The other two are "hot" (or, as they are sometimes called, "live"). If you touch a live wire, the shock can be fatal.

(b)

The three wires that enter your home first pass through the meter (a), and then connect to the service entrance panel (b).

(a)

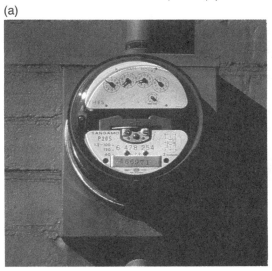

The three wires make it possible to obtain either 120 V or 240 V of electrical energy. If a voltmeter were connected to the neutral wire and one hot wire, it would give a reading of 120 V, the maximum voltage required by most household appliances. If a voltmeter were connected to both hot wires, it would give a reading of 240 V, the maximum voltage required by larger appliances such as clothes dryers and stoves. For extra safety, such appliances must be plugged into special outlets.

The master panel distributes electricity throughout the household circuits. Each circuit is designed to carry only a certain amount of current; too much current could cause the wires to overheat and could start a fire. For safety, fuses or circuit breakers are installed in the panel to limit the amount of current in each circuit.

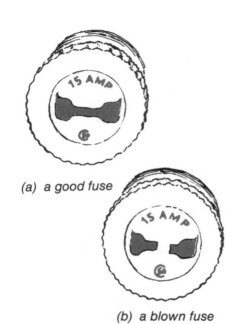

(a) a good fuse

(b) a blown fuse

Fuses

A fuse is a thin piece of a special metal alloy that melts rapidly if the current passing through it exceeds a certain limit. The melting of the metal breaks the circuit and thus stops current from flowing.

If the wiring of a circuit can safely handle a current of up to 15 A, then a 15 A fuse is placed in the circuit. Fuses of 20 A and 30 A are used in circuits able to handle higher currents. Once fuses are melted or "blown," they must be replaced with a fresh fuse. *Never* place a 20 A or 30 A fuse in a circuit designed for a current limit of 15 A. Doing so can allow too much current in the circuit, causing the wires to overheat and even catch fire.

Circuit Breakers

Instead of a fuse, some circuits use a circuit breaker. One type of circuit breaker consists of a bimetallic strip connected to a switch. When the current goes through the circuit, this strip is heated. One metal expands more than the other, causing the strip to bend. The higher the current, the greater is the bending. Too high a current in the circuit will cause the strip to bend enough to "trip" the switch, thus breaking the circuit. Once the overload is corrected, the circuit breaker can be reset. Unlike fuses, circuit breakers do not have to be replaced every time a circuit is broken.

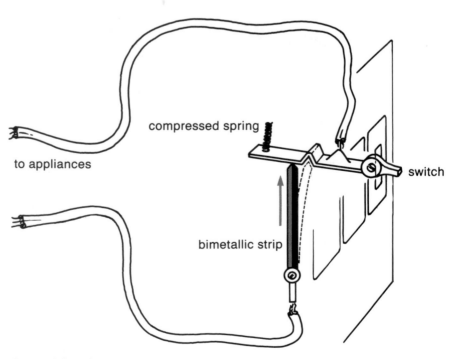

to appliances

compressed spring

switch

bimetallic strip

A circuit breaker.

Four dangerous situations. Water is a good conductor; you could receive a painful electric shock if you use an appliance while you're wet. Overloaded circuits can heat up, causing a fire. Fire can also result from a "short circuit" in frayed wiring. Power lines are kept well away from trees and buildings; you could be electrocuted if you touch them.

Electricity and Safety in the Home   **215**

Working with Telephone Systems

Electricity is used in our society for many applications. Featured here are just two of the numerous careers that apply the physics and mathematics of electricity.

A Traffic Engineer

Gillian Woodruff is a member of the scientific staff at Bell Northern Research. She works as a traffic engineer in the telephone industry. In her career, traffic refers to telephone calls. Her department evaluates new equipment to be sure it will carry the expected load of telephone traffic without unreasonable blocking and delay. To do this, they use mathematical modelling or computer simulations.

What made Gillian decide on a career in telephone traffic engineering? "I enjoyed mathematics, statistics, and physics in high school. I followed a B.Sc. (Engineering) program, in which I took both pure mathematics and electrical engineering courses. This prepared me for continuing my studies in either mathematics or engineering. I chose to continue with an M.Sc. in electrical engineering. Engineering degrees generally open doors to more employment opportunities in industry than do the more theoretical subjects."

An Electronics Technologist

Electronics technologist, Julia Pederson, makes sure that equipment is operating correctly before it leaves the factory. The company she works for manufactures multiplex systems, the electronics behind the telephone system.

The many panels of control equipment that make up multiplex systems all have to be tested, and Julia tests each one. In testing an oscillator panel, which produces waves, she first uses an oscilloscope to check the shape of the waves. She then compares the wave on the oscilloscope screen to that in the engineer's drawings. Next, she checks the wave's frequency using a counter. Finally, she uses a power meter to test the strength of the signal. What Julia finds most interesting is trying to pinpoint why a component isn't working properly. Once she locates the problem, it's her job to fix it.

Julia took math and science in high school and was able to go directly into the two-year Biomedical Electronics program at an institute of technology. "Although I took the biomedical program, I have never worked in that field. After graduation I went directly into telecommunications."

Gillian Woodruff

Julia Pederson

Checkpoint

1. The diagrams (a), (b), (c), and (d) show four circuits. In which one(s) will the bulb(s) light? Redraw the other(s) to show how the circuits could be changed so the bulbs will light.

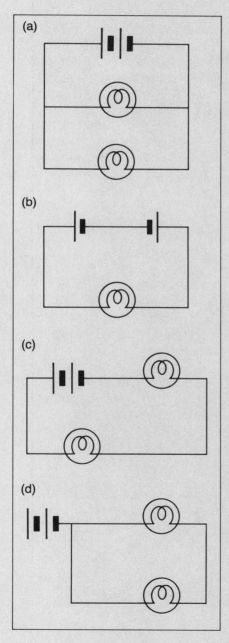

2. (a) Explain the difference between a parallel circuit and a series circuit.
   (b) Draw a diagram for a circuit that has a battery of two cells, a variable resistor, and a motor, all in series.
   (c) Draw a diagram for a circuit that has a battery of four cells, with a motor and a lamp in parallel.

3. Explain why one bulb in a chandelier may burn out, yet the others continue to shine.

4. Consider the circuit shown below. In which of the following situations would current flow to the bulbs and the bell?
   (a) Switch 1 is open, but switches 2 and 3 are closed.
   (b) Switch 2 is open, but switches 1 and 3 are closed.
   (c) Switch 3 is open, but switches 1 and 2 are closed.

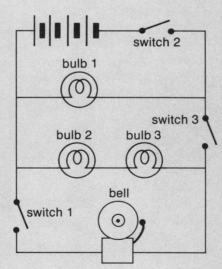

5. Draw a circuit diagram for a parallel circuit that has two motors and a lamp. There should be switches to turn each device on and off, and a switch that turns both motors on but doesn't affect the lamp.

6. On one day, a household used a dryer for 2 h, a colour television for 4 h, a vacuum cleaner for 1 h, and a toaster for 0.2 h.
   (a) What was the total energy these appliances consumed that day? Express your answer in kW·h. (Use Table 4-2 on page 212 to determine the power ratings.)
   (b) If electrical energy costs 7¢/(kW·h), what was the total cost of the energy used by these appliances on that day?

7. Explain why each of the following is unsafe:
   (a) a 20 A fuse inserted in a fusebox socket designed for a 15 A fuse;
   (b) an electric drill operated in a room where the floor is wet;
   (c) a nail used to pick away dirt that has collected in a wall socket.

Focus

- Static electricity is the presence of an electric charge that is not moving.
- Electric current exists when electrons flow and is measured in amperes (A).
- Chemical cells convert chemical energy into electrical energy. Two or more cells joined together make up a battery.
- All cells contain two different metal electrodes and an electrolyte solution.
- The energy of electric charges in a cell is measured in volts.
- Different substances offer different amounts of resistance to electric current.
- A generator is a device that converts mechanical energy into electrical energy. It is made from a coil of wire and a magnet that move relative to one another.
- A generator produces more current when the magnet is stronger, when the coil has more turns of wire, and when the motion of the magnet relative to the coil is speeded up.
- Electric motors convert electrical energy into mechanical energy.
- Electrical energy can be converted into other forms of energy, and they can be converted to electrical energy.
- An electric circuit is a complete pathway for electric current.
- A series circuit has only one conducting path for electrons, whereas a parallel circuit has two or more alternative paths.
- The amount of energy used by electrical devices in homes is measured in kilowatt hours (kW·h).
- Devices such as fuses and circuit breakers help ensure that electric circuits are safe.

Backtrack

1. Explain the difference between:
   (a) static electricity and current electricity.
   (b) a cell and a battery.
   (c) an electrode and an electrolyte.
   (d) a generator and a motor.
   (e) an open circuit and a closed circuit.
   (f) a parallel circuit and a series circuit.
   (g) a fuse and a circuit breaker.
2. In what units do you measure
   (a) current?
   (b) voltage?
   (c) energy used by electrical devices in your home?
3. You have been stroking your cat on a winter day. When you reach to turn on the light, you receive a shock. Using what you know about static electricity, explain why you got a shock.
4. Of the following, which would produce the most effective chemical cell? Explain your answer.
   (a) two copper strips placed in a beaker of salt water
   (b) two zinc strips placed in a beaker of pure water
   (c) a copper strip and a zinc strip placed in a beaker of pure water
   (d) a copper strip and a zinc strip placed in a beaker of salt water
5. A certain metal has much higher resistance than nichrome. If someone uses wire made of this metal to replace nichrome wire in a toaster, what would you expect to happen to the heat produced by the toaster?
6. A lamp is powered by a generator. What will happen to the brightness of the lamp if
   (a) more turns of wire are added to the generator's coil?
   (b) the generator's shaft is turned more slowly?
   (c) a stronger magnet is used in the generator?
7. How can you increase the lifting strength of an electromagnet?
8. (a) List the major parts of a typical electric motor.
   (b) Explain the function of each part.
9. What function does a switch have in an electric circuit?

10. (a) Draw a diagram of an electric circuit that contains a battery of three cells and two lamps in series.
(b) Draw a circuit diagram to represent a string of decorative lights with eight lamps connected in parallel.

11. In the circuit diagram shown below, insert
(a) one switch symbol to show how the light can be turned on and off without affecting the motor;
(b) a second switch symbol to show how the motor and the lamp can be controlled at the same time.

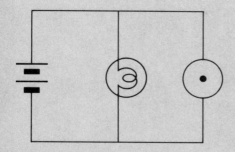

12. It's house-cleaning day. You spend 3 h running the vacuum cleaner and 4 h running the clothes dryer.
(a) How much energy, in kilowatt hours, did you consume? (Use the power ratings in Table 4-2, page 212.)
(b) If energy costs 7¢/(kW·h), how much did your day of house cleaning cost?

Synthesizer

13. (a) If you comb your hair with a plastic comb and then bring it near some small pieces of paper, the paper will cling to the comb. Explain this phenomenon.
(b) After a few seconds, the paper doesn't cling to the comb any more, but falls off. Why?

14. A "motocross" is a race in which lightweight motorcycles are driven over a cross-country obstacle course. Suppose a battery-manufacturing company hires you to design a battery for a motocross motorcycle. What design feature would you include? Why?

15. Small generators are sometimes used to power lights on bicycles.
(a) How is the coil in this kind of generator made to turn?
(b) When the generator is in use, why might it take a bit more effort to ride the bicycle?

16. (a) Which components do a motor and generator have in common?
(b) Explain how turning the shaft of the motor by hand makes it work like a generator.

17. Thermocouples are used in ovens to make electric thermometers that measure the temperature of the oven. What is the advantage here of using a thermocouple rather than an ordinary thermometer?

18. Sketch a design for a burglar alarm. You may use any or all of the following: a flashlight with batteries in it, a long roll of insulated copper wire, an iron spike, a frying pan, an empty tin can, and scissors.

19. (a) Draw a diagram to represent a circuit with three motors in parallel, connected to a battery of three cells.
(b) Design a switch that will, at the same time, turn one motor off and another motor on.

A motocross motorcycle.

Diversity of Living Things

Imagine a penguin swinging through the trees of a tropical forest. How about a monkey diving below the Antarctic seas in search of fish? Both seem like crazy ideas, but why? In fact, penguins and chimpanzees have much in common. Both animals have internal skeletons and similar organs. Both grow, move, obtain food, and reproduce. However, they carry out their activities in very different ways and in very different environments. A penguin's flippers and webbed feet are designed for moving it underwater, but not along the branches of a tree. Its beak is shaped for catching fish, but not for eating fruit. Its structure and behaviour allow it to survive in the polar seas, but not, as with the chimpanzee, in a tropical forest.

In this Unit, you will investigate why there are millions of different kinds of organisms, each with its own way of life and place to live. You will explore why a particular species has the shape and structures it does. You will also see that diversity exists among individuals of the same species, as well as among different species. You will look at evidence that shows how species may change in appearance over several generations. As well, you will discover how scientists use similarities and differences to classify living things, and how classification helps people to understand the relationships among types of organisms.

Diversity and Adaptation

In and around your community, you can probably identify many different living things, such as trees, dandelions, grass, cats, ants, pigeons, spiders, and people. Because these organisms look so different from one another, you might find it hard to believe that they have anything in common. Why does each type of organism look the way it does? For example, why does a fish have fins, gills, and a streamlined shape? Why does a tree have branches and leaves? Why do people have legs?

Why do all these animals have fins?

By comparing the shapes and structures of many different organisms, you may begin to see a pattern. Take a structure such as fins. Thousands of different kinds of animals have fins, from scorpion fish to eels, and from sea horses to dolphins. All of these animals live in water. From observations such as these, you may infer that fins have functions that help animals survive in water. From more observations and experiments, you may conclude that most animals with fins use these structures to stabilize, steer, and move themselves through the water.

Adaptations are features that increase an organism's chances of surviving and reproducing in a particular environment. Fins are an adaptation for locomotion underwater. Organisms also have adaptations that allow them to obtain food, digest food, protect themselves, reproduce, and so on. Adaptations include an organism's behaviour, internal structures, and **life cycle** (its stages of development) as well as its external appearance. Taken together, adaptations help fit each organism to a particular environment and way of life.

Adaptations of this seaweed for life under water are very different from the adaptations of the cactus.

To confirm that each organism is adapted to the particular environment where it lives, you could try moving an organism to an environment in which it is not normally found and then observe whether it survives there. A cactus plant that grows in dry areas on land would not survive in the sea. On the other hand, a seaweed that thrives in the sea would die on land. Each of these organisms has structures that adapt it to one environment but not another.

Suppose you visit a museum and see an organism that you have never seen before. Its label is missing. Using the concept of adaptation, can you infer what environment the organism lives in and how it survives? If the organism has gills, you may infer that it lives in water. If it has large eyes, you may infer that it is active in the dark. If it has teeth, their size and shape may tell you what it feeds on.

An organism may have several structures adapted for one general function. For example, the strong, curved claws and the large canine teeth of a lion are both adaptations for obtaining food. As well, one structure may have more than one function. For example, the roots of a plant serve both to anchor it in the ground and to obtain water and dissolved nutrients.

In the next Activity, your class will look at a variety of organisms and match some structures to particular functions or environments.

Did You Know?

The English naturalist Charles Darwin once observed a large orchid that grows on the island of Madagascar. Its flowers produce a sweet-tasting nectar at the base of 20 to 30 cm long tubes. Observing this, Darwin predicted that there must be a certain insect with a sucking tube, or proboscis, long enough to reach the nectar of this flower. When the orchid was discovered, there was no known insect that had a proboscis long enough to do the job. Remarkably, a nectar-feeding moth with a 30 cm long proboscis was discovered on Madagascar 40 years later!

Based on his knowledge of flower structure and function, Darwin had been confident that such an insect must exist in the area. He was right.

The nectar is stored here, in a tubular extension of the flower.

Structures for Survival

Problem

How do various structures help organisms to survive?

Materials

large sheet of paper
reference materials about plants and animals

Procedure

1. Choose a particular plant or animal to research. Make a sketch or obtain a picture of it. Fasten your sketch or picture to the middle of the large sheet of paper.
2. Record the name of the organism and the environment it lives in. Alongside the sketch or picture, list some functions that are important to the survival of the organism. For example:
   - locomotion,
   - obtaining food,
   - protection from other animals or plants,
   - protection against physical factors such as dryness, wind, snow, heat or light.
3. Beneath each function, record the adaptations that are associated with it. Draw a line to connect each adaptation to the appropriate part of the organism.

4. If necessary, make separate detailed illustrations to show structures that are small or are inside the organism.
5. Once you have completed your research and prepared your report, compare the various structures and functions of your organism with those of your classmates.

Analysis

1. Choose one function, such as protection from predators. List all the structures found by your class that carry out this function for different kinds of organisms.
2. Give examples of two kinds of organisms living in the same environment that have
   (a) no structures in common for locomotion or obtaining food.
   (b) one or two structures in common for locomotion or obtaining food.

3. Give examples of two kinds of organisms living in different environments that have
   (a) no structures in common for protection from predators.
   (b) one or two structures in common for protection from predators.

Further Analysis

4. Imagine you could take a structure from one kind of organism and use it to replace part of, or add onto, another kind of organism. Describe the structural change and explain why the new structure might
   (a) increase the organism's chance of survival in its environment.
   (b) decrease the organism's chance of survival in its environment.
5. (a) Describe or sketch the appearance of a structure you might expect to find on each of the following and explain the function of that structure:
   • an animal that has many predators;
   • an animal that lives in water;
   • an animal that feeds on plant roots.
   (b) Compare the different structures your class came up with for each organism above.

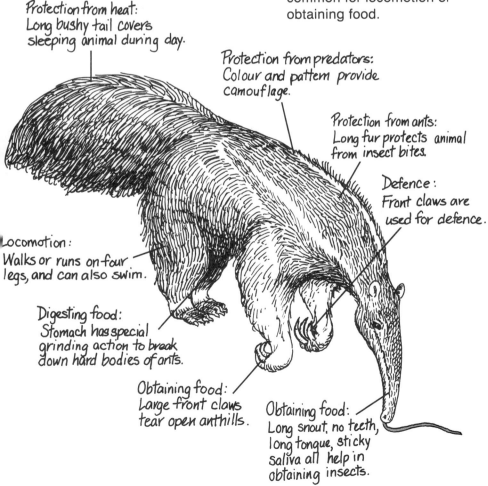

Protection from heat:
Long bushy tail covers sleeping animal during day.

Protection from predators:
Colour and pattern provide camouflage.

Protection from ants:
Long fur protects animal from insect bites.

Defence:
Front claws are used for defence.

Locomotion:
Walks or runs on four legs, and can also swim.

Digesting food:
Stomach has special grinding action to break down hard bodies of ants.

Obtaining food:
Large front claws tear open anthills.

Obtaining food:
Long snout, no teeth, long tongue, sticky saliva all help in obtaining insects.

The Same But Different

Using structures as a guide, you can look for similarities among the huge variety of living things. By sorting organisms with similar structures into groups and then comparing the groups, you can test the idea that each organism's appearance is related to a particular environment or way of life.

Consider a group that includes all organisms with leaves. Leaves are adapted in various ways to carry out photosynthesis and have a number of features that help them do this. The green pigment chlorophyll, present in leaves, captures sunlight. The veins in leaves transport materials to and from the leaf tissues. The pores in the surfaces of leaves allow gases to pass in and out. In these and other ways, all leaves are the same. Yet organisms with leaves vary greatly in appearance from one another, and leaves themselves come in many different shapes and sizes. Can these differences also be interpreted as adaptations?

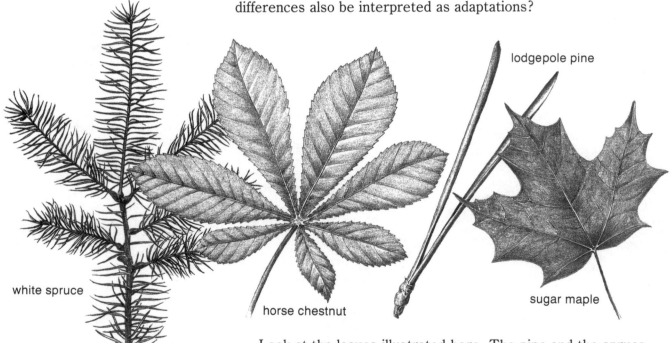

white spruce

horse chestnut

lodgepole pine

sugar maple

How are these leaves the same? How are they different?

Look at the leaves illustrated here. The pine and the spruce leaves are both needle-shaped, and the maple and the horse chestnut leaves are both broad and flat. Pine and spruce commonly grow in cold, dry conditions; maple and chestnut grow best in milder, wetter areas. The different shapes may therefore be related to survival in different climates. A needle-like shape reduces evaporation and helps conserve moisture. Leaves with this shape are found on many kinds of desert plants as well as on those growing in the cold north. In climates where water loss is not as much of a problem, it is an advantage for a leaf to have a larger surface area to absorb sunlight.

As this example shows, the concept of adaptation can be applied at several different levels. There are broad adaptations shared by many organisms, such as the adaptations of fins for swimming, of wings for flight, and of leaves for photosynthesis. Within these large groups there are narrower adaptations, shared by fewer organisms. They may be adaptations for different climates, or for different ways of obtaining food, or for different methods of locomotion, protection, or reproduction. The smallest group, in which all the organisms have a very similar set of adaptations, is the **species**. The members of one species differ from other species in at least one characteristic, and they usually reproduce only among themselves (that is, with other members of their own species).

In the following Activity, you'll be measuring the diversity of plants that grow in your schoolyard or in a nearby park. In what ways are these plants similar? In what ways are they different? How might each variation be an adaptation for a particular condition or way of life? How many different species of plants can you distinguish?

Probing

Using an appropriate reference, investigate how each of these birds lives and how large each one is. How does the structure of each bird's type of feet help adapt it to its way of life? (These are not drawn to scale.)

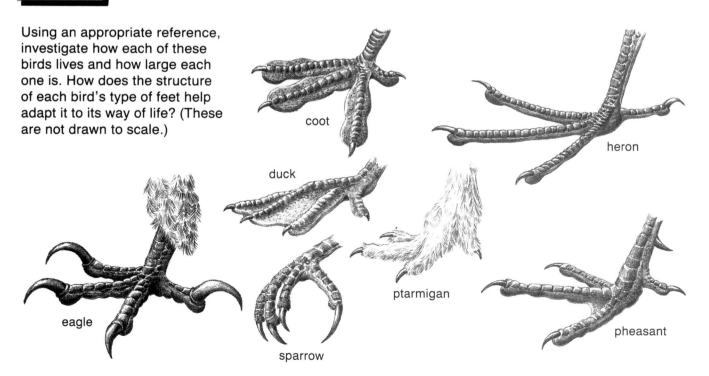

coot

duck

heron

eagle

ptarmigan

pheasant

sparrow

Measuring Diversity

Problem

What differences and similarities are there among the plants growing in your area?

Materials

hand lens
metre stick
notebook
field guides

Procedure

1. Prepare a table to describe the appearance of each species found by your class. The table should allow you to record answers to questions such as:
   - Is it a small plant, a bush, or a tree?
   - Is it upright, climbing, or crawling along the ground?
   - How tall is it?
   - What are the shape and colour of its leaves?
   - Does it have flowers?
   - Does it bear fruit?

2. Work in groups, with each group studying one particular area of the schoolyard or park.
3. Describe the environment of your group's area. Is it sunny or shady? Exposed or sheltered? Damp or dry?
4. Begin at one side of your area and work slowly through to the other side, recording each new species of plant you encounter. Record the names of any species you can identify. Sketch the others, noting special features that will help you identify them back in your classroom. Note any special adaptations that each of the plant's structures may have, such as thick, waxy leaves, tough stems, hairs, or spiny projections.

CAUTION: Do not break or pick any plants without your teacher's approval.

5. In the classroom, combine your results with those of the other groups. Record the total number of different plant species found by your class.

Diversity Within Species

Analysis

1. Name three structures that all of the plants have in common.
2. Choose two plants from your own study area. Describe or sketch one structure that:
   (a) looks similar in both plants.
   (b) looks different in the two plants.
3. Describe three adaptations from among the plants observed by your class. Infer how each helps the plant survive or reproduce.

Further Analysis

4. All the species of plants that grow in your area are surviving in a particular environment. In your library, find illustrations of plants that grow in very different environments, such as a tropical forest, a desert, or a coniferous forest. Select three plants, each one from a different environment.
   (a) How does each plant appear different from those that grow in your area?
   (b) How does each appear similar to those that grow in your area?

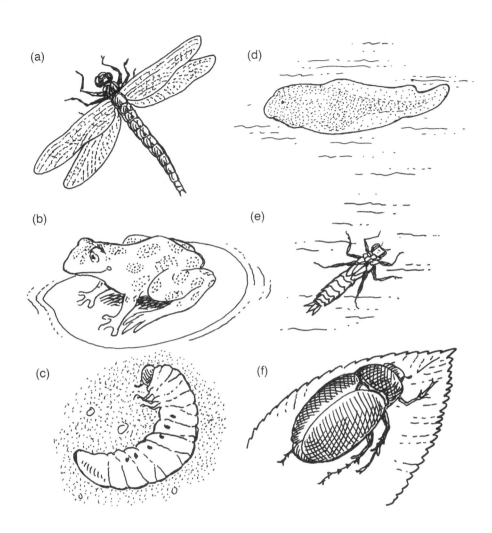

(a)   (d)

(b)   (e)

(c)   (f)

How many different species are shown here? At first glance, there appear to be six species, because each individual organism looks quite different from all the others. In fact, there are only three species shown. Considerable diversity may be found within the *same* species, just as it is found among different species. Each different form within a species has adaptations for a different function or a different environment.

For example, the young of some species look very different from the adults. In the illustration above, (d) is a tadpole, the young form of a frog (b); (c) is a grub, the young form of a June beetle (f); (e) is a nymph, the young form of a dragonfly (a). Such dramatic changes in appearance and habits during an individual's life cycle are called **metamorphosis**. Frogs, June beetles, and dragonflies undergo metamorphosis during their life cycle.

Butterflies as well undergo metamorphosis. A caterpillar (the **larva**, or young form, of a butterfly) has adaptations for obtaining food. Most of its long body consists of a digestive tract, and it has large, biting mouth parts. It spends most of its time eating leaves and growing larger. In the next stage, the pupa, the caterpillar's tissues are re-specialized to form the adult butterfly. The adult butterfly has adaptations mainly for reproduction. Many adult butterflies do not feed at all, or feed only on nectar. They do not grow in size. Their role in the life cycle is to find a mate and reproduce. After mating with a male, the female butterfly lays the eggs that hatch into the next generation of caterpillars.

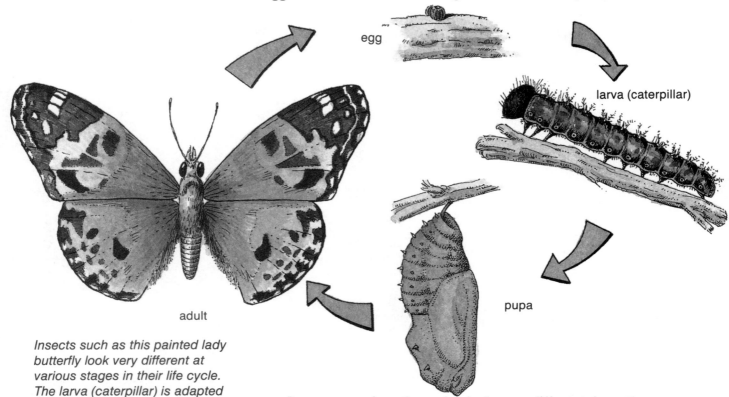

egg

larva (caterpillar)

pupa

adult

Insects such as this painted lady butterfly look very different at various stages in their life cycle. The larva (caterpillar) is adapted for one way of life; the adult butterfly, for another.

In some species, the males look very different from the females. This variation in the appearance of the sexes is called **sexual dimorphism**. For example, the plain brown female peafowl shown on the next page looks very different from the elegantly coloured, long-tailed male of the species. The bright colours of the male help him to attract a mate. After mating, the female bird sits on the eggs and later looks after the young that hatch from them. The female's drab colours are an adaptation that helps to conceal her from predators.

In the other illustration, you can see that a female paper nautilus (an animal similar to the octopus) not only differs in appearance from the male but also is nearly ten times bigger! Can you think of a hypothesis to explain this difference?

Male and female peafowl. The peacock (male) and peahen (female) look so different, you'd hardly think them the same species.

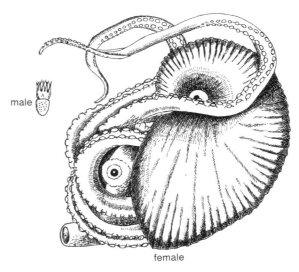

male

female

The paper nautilus shows an extreme sexual dimorphism in both size and appearance.

Among species of social insects (those insects living together in colonies), such as bees, ants, and termites, there may be several different-looking forms of both males and females. The existence of several distinct forms within the same species is called **polymorphism**. Each form has adaptations for a particular function, such as reproduction, protection, dispersal, or obtaining food. For example, in termites like the ones shown below, the workers build the nest and keep it in operation. The soldiers have large jaws that they use to defend the nest against attackers. The "king" and "queen" are mated for life, adapted to perform one function — reproduction.

The words metamorphism, dimorphism, and polymorphism all use a Greek word *morph*, meaning "form." The first part of each word also comes from Greek. *Meta* means "change"; *di* means "two"; and *poly* means "many." If you understand what each part of the word means, you can better understand the meaning of the whole word.

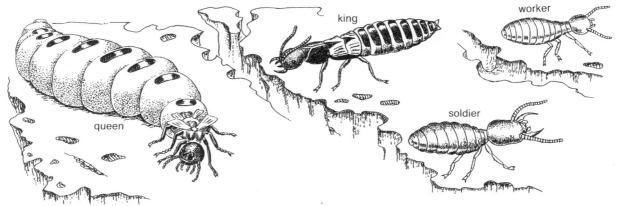

queen

king

worker

soldier

Other Variations Within Species

In addition to these dramatic examples of diversity within species, there are numerous other variations in form that are less extreme. Many animals have young that look moderately different from themselves. Ostriches, wild pigs, tapirs, and several species of deer have young that are striped or spotted. The young lose these patterns when they mature. Young arctic hares are darker than their parents and young harp seals are lighter than their parents. Other variations are related to changing seasons. Several northern species, such as the ptarmigan, ermine, and snowshoe hare, grow lighter in colour every winter. Many trees lose their leaves each winter, making them look very different from their summer appearance. All of these variations may be seen as adaptations for survival.

A birch tree adapts to winter conditions by losing its leaves.

Young deer and young seals look different from their parents, but are recognizable as the same species.

The arctic hare (shown here in its summer adaptation) and the weasel change colour to remain camouflaged in both winter and summer.

Reproduction and Survival

Would you expect a cat to give birth to puppies, or an apple seed to grow into a rose bush? Despite the sometimes large variations within species, observation shows that cats only produce more cats, and apple trees give rise to more apple trees. By reproducing versions of themselves, organisms can pass on adaptations that allow their young to survive and reproduce in their turn.

In many organisms, offspring may be produced from a single individual. For example, plants like the one shown below produce runners. At intervals along each runner, tiny new plants sprout roots and leaves. This is one type of **asexual reproduction**: reproduction that doesn't require two individuals to mate before young can be produced. Asexual reproduction can occur in a variety of ways. It is most common among plants and micro-organisms such as bacteria and fungi but also occurs in some animals, like the hydra shown below. Every hydra can produce more hydra by budding off tiny replicas of itself. When fully grown, each offspring produced by asexual reproduction is an exact copy of the parent from which it developed. As long as the environment remains the same, offspring produced in this way will be able to survive and reproduce as well as their parents did.

Most organisms, like the cats shown in the cartoon, produce offspring by **sexual reproduction**. This process requires two parents. Each parent produces special reproductive cells, called **gametes**, that combine at fertilization. (The sperm and egg are gametes.) A new individual grows from the fertilized cell, or **zygote**.

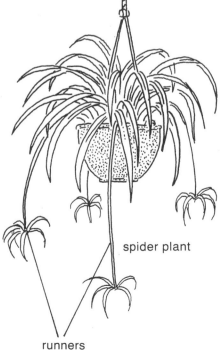

spider plant

runners

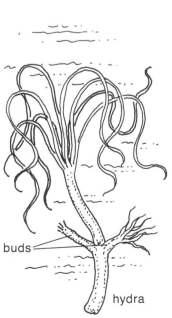

buds

hydra

The spider plant and the tiny pond-living hydra (an animal) can both reproduce asexually.

Inheriting Traits

Whereas offspring produced asexually are very like the individual from which they came, offspring produced sexually resemble both parents, but are not exactly like either one. The characteristics that an organism inherits from its parents are called **traits**.

These puppies have the same parents, but each one is different from all its siblings and from both its parents.

Some inherited traits can be changed or modified by the influence of the environment, whereas others cannot. If you inherit Type A blood, for example, your blood type will always be, and can only be, Type A. On the other hand, you might inherit a tendency to be tall. How fast you grow, and what height you eventually reach, however, will be affected by a variety of factors, including how much and what you eat while you are still growing.

In humans, traits include such things as the colour of a person's hair, skin, and eyes, the blood group, the shape of nose and lips, and the tendency to be short-sighted or to become bald. Some of these traits can be seen in the newborn infant, but others do not appear until later in life. What traits do you have that are similar to those of someone in your family? Although family members may resemble one another closely, no individual is exactly like another —with the exception of identical twins. They are produced from a single zygote that separated into two shortly after fertilization. Identical twins therefore have exactly the same traits.

If you ever become a parent yourself, can you predict the chances that your children will inherit the same forms of some trait that you have? In the next Activity, you will find that some forms of traits are more common than others. You should note that having a more common form of a trait is neither better nor worse than having a less common form. The traits you will be observing are simply ones that are fairly easy to distinguish.

Tracking Traits

Problem

Are some forms of traits more common than others?

tongue roller

non-roller

free ear lobe

attached ear lobe

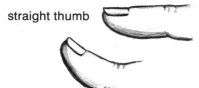
straight thumb

curved thumb

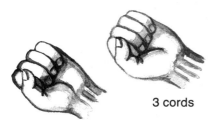
2 cords

3 cords

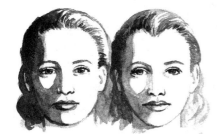
straight hairline   widow's peak

Materials

information in Table 5-1
mirror

Procedure

1. Prepare a table similar to Table 5-1. In it, record under "My Form of the Trait" the appropriate description.
2. See whether you can roll your tongue into a U-shape as shown. Record whether you are a roller or non-roller.
3. Record the shape of your ear lobes. (You may need to look in a mirror, or have a friend help you decide.)
4. Hold up your thumb and note whether it is fairly straight or curved back.
5. Hold one hand in front of you with the palm facing you. Roll your fingers in to make a tight fist, then bend your hand back at the wrist as shown. Observe how many cords stand out below your wrist.
6. See whether the hairline at your forehead goes straight across or forms a point (called a "widow's peak") at the centre. (Again, you may need to look in a mirror, or have a friend help you decide.)
7. Combine the class results and record totals for each form of each trait.
8. Calculate the percentage of the class having each form of each trait.
9. Answer the questions on the following page.

Table 5-1

TRAIT	MY FORM OF THE TRAIT	FORMS OF THE TRAIT	CLASS TOTAL	PERCENTAGE OF CLASS
tongue roller		roller		
		non-roller		
ear lobe attachment		free		
		attached		
thumb shape		curved		
		straight		
number of cords in wrist		two		
		three		
hairline shape		straight		
		widow's peak		

Analysis

1. Based on the data from your class, are some forms of traits more common than others? If so, describe them.
2. (a) How many students in your class have exactly the same combination of all forms of the five traits investigated?
   (b) Make a general statement about variation in humans, based on your results.

Further Analysis

3. All the traits you have looked at have only two possible forms. Everybody has either one form or the other. Give an example of a human trait that has more than two forms.
4. A person's height is not determined as a choice of two or three alternative heights. Rather, it may fall anywhere on the scale between the shortest and the tallest human. Height is therefore said to be *continuously variable*; that is, it varies along a continuum. Give another example of a trait that is continuously variable.
5. If there was no variation among inherited traits, would you be able to tell the difference between individuals of the same species? Explain.

Extension

6. Find out the form of the trait a person is likely to inherit if:
   (a) both parents have attached ear lobes.
   (b) both parents have free ear lobes.
   (c) one parent has attached and one parent has free ear lobes.
7. Find out how non-identical (fraternal) twins are produced.

Did You Know?

A pattern in the inheritance of traits was first recorded in detail in 1865 by Gregor Mendel, an Austrian monk. He produced several generations of pea plants and observed the occurrence of different traits in each generation. Mendel measured two alternative forms of each trait (just as you did in Activity 5-3). He found that one form of each trait occurred in about 75% of individuals while the other form occurred in only 25%. From Mendel's work developed the science of heredity, or genetics.

Gregor Mendel (1812–1884)

Genes, Inheritance, and Sex

What does it mean when we say a girl got her brown eyes from her father and her black hair from her mother? What did she receive from each parent? At fertilization, when the development of a human begins, each parent contributes one gamete: a sperm cell from the father and an egg cell from the mother. Inside these cells are units called **genes**, which contain instructions that control the development of inherited traits. Each trait is controlled by a different gene or by several genes working together. Thus, there are genes for eye colour and genes for blood group, and so on. Studies on the inheritance of traits indicate that genes have two or more forms, called **alleles**. Whether you have attached ear lobes or free ear lobes, for example, depends on which alleles are present. Many different combinations of alleles are possible in each offspring.

Consider why, in the inheritance of traits, sexual reproduction is important to many kinds of organisms. You know that offspring produced by asexual reproduction are copies of their single parent and therefore have the same adaptations. What might happen to these organisms if the environment in which they live changes? Obviously, if the environment changes over time and the organisms have not changed, the offspring will be less well adapted. On the other hand, in species that reproduce sexually, forms of traits are shuffled with each generation. Because there are so many combinations of traits produced in the offspring, some individuals in a large population may be better adapted than others to a changed environment. At least some members of the species will thus probably survive (and continue reproducing) in the new environment.

The potential for variation made possible by sexual reproduction is enormous. A male and a female of an imaginary species having only 100 traits could produce more than 1 000 000 000 000 000 000 000 000 000 000 (1×10^{30}) different genetic combinations among their offspring! (That is not to say they *could* have that many offspring, of course; but if they did, none of them need be identical!)

In the early 1900s, Canadian scientists experimented to develop new varieties of wheat that would be better adapted to the conditions of the Prairies. A scientist named Charles Saunders crossbred hundreds of varieties of wheat from all parts of the world to produce plants with different combinations of traits. After years of experiments, Saunders developed one variety that ripened earlier than others, withstood early frosts, and produced high quality flour. Within a few years, this variety of wheat, named Marquis, was being planted throughout the Prairies.

Artificial Selection

Long before the discovery of genes, people knew that certain characteristics are inherited. Most of this knowledge came from the experience of farming. Thousands of years ago, people learned that if they saved the seeds from their crops and planted them, they would get a new crop similar to the last one. Over time, some people realized that they could improve their domestic plants by choosing which ones to breed. For example, a farmer might choose to save and plant the seeds from only the largest corn cobs. The other corn would be eaten or traded. When this process was repeated year after year, the farmer would gradually develop corn plants that, on average, grew larger corn cobs than did the previous ones. Such deliberate alteration of particular traits is called selective breeding or **artificial selection**.

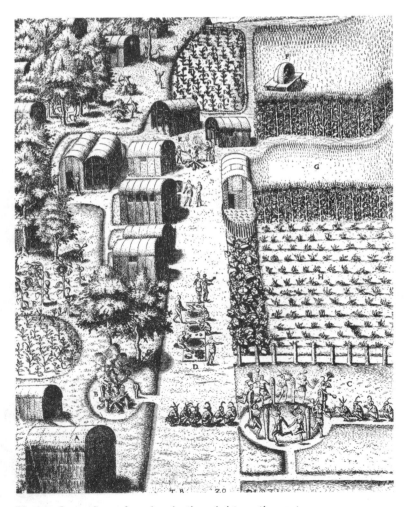

Native Canadians farming in the eighteenth century.

(a) skull of greyhound

(b) skull of European wolf

Although most dogs do not look much like wolves, their skulls have many similarities.

All of our domestic varieties of animals and plants, such as cattle, sheep, pigs, dogs, hens, wheat, rice, tomatoes, and apples, were developed from wild species by the process of artificial selection. All domestic dogs, for example, are thought to be descended from wolves. Although some dogs look very different from wolves, their internal structure and behaviour are quite similar. Dogs have the same basic shape and arrangement of bones, teeth, skull, brain, and digestive system as the wolf has. The relationship between dogs and wolves is so close that individuals of many larger varieties of dogs may be crossbred with wolves.

As you know, there are a great many varieties of dogs. Each variety was gradually produced by crossbreeding only those offspring with desired forms of traits. Poodles, for example, were developed by selecting dogs that had features such as curly hair and floppy ears. These dogs were bred with one another; some of the resulting puppies inherited these traits. They, in turn, were bred only with other dogs with the same traits. After many generations, and perhaps hundreds of years, this process of artificial selection produced a distinct variety of dog. Other varieties of dog were selected for other traits, such as small size, long hair, ability to retrieve, or a good sense of smell.

Many breeds of dogs have been produced by selective breeding.

In 1933, a typical North American hen laid an average of 126 eggs per year. Using the process of artificial selection, animal breeders slowly developed a variety of hen that laid more eggs. Today, the average number of eggs produced by a hen in a year is 230.

While shuffling genes to get the results they want, breeders may also produce some unwanted traits in their animals or plants. Several breeds of dog, for example, have medical problems connected with their appearance. Breeds with flattened faces, like the Pekingese, often have breathing problems. Dogs with floppy ears often have ear infections. Other breeds have problems with their hips or eyes as a result of inherited characteristics.

In the next Activity, you'll have a chance to consider what forms of traits could be selected through artificial breeding of several different types of organisms.

Activity 5-4

Artificial Selection

1. For each of the plants and animals listed below, write in your notebook one or more characteristics that a breeder might artificially select for breeding. Explain why each trait would be regarded as advantageous by the breeder.

racehorses	beef cattle
apple trees	dairy cattle
turkeys	rodeo cattle
hunting dogs	sheep
guard dogs	corn
roses	wheat

2. Could any of these organisms survive on its own in the wild, without being cared for by humans? Explain why or why not.

3. All the vegetables shown here were produced by artificial selection from one species, the wild cabbage. Which traits do you think the breeder wanted to develop in each type, and why?

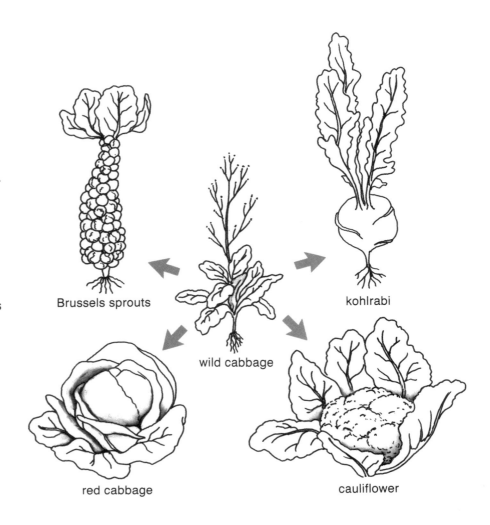

Brussels sprouts

kohlrabi

wild cabbage

red cabbage

cauliflower

Checkpoint

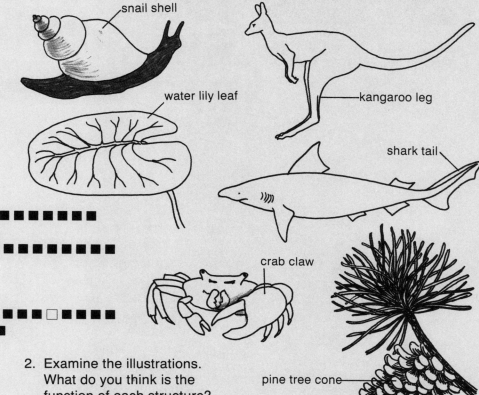

snail shell

water lily leaf

kangaroo leg

shark tail

crab claw

pine tree cone

1. Use the definitions to find the correct words for the word puzzle. When you have completed the puzzle, the highlighted letters will spell two words that have to do with an organism's development between birth and reproduction.

(a) ■■■■■□ ■■■■■■■■■■■
(b) ■■■■■■■□■■■
(c) ■■■■□■■■■■ ■■■■■■■■
(d) ■■■■■□
(e) ■■■□■■■■
(f) ■□■■■■■
(g) ■■■■■■■■ ■■■■■■■■■□■■■
(h) ■■□■■■■■■■■■■■
(i) ■■■□■

(a) a process that allows for the "shuffling" of genes (two words)
(b) features that increase an organism's chances of survival and reproduction
(c) also known as selective breeding (two words)
(d) reproductive cell
(e) a group of organisms that can interbreed with one another
(f) fertilized egg
(g) results in offspring that are an exact copy of the parent (two words)
(h) the existence of several different forms within the same species
(i) small units within cells with instructions that control the inheritance of traits

2. Examine the illustrations. What do you think is the function of each structure?

3. Suggest a structure that each of these organisms uses to carry out the function mentioned:
   (a) a bean plant reproduces;
   (b) a goldfish moves from place to place;
   (c) a cactus plant conserves water;
   (d) a hawk builds a nest.

4. (a) Explain the meaning of the terms metamorphosis, sexual dimorphism, and polymorphism.
   (b) Give an example of an organism in which each of these occurs.
   (c) For each example, suggest how the different forms of the species are adapted for special functions or environments.

5. Horses and zebras are similar in many ways. What advantage might stripes be to a zebra in its environment?

6. What are two differences between sexual reproduction and asexual reproduction?

7. You are growing several red rose plants. Some are bright red, others are pale; some have a strong scent, others a weak one.
   (a) Name at least two traits of these roses.
   (b) How would you go about trying to produce a group of roses, all of which were bright red and strongly scented?

How Species Change

When an organism such as a bird produces offspring it helps to ensure that the species to which it belongs will continue to exist for another generation. Adaptations that allowed the parent birds to survive and reproduce are inherited by the offspring. But not all of these young will survive to have offspring of their own. Some of them may be eaten by predators; others may die from starvation, disease, or other causes. As you have observed, offspring produced by sexual reproduction differ slightly from one another and from their parents. Is there any way you might be able to predict which offspring are more likely to survive, and which are not? Or is survival simply a matter of chance?

The following Activity is a model to show how variations in a trait might affect the survival of a population of moths that tend to spend the day resting on the trunks of trees. In the Activity, black and white paper squares represent the two forms of an insect called the peppered moth. You yourself will be the predator, acting as a bird that eats the moths. Large sheets of black and white paper represent the tree trunks in the moths' habitat.

Many young birds, such as these robin nestlings, do not survive to adulthood. Predators, accidents, a poor berry crop—any of these may take their toll.

A Matter of Life or Death

Problem

How might a predator affect a population of prey?

Materials

50 white paper squares (about the size of a dime)

50 black paper squares (the same size)

drinking straw

large sheet of white paper (about 1 m²)

large sheet of black paper (about 1 m²)

timing device

bowl

Procedure

1. Work in a group with at least four other students. Your group will need *either* one black sheet *or* one white sheet, and squares of both colours. So that the class can compare results, the number of groups using a black sheet should be the same as that using a white sheet.

2. Prepare a table similar to Table 5-2 in which to record your data.

3. Place the sheet of paper on the floor. A white sheet represents the surface of a light-coloured tree trunk; a dark sheet, the surface of a dark-coloured tree trunk.

4. Randomly scatter all the paper squares of both colours on the sheet. Make sure the squares are all separated so that one does not lie on top of another. The white squares represent white moths; the black squares, black moths.

5. In your group, choose a student to represent a bird that feeds on moths and another student to act as timekeeper. The "bird" must catch "moths" by sucking on one end of the straw, picking up a square, and continuing to use the straw to transfer the square to a bowl. Practise this technique a few times before going on to the next step.

Table 5-2

	NO. OF WHITE MOTHS AT START	NO. OF BLACK MOTHS AT START	NO. OF WHITE MOTHS AT END	NO. OF BLACK MOTHS AT END	NO. OF WHITE MOTHS +50%	NO. OF BLACK MOTHS +50%
1st generation	50	50				
2nd generation						
3rd generation						
4th generation						
5th generation						

6. At a signal from the timekeeper, the bird begins to catch moths as quickly as possible, continuing to do so until the timekeeper signals that 30 s have passed.

7. After the 30 s, count the black and white moths that *remain* on the sheet. Record these numbers in your table.

8. Have the surviving moths now "reproduce." To do so, add 50% to the surviving total of each colour on the sheet. (For example, suppose 20 black moths remain; 50% of 20 is 10. Add 10 more black squares at random to the sheet. Round up fractions to the next whole number.) Record the new totals in your chart.

9. Repeat Steps 6 to 8 for four more "generations" of moths. For each "generation," have other students take the part of the bird.

10. After the fifth generation, average your group's final results and the results of other groups that used a sheet of the *same* colour.

Analysis

1. At the beginning of the experiment, the population consisted of 50% black moths and 50% white moths. How has this ratio changed after five generations for moths living
   (a) against the white background?
   (b) against the black background?

2. Is this result what you might have predicted? If not, why?

3. Was either colour of moth completely safe from the bird?

Further Analysis

4. After several more generations, what might happen to the population of moths living
   (a) against the white background?
   (b) against the black background?

5. Suppose there were a predator that hunted by using its sense of smell, rather than sight. What might happen to the original moth population if it were hunted by this predator?

6. In the model represented by this Activity, were there some factors that were not considered? If so, what were they, and how might they have affected the results?

Monarch butterflies and Viceroy butterflies are two different species, yet they look so much alike it is difficult to distinguish between them. The appearance of one *mimics* the appearance of the other. In another case of *mimicry*, the *Ophrys* orchid (a flower) looks very much like the wasp that pollinates it. Use a reference book to discover an explanation for these and other cases of mimicry in the natural world.

The Case of the Peppered Moth

The Activity you have just completed is based on real-life events that took place in 19th-century England. Before 1848, every individual peppered moth that biologists had observed had light-coloured wings. After 1848, however, people began to notice a dark-coloured form of the moth. From that time on, they noticed that the proportion of dark moths steadily increased. By 1898, dark moths outnumbered the previously more common light-coloured moths in some areas by 99 to 1.

Looking for an explanation for this change, the biologists noticed that dark moths were commonest around cities, where the tree trunks had been darkened by air pollution from newly developed industries. In the countryside, most tree trunks were covered with light-coloured lichens. A scientist named H. D. B. Kettlewell hypothesized that the change in the moth population had come about because dark moths were better adapted to the industrial environment; that is, they had a better chance of survival because birds that eat moths would find it harder to see them against the dark trunks. On the other hand, the light-coloured moths, which had an advantage on light-coloured trunks, would be easily picked out against the darker trunks.

To test his idea, Kettlewell released equal numbers of light and dark moths in two different areas, one with light tree trunks and the other with blackened trunks. Some time later, he recaptured moths in both areas and calculated the proportions of light to dark individuals. His results (see Table 5-3) indicated that the chances of survival were better for dark moths in areas with dark trunks and for light moths in areas with light trunks. To confirm that the birds relied on their sense of vision, Kettlewell filmed them hunting moths. In some cases, birds passed right over a moth that was camouflaged against its background!

Against a dark tree trunk, a dark moth is camouflaged and a light moth stands out. Against a light trunk the opposite is true.

Table 5-3 *Kettlewell's Results*

unpolluted wood (light trees)		WHITE MOTHS	BLACK MOTHS
	released	496	473
	recaptured	62	30
	% recaptured	12.5%	6.3%
polluted wood (dark trees)	released	137	447
	recaptured	18	123
	% recaptured	13.1%	27.5%

Theory of Natural Selection

The process that produced the change in the moth population is called **natural selection**. It works in the same way as artificial selection except that conditions in the environment, rather than deliberate intervention by people, determine which organisms survive and reproduce. Two English naturalists, Alfred Russel Wallace and Charles Darwin, first proposed the theory of natural selection during the last century. After studying organisms in different parts of the world, both men came to the same conclusions independently of each other. The theory was explained in detail in Darwin's famous book, *On the Origin of Species*, first published in 1859. It describes not only how organisms become adapted to their environment, but also how species may gradually change over time and develop into new species.

The idea that species may change over time, or **evolve**, was not new. But Darwin and Wallace came up with an explanation of how this process might happen. The key observations that led to their theory are:

- there is variation in traits among the individuals in a species;
- individuals pass on forms of traits to their offspring by reproducing;
- not all individuals reproduce.

Darwin spent five years on a trip around the world in a sailing ship, hired as the ship's naturalist. His observations of fossils, landforms, plants, and animals formed the foundation of his life's work.

You have already made these observations for yourself. Darwin realized that some of the variations among individual offspring might determine which of them survived and reproduced in a particular environment and which did not. For example, some individuals might have forms of traits that make them better at avoiding a predator, obtaining food, surviving cold or dry conditions, or attracting a mate. These advantages might be due to a structure, such as a thicker coat, deeper roots, or smaller ears. Or they might involve a behaviour, such as ''freezing'' to hide from a predator, or displaying to attract a mate. Whatever the reason, only the organisms that survive and reproduce pass on their characteristics to the next generation. In this way, the environment continually ''selects'' the individuals best suited to it, that is, the individuals best at surviving and reproducing in it.

These male Common Grackles are displaying to each other by pointing their bills upward. This aggressive behaviour serves to threaten a competitor. The "winner" in this competitive display is more likely to be selected as a mate by a female grackle.

Whether a particular characteristic is helpful or harmful to an organism depends not only on the trait itself, but also on the organism's environment, including other individuals with whom it competes. For example, is it an advantage for a plant to be tall or short? As you can see in the illustration, a tall plant may have an advantage in one environment, and a short plant may have an advantage in a different environment. In some cases, the fate of an individual may not be related to its traits at all. Its fate could be determined by chance alone. For example, if an area becomes flooded under 10 m of water, then neither tallness nor shortness makes much difference to a plant's survival.

windy environment

sunny environment

In each situation, which plant— the tall one or the short one—has a better chance of survival? What features of the environment are important to consider?

Many traits may provide neither an advantage nor a disadvantage. For example, the colour of your eyes is unlikely to affect your chances of surviving or reproducing. Nevertheless, it is beneficial for a species to have a wide range of forms of traits among its members. If the environment changes rapidly, a form of trait that once made little difference to an individual's survival may suddenly save its life.

The result of natural selection in Darwin's theory has sometimes been called "survival of the fittest." Many people wrongly think this phrase means that only the biggest and strongest individuals survive. However, "fittest" here means *those that are best fitted to their environment.* The fittest individuals may be smaller, or bigger, or slower, or quicker, than others. They may have any combination of characteristics that give them an advantage over other individuals in a particular environment.

The Appearance of New Species

You have seen that in a species such as the peppered moth, new adaptations that alter its appearance can arise. Can adaptive changes in populations eventually produce new species?

Recall that a species may be defined as a group of organisms that can interbreed with one another. In nature, however, the members of a species do not have an equal chance of breeding with all other members of their species. Organisms tend to live in distinct populations that may have little or no contact with one another. For example, consider a population of fish that has become isolated in a cave. Over time, these fish become more and more different from the members of their species outside the cave. In a dark environment there is no advantage to having eyes; the energy used to develop eyes, not needed in this environment, is used in other ways. Over the generations, fish without eyes have an advantage over those with eyes until, eventually, the entire population is blind. In this and undoubtedly other ways, the isolated fish have become so different from their ancestors that they can be considered a new species. This is how scientists think various species of blind cave fish developed.

Blind cave fish.

As another example, suppose that a large lake has one population of trout living at one side of the lake and another population of the same species living at the other side of the lake. Both populations, as members of the same species, share a similar appearance and behaviour. However, individuals from the two populations may be unlikely to meet and interbreed. In the next Activity, you will study a situation that might cause two such populations to become two different species.

Two Species from One

The three illustrations show changes in a lake environment over thousands of years. Study the illustrations carefully and explain the process that appears to be taking place among the fish that inhabit the lake.

Analysis

1. Why did the two populations of fish develop different forms of traits?

2. Suppose the water level in the lake rose once more, allowing the two groups of fish to mix as they had been able to do previously. What factors might prevent fish in the two populations from interbreeding?

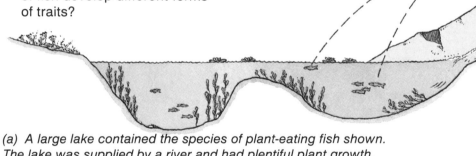

(a) A large lake contained the species of plant-eating fish shown. The lake was supplied by a river and had plentiful plant growth.

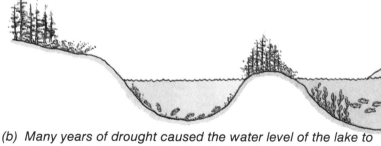

(b) Many years of drought caused the water level of the lake to drop drastically. A ridge extending the length of the lake then separated one portion of the lake from the part supplied by the river. It became two separate lakes.

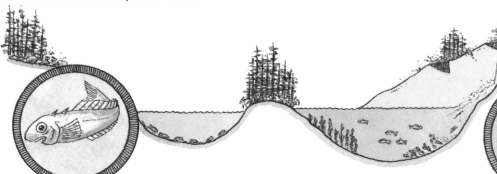

(c) Over a long period of time, this is what happened.

How Species Change **249**

Relationships among Different Species

Experiments with animals in captivity, as well as evidence from populations in the wild, show that organisms considered to be different species can, in fact, mate with each other. Zebras and horses have mated and produced offspring, as have lions and tigers. Horses and donkeys are often crossbred to produce mules, and wolves and dogs can also interbreed. Such experiments demonstrate that the boundaries between similar species are not rigid.

The fact that species such as horses and zebras share many traits is evidence that they may have developed from two populations of a single species. In other words, species such as horses and zebras may have inherited traits from a shared ancestor. Their differences have probably evolved because they live in different environments.

Many biologists are interested in learning about the relationships among species. Similarities and differences in the appearance and behaviour of organisms may be clues to genetic relationships. On the basis of resemblances among them, different species can be grouped into "families." In the next Topic, you'll look at the problems of identifying species and establishing relationships among them.

Relatives: a zebra, a hinny, and a donkey. (A hinny is the offspring of a donkey and a pony.)

Classifying Living Things

There is such an overwhelming number of organisms that no one could possibly know all of them. Biologists estimate that there may be as many as 50 million different species of organisms on Earth today. Anytime human beings encounter a large number of objects, they tend to group them together into similar categories. It has been only natural for people to try to make sense of the multitude of living things by devising systems to classify them.

Five Kingdoms

What characteristics would you use to classify living things? More than 2000 years ago the Greek philosopher Aristotle divided all known living things into two groups, plants and animals. He further divided each of those two groups into three others. He grouped plants as trees, shrubs, or herbs, according to the structure of their stems. He grouped animals according to where they lived: in the water, on land, or in the air.

Can you see any problems with Aristotle's system of classification?

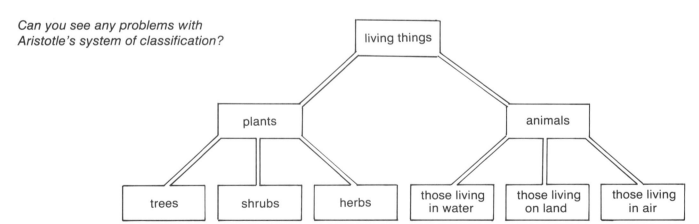

At the time when Aristotle was alive, the Greeks knew of only about 1000 different kinds of organisms. Today, scientists have named and described more than a million species. As Aristotle did, modern scientists still recognize animals and plants as two major groups. But there are many living things that do not fit easily into either of these groups.

Microscopic organisms, such as bacteria, fungi, and most single-celled organisms, have been known and studied only since the invention of the microscope in the 17th century. Scientists at first grouped some of these organisms with animals and others with plants. Yet as they further studied the structure and characteristics of these organisms, they had to extend the classification system and provide different groups for them.

Most scientists now divide all living things into five major groups, based mainly on differences in the organisms' structures. These groups are called **kingdoms**. As you examine the system of classification used today, remember that it is only a catalogue of the world's diversity of life. Scientists developed the system to help them understand the relationships among living things. The system is a useful device, but it may change as this understanding changes.

The five kingdoms generally recognized today are shown below. How are the organisms in each kingdom alike? How much diversity can be found in each kingdom? A brief survey will give you an introduction to the creatures — large, small, beautiful, and strange to the human eye — that exist on Earth.

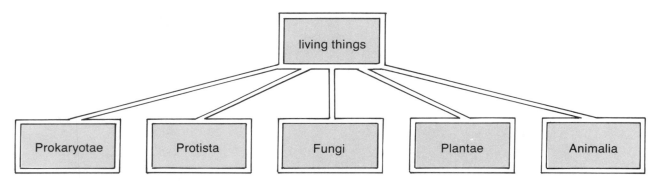

```
                    living things
        ┌──────┬────────┬────────┬────────┐
   Prokaryotae  Protista  Fungi  Plantae  Animalia
```

Prokaryotes (bacteria) are extremely small. Even with the best light microscope they are barely visible. There are many kinds of bacteria, found in every environment on Earth, with three general shapes: round (coccus), rod-like (bacillus), and spiral (spirillum).

Prokaryotae (or Monera)

This kingdom contains the bacteria. They are simple **unicellular** (single-celled) organisms that may live singly or in colonies. Unlike all other organisms, bacterial cells do not have a separate cell nucleus. (For a review of cell structure, see Skillbuilder Five, *Using the Microscope*, on page 359). The name of this kingdom is made up from the words *pro*, "before" and *karyon*, "nucleus." In many books, the kingdom is called Monera (from the Greek *moneres*, "single"). More than 10 000 species of bacteria have been described and named.

Protista

The members of this kingdom are more diverse in appearance than those of any other kingdom. They are most easily defined by what they are not: they are not bacteria, fungi, plants, or animals. Most protista are unicellular organisms; however, some scientists also include unicellular and **multicellular** (many-celled) algae in this kingdom. The amoeba, euglena, and paramecium are some of the most familiar protists. All protists have cell nuclei and other cell structures.

Some of the many kinds of protists. Like the bacteria, some of these are disease-causing, others are not. Most are microscopic. (These are not drawn to scale.)

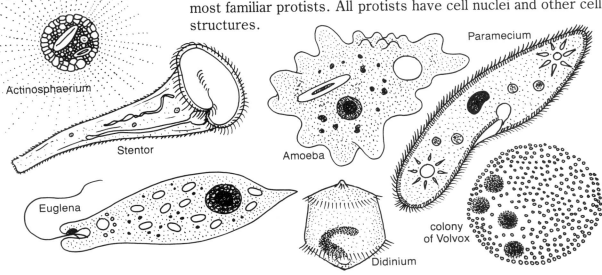

Fungi

Yeasts, moulds, and mushrooms are perhaps the best known of the numerous types of fungi. Some fungi are unicellular and others are multicellular. Fungi were once grouped with plants but, unlike plants, they are not capable of photosynthesis. Instead, they must obtain their food from other organisms. They digest this food outside themselves and then absorb it. Fungi have stiff cell walls made of chitin. It is estimated that there are about 100 000 species of fungi.

Several kinds of fungi. From whatever source—rotting wood, damp bread, or an old orange—all fungi must obtain their food from other organisms.

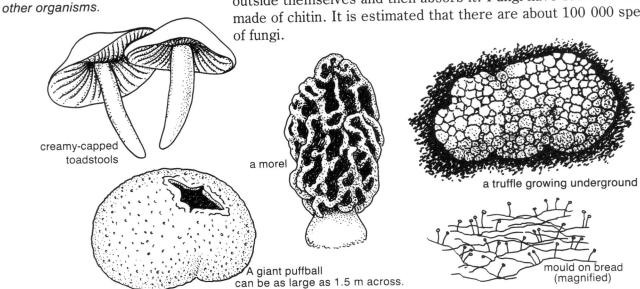

Plantae

This kingdom groups together multicellular organisms that make their own food by photosynthesis. Around the outside of each cell, there is a rigid cell wall. Plants include such organisms as ferns, mosses, trees, flowering shrubs, and herbs. Some scientists include algae in this kingdom.

For sheer beauty, nothing can beat Kingdom Plantae. This diverse group provides the world with food and oxygen.

Animalia

It's difficult to imagine that organisms as different as a jellyfish, a sea urchin, a kangaroo, a snail, and an ant are similar, but they are. They are all members of Kingdom Animalia. This kingdom consists of multicellular organisms, most of which are capable of locomotion. All are dependent on other organisms for their food. Their cells are bounded by flexible membranes only. Animals include sponges, jellyfish, worms, insects, fish, amphibians, reptiles, birds, and mammals, as well as many other groups. Later in this Unit, you will look more closely at the tremendous diversity within the plant and animal kingdoms.

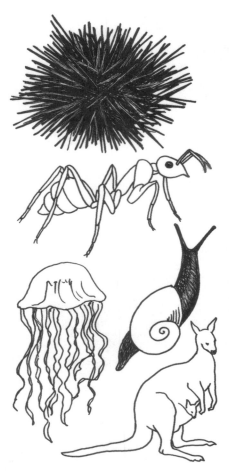

Kingdom Animalia includes a lot more than just the familiar large animals. Insects are some of the most numerous; scientists have listed nearly 8000 species of ants and 290 000 species of beetles! All the large mammals, taken together, total only 3500 species.

A Hierarchical System

The classification of all living things into five kingdoms allows people to distinguish important differences between, say, a grasshopper and a pine tree, or a mushroom and an amoeba. But each kingdom includes thousands of species, which differ greatly among themselves. To distinguish among the different species in the plant or animal kingdoms, for example, each of these large groups must be subdivided several times. After each division, there are fewer species within each subgroup. The smaller the group, the more closely related are the species in the group. Eventually, the process of division produces a very large number of "groups," each containing only a single species.

Such a system is said to be **hierarchical**. Biologists today use a classification system with seven major levels:
- kingdom,
- phylum (plural form: phyla),
- class,
- order,
- family,
- genus (plural form: genera),
- species.

Every organism is classified at each level. For example, as shown in Table 5-4, the Kingdom Animalia includes multicellular organisms, most of which can move from place to place. Phylum Chordata, the level below, includes only those organisms that have a flexible rod, called a notochord, along their backs. Class Mammalia contains only those chordates that suckle their young. Order Carnivora includes only meat-eating mammals with similarly specialized teeth. Within this order are several families, including Family Canidae (dogs, wolves, and foxes), Family Ursidae (bears), and Family Felidae (all cat-like carnivores, including lions and tigers). Within Family Felidae are several genera, including *Felis* (cats and tigers), *Lynx* (lynxes and bobcats), and *Acinonyx* (cheetahs). Table 5-4 shows how a house cat would be classified at all levels in the hierarchy.

Table 5-4 *An Example of Hierarchical Classification*

LEVEL	EXAMPLE	SOME ANIMALS INCLUDED IN THIS LEVEL
Kingdom	Animalia	earthworm horse lynx tiger house cat bobcat dog lizard snail
Phylum	Chordata	horse lynx tiger house cat bobcat dog lizard
Class	Mammalia	horse lynx tiger house cat bobcat dog
Order	Carnivora	lynx tiger house cat bobcat dog
Family	Felidae	lynx tiger house cat bobcat
Genus	*Felis*	tiger house cat
Species	*Felis domesticus*	house cat

Different members of the cat family. Which ones are most closely related to the house cat, Felis domesticus?

lynx
Lynx canadensis

European wildcat
Felis sylvestris

bobcat
Lynx rufus

black-footed wildcat
Felis nigripes

mountain lion
Puma concolor

jaguar
Panthera onca

lion
Panthera leo

tiger
Panthera tigris

Linnaeus and the "Two-Name" System

You may wonder why individual species of organisms are given names such as *Felis domesticus*, or why the larger groups have names such as Felidae or Carnivora. These names are derived from Latin. The system of classification that uses such names was first developed in the 1700s by a Swedish scientist named Carl von Linné. During Linné's time, most scientific writing in Europe was done in Latin. (In fact, Linné generally signed his name in its Latinized version, Linnaeus.) The use of Latin has been kept for naming living things because it allows all scientists, regardless of the language they speak, to refer to an organism by the same name. Without such a system, it would be very confusing to talk about organisms. A common animal or plant may have hundreds of different names in different languages. Even in English, the same organism may have several different names. For example, the "groundhog," "marmot," and "woodchuck" are different names for the same animal. Its Latin, or scientific, name is *Marmota monax*. Everywhere in the world, biologists refer to it by that name.

The scientific name of each species always consists of two words. The first word is the name of the genus in which the organism is classified. It is written with a capital letter. The second name is never capitalized. Because there are often several species in one genus, there may be several species with the same first name. For example, *Canis lupus* is the wolf, *Canis latrans* is the coyote, and *Canis familiaris* is the domestic dog. The fact that these three animals have the same first name tells you that they are in the same genus (*Canis*) and therefore that they are very similar to one another. It also tells you that any other animal with the first name *Canis* must be very similar to them and not, for example, a bird or a cat. In the next Activity, you will put yourself in the place of a scientist who has just discovered a new species. You need to give the species a scientific name so that anyone anywhere can use it when speaking or writing about the organism.

Name That Species

Imagine you are a scientist and need to give a scientific name to each of the species shown here. The names should describe some characteristic of the organism. You might also choose to put very similar organisms in the same genus. Use some of the Latin or Greek names in Table 5-5 or make up names of your own (in Latinized form). Here is an example to get you started. What might you name a sawfly that lives in Canada and has striped larvae that feed on trees? You might name it *Arbivorous lineatus*, or *Dendrophilos canadensis*. Check the list to see what these names tell someone about the insect.

Analysis

1. Trade your list with a partner. Match each of your partner's names to one of the organisms.
2. Did your partner's scientific names help you identify the organisms? Explain.

Extension

3. Look up the scientific names of five species similar to some shown here and try to interpret their meaning.

Table 5-5 *Common Latin and Greek Terms Used in Biological Classification*

ala (L), wing
anser (L), goose
aquila (L), eagle
arbor (L), tree
arctos (G), bear
avis (L), bird
campestris (L), from the plains
cephal (G), head
chen (G), goose
dendron (G), tree
demos (G), house
equus (L), horse
ensis (L), belonging to; "from"
erythros (G), red
lacerta (L), lizard
leukos (G), white
lineatus (L), lined; striped
maritimus (L), from the sea
mephitis (L), bad odour
mus (L), mouse
musca (L), a fly
myos (G), mouse
ophidion (G), serpent
pallens (L), pale
philos (G), loving; having an attraction to something
pteridion (G), wing
quinque (L), five
rubra (L), red
serpis (L), serpent
terra (L), ground
thalassa (G), sea
vermis (L), worm
volans (L), flying
voro (L), to devour

The Usefulness of Classification

The system of scientific names is more than an exercise in order and neatness. It allows scientists to communicate about species without confusion and to identify relationships among species. For example, look at the illustration of these three marine animals. On the basis of their resemblances and differences, should these three species be classified in the same genus, family, order, or class?

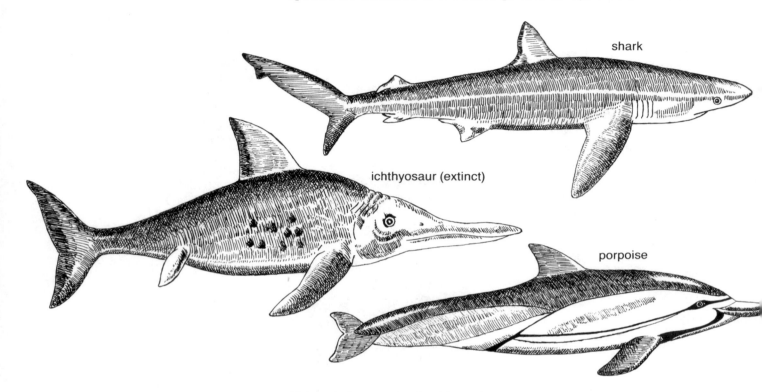

shark

ichthyosaur (extinct)

porpoise

Although these animals all have fins and a similar body shape, they are not closely related.

The similarities in appearance at first might suggest that these animals are closely related. A closer examination, however, shows major differences. For example, the shark has a rough, sandpaper-like skin, has a skeleton composed of a light, soft material called cartilage, and breathes using gills. It is a ''cold-blooded'' animal, or **ectotherm**; that is, its body temperature depends on the temperature of the surrounding environment. The ichthyosaur, well known from fossils, had a scaly skin, had a bony skeleton, and respired using lungs. The porpoise also has a bony skeleton and lungs, but a smooth skin. It is a ''warm-blooded'' animal or **endotherm**; that is, it can maintain its body temperature at a constant level independent of the temperature of its environment. Unlike either the shark or the ichthyosaur, the porpoise has mammary glands that produce milk for its young.

Because of these differences and others, the three animals are not grouped by scientists in the same genus, family, order, or even class! They are all members of the same phylum, however — the chordata. Within this phylum, the shark is classified as a fish, the ichthyosaur as a reptile, and the porpoise as a mammal. This classification indicates that, despite appearances, these animals have a lot more differences than similarities. Their superficial resemblance is a result of having similar adaptations to the same environment (the sea) and way of life (capturing other animals for food).

As you have seen, a system of classification has very practical value. In the same way that classifying books or records into categories can help you find one particular item on a shelf, the classification of living things can help you identify an "unknown" organism and find the group of organisms to which it belongs.

Keys to Identification

In order to discover which group and which species an unknown organism belongs to, it is useful to use a *key* — a checklist that describes various characteristics that an organism may or may not have. A **dichotomous key** is a list of alternatives arranged so that you can track down the group to which an organism belongs, and then find the name of the species itself. To use a dichotomous key, begin with the first choice and read *both* descriptions at that number. At the end of each description will be either a number that leads you to another pair of alternatives, or the name of the organism that fits the description. If you find the name of the organism, then you have identified it, and you need go no further in the key. If you find a "go to" statement, continue with further choices as indicated.

Using the animals on page 258 as an example, you could use a dichotomous key to identify them as shown in Table 5-6. Most dichotomous keys require far more choices in order to identify organisms, but this example and the one on the next page illustrate the idea, using just a few characteristics.

Table 5-6 *Sample Dichotomous Key*

CHARACTERISTICS	IDENTIFICATION
1. (a) has gill slits behind its head	shark
(b) has no gill slits behind its head	go to 2
2. (a) upper jaw extends far beyond lower jaw	ichthyosaur
(b) upper and lower jaws about the same length	porpoise

Using a Dichotomous Key

The illustration shows some common North American ungulates (animals with hooved feet). Some mammals in the Order Ungulata have horns and some have antlers. Scientists distinguish horns from antlers by looking closely at the structure of the material. However, the horns of common North American ungulates are relatively small and unbranched, so they are easily distinguished from antlers. Thus, for North American ungulates you don't need to examine the material in order to tell one from the other.

How many of the ungulates shown here can you identify? Even with some general knowledge, this could be a complicated task. A dichotomous key can help you solve this question of identity.

Activity 5–8

What Animal Am I?

1. In your notebook, list the letters (a) to (h).
2. Select one of the heads in the illustration. To identify it, work through the dichotomous key in Table 5-7 below, and record your answer beside the appropriate letter in your notebook.
3. Identify the other species shown, and write the name of each animal beside the appropriate letter.

Table 5-7 *Dichotomous Key for Some Common Ungulates*

CHARACTERISTICS	IDENTIFICATION
1. (a) animal with horns	go to 2
(b) animal with antlers	go to 3
2. (a) horns with prominent backward spiral curl	bighorn sheep
(b) horns short and dagger-like with slight backward curl	mountain goat
3. (a) antlers very broad and flat	moose
(b) antlers slender and branched	go to 4
4. (a) antlers thin, not prominent	go to 5
(b) antlers prominent, heavy, sweeping back, with several branches	wapiti (American elk)
5. (a) antlers with forward sweep	white-tailed deer
(b) antlers upright	go to 6
6. (a) antlers forked or V-shaped	mule deer
(b) antlers flatten and widen, with one branch extending over animal's brow	caribou

Some North American ungulates.

Making a Dichotomous Key

By following a few simple steps, you can make a dichotomous key that can be used by others to identify organisms. In the same way that you find it useful to produce an outline before doing a project, you will find it easier to produce a dichotomous key if you first draw a diagram to separate the characteristics you decide to use.

- Divide the group of organisms in two on the basis of a single difference. Make sure the difference allows you to categorize any member of the group into one of the two smaller groups.
- Divide each subgroup in two on the basis of another difference. Continue dividing each subgroup until you obtain groups, containing only a single organism. Draw a diagram to show the connections between the groups, as shown in the example of insect classification.

A diagram to help produce a key to identify insects.

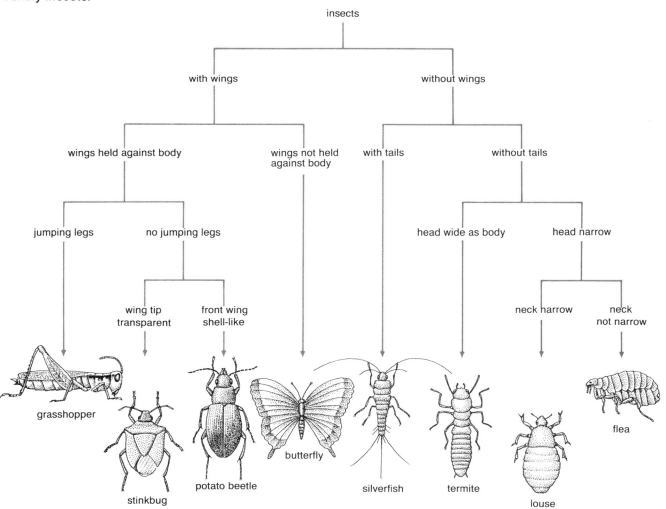

- To prepare a key from the information on your diagram, write down each pair of alternative characteristics in a numbered list. An example is shown in Table 5-8.

Table 5-8 *Dichotomous Key for Insects*

CHARACTERISTICS	IDENTIFICATION
1. (a) with wings	go to 2
(b) without wings	go to 3
2. (a) tails	silverfish
(b) no tails	go to 3
3. (a) head wide as body	termite
(b) head narrow	go to 4
4. (a)	
(b)	

- Finally, fill in the remaining steps. Note that in any dichotomous key, the number of steps will be one less than the total number of items to be classified. In this example, there are eight items to classify; therefore, there are seven steps in the dichotomous key. In the following Activity, you'll make your own dichotomous key for another group of organisms.

Activity 5–9

A Key for Leaves

Prepare a dichotomous key to identify each leaf in the group of leaves provided by your teacher. Start by organizing pairs of groups into a diagram as described on page 261. Use easily observed traits such as leaf shape, pattern of veins, and shape of leaf edge to create your groups. Using your diagram, produce a key.

Trade your dichotomous key with another student and try to identify the leaves by using it. Does the key that you prepared work? If not, how can you improve it?

Checkpoint

1. (a) List five kingdoms into which all living things are grouped, giving the main identifying characteristics of each.
   (b) Give examples of five living things that are in different kingdoms and name the kingdom to which each organism belongs.
2. What kind of environment might favour the survival of a plant with
   (a) long roots extending deep into the ground?
   (b) short roots?
   (c) long roots spread out close to the surface?
3. When first used, a type of rat poison proved very effective in killing rats. But after several years of its use, rats began to survive in places where the poison was present. Explain in a few sentences how and why the rats might have survived.
4. A pair of mice has five offspring born in an underground nest. After several weeks, the young mice venture out of the nest in search of food. Within a week, three of them are caught by owls, which hunt by sound at night. However, the other two offspring avoid capture and grow to adulthood. Suggest two traits that the surviving mice might have had that the captured mice did not have.

Table 5-9 *Some Related Species*

ENGLISH NAME	SCIENTIFIC NAME	FAMILY
house mouse	*Mus musculus*	Muridae
grey squirrel	*Sciurus carolinensis*	Sciuridae
red squirrel	*Tamiasciurus hudsonicus*	Sciuridae
eastern chipmunk	*Tamias striatus*	Sciuridae
black rat	*Rattus rattus*	Muridae
fox squirrel	*Sciurus niger*	Sciuridae

5. Why does the process of natural selection require the following in order to occur:
   (a) variation within a species?
   (b) inheritance of traits?
6. Salamanders like the ones shown here are washed by a flood into an underground cave and trapped there by a landslide. A stream flowing through the cave supplies enough food for the animals to survive and reproduce. Suggest two traits that might change among the population of salamanders over several generations as a result of natural selection.

7. Using the information in Table 5-9, answer the following questions. Give reasons for your answers.
   (a) Which two species are most similar to one another?
   (b) Which species is most similar to the house mouse?
   (c) Is the eastern chipmunk more like the red squirrel or the black rat?

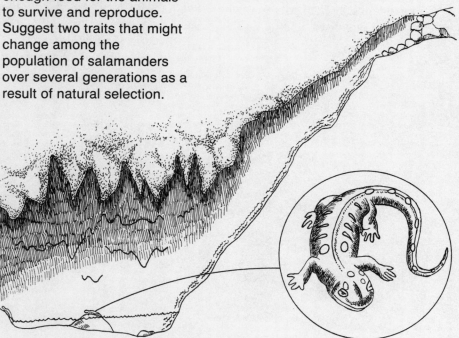

Science and Technology in Society

Why Save Species?

When you hear the word "sharks," you are unlikely to think of an animal that may save human lives. But recent studies on the blood and tissues of sharks have revealed amazing properties that may have many applications in medicine. Sharks are one of the few animals that seem immune to cancer. In nature, they rarely get sick; when wounded, they quickly heal. Curious to discover how sharks keep so healthy, scientists in Florida examined their blood. They found that it contains substances that destroy a large variety of harmful micro-organisms and disease-causing chemicals. Further studies of these substances from sharks may eventually help patients fight off infectious diseases.

With discoveries such as these, scientists learn that almost every species has something valuable to teach us. Although some organisms have been studied in great detail, scientists know little or nothing about most of the millions of species with which we share this planet. Many of these species may hold useful secrets.

Sharks are not the only organisms that have medical value. Most of the wonder drugs of the past half century have come from plants. There are numerous examples of plant chemicals (i.e., substances produced by particular plants) that we use as drugs, such as digitalis for heart failure, curare as a muscle relaxant, and quinine for malaria. Aspirin contains a chemical derived from the willow tree.

Another value of plants is as food. Today we depend on only 20 species of plants for 90% of the world's food supply. Yet about 70 000 species of plants are known to have edible parts, and at least some of these have a much greater nutritional value than plants we currently eat.

Other species of plants supply people with valuable materials. For example, the plant *Euphorbia lathrys* is rich in hydrocarbons, from which oils and gases can be extracted. This plant is being studied as a possible source of energy for the future. You may have seen another plant oil, called jojoba, listed on the ingredients of your shampoo. The seed of the jojoba plant is very rich in a clear, golden oil that is used as a top quality lubricant as well as in cosmetics, paints, cooking oils, and drugs. Moreover, this useful plant can grow in dry and salty conditions where few other species of plant can survive.

Some shark species are now in danger of extinction.

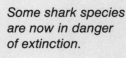

Seeds of the jojoba plant.

However, at the same time scientists are discovering the value of species, the growing human population is forcing many species into extinction. The greatest diversity of species is found in tropical forests, which are being cut and burned at an alarming rate.

Scientists are also concerned about the reduction in variation *within* some domesticated species. For example, the wheat plants in most fields are very similar to one another. Through selective breeding, most of them have inherited the same form of each trait. This makes it easier to harvest the field to ensure very productive wheat plants, but what happens when a disease attacks? What if none of the wheat is immune to the disease? Is there enough variability in the wheat so that at least some plants will have inherited an immunity to the infection, and thus survive? These are the kinds of questions that are leading scientists to recommend a return to greater variation within crop species.

It is not too late to help save the world's rich diversity of living things. Many species of plants are being preserved for the future by collections of their seeds. Some rare species of animals are being helped by breeding programs in zoos and parks. For example, the Reptile Breeding Foundation in Picton, Ontario, is a world leader in breeding rare reptiles and amphibians.

Collections and zoos alone, however, are not enough. Large areas of natural habitat, such as rainforests, must be kept free from human activity to act as "reservoirs" of species, so that future generations can discover and enjoy the values of all living things.

Think About It

1. You have just read about some of the direct and practical values of different species as sources of food, medicine, and raw materials. This is not the only argument for saving species from extinction. What other reasons can you think of why it is important to preserve the widest possible variety of living things?

2. Select a plant or an animal that is in danger of extinction. Research and report on the present efforts being made to help it survive.

A Jamaican boa at the Reptile Breeding Foundation in Ontario.

There is very little variation among these wheat kernels.

Pandas and rhinos are in danger of extinction, pandas mainly because of loss of habitat, and rhinos because of excessive hunting.

Similarities and Differences Among Plants

All of the organisms classified as plants are **autotrophs**—that is, they are able to make their own food through the process of photosynthesis. Most plants live on land, and most are unable to move from place to place. Many plants can reproduce both asexually and sexually. In this Topic, you'll survey some of the major groups into which plants are classified and examine differences in structure among and within different groups.

Table 5-10 *Major Plant Phyla*

PHYLUM	EXAMPLES
Bryophyta	mosses, liverworts
Filicinophyta	ferns
Coniferophyta	pines, spruces, cedars
Angiospermophyta	flowers, grasses, oaks, elms

Mosses and Liverworts

The simplest land plants are **mosses** and **liverworts**, classified in the phylum Bryophyta. These plants are restricted to relatively damp areas such as forest floors, stream edges, wet rocks, and shaded tree surfaces.

Mosses and liverworts are simple plants. Liverworts have a flat, lobed shape that resembles the shape of a liver.

These **bryophytes** are unable to survive in drier areas for a number of reasons. They do not have true roots for absorbing water, nor do they have **vascular tissue**, the network of conducting vessels that transport water and dissolved materials in other land plants. Instead, bryophytes absorb water into their outer cells, from where it diffuses to other parts of the plant. This process is slow and does not allow the plants to grow to a large size. Some bryophytes have small, root-shaped cells on their undersurface. These cells take in water and help to support the plant.

Mosses and liverworts reproduce both sexually and asexually. From small stalks, they produce asexually by releasing microscopic **spores** that become new plants. At another stage in their life cycle, mosses and liverworts produce sperm and eggs, thereby reproducing sexually. The male plant produces sperm, which must swim to reach and fertilize an egg cell. This is another reason mosses and liverworts must live in moist areas — because they need moisture for sexual reproduction. Sexual reproduction occurs after rainfall leaves a film of water on the plant surface.

Bryophytes are adapted to survive periods of dryness by becoming dormant. They quickly revive and grow again when suitable conditions return.

Did You Know?

Peat moss has such a high energy content that it makes a superb fuel. In Ireland, blocks of it are cut from the landscape, dried in the sun, and burned for home heating.

You might think of this peat as "peat being on its way to becoming coal." When the plants die, they often sink into the water. Because it is not exposed to the air, and because its acid environment discourages bacteria, it does not readily decompose. Thus, peat is preserved and compressed over time, becoming higher and higher in its stored energy content just as coal has, over an even longer period of time.

Peat moss (Sphagnum) *forms huge bogs throughout much of North America's northern forests and tundra. Peat moss can hold as much as 20 times its dry mass of water. (Most plants can hold only about four or five times their dry mass.) For this reason, people have long used peat moss to absorb liquids. Today, they use it in gardening because of these water-holding qualities.*

Ferns

Ferns, belonging to the phylum Filicinophyta, are better adapted than bryophytes to the land environment. They have underground stems called **rhizomes** and develop true roots. They grow large leaves called **fronds**.

Ferns have cells that are specialized to form tubes. This vascular tissue in ferns can rapidly transport water and food to all parts of the plant; it also helps keep the plant stems and leaves rigid. For these reasons, ferns can grow to much larger sizes than bryophytes. In tropical forests, some fern species may reach a height of 15 m.

The familiar green fern fronds seen in forests produce spore cases that appear as raised brown dots on the underside of their leaves. When the spore cases burst open, dust-like spores (asexual reproductive structures) are shed onto the ground. Each spore then develops into a tiny, heart-shaped structure that produces both sperm and egg cells on its undersurface. These cells mature at different times, so the sperm must swim to the eggs; therefore, like the bryophytes, ferns must grow in moist places. In the next Activity, you'll examine some adaptations of the fern for life on land.

Activity 5 – 10

The Fern—A Simple Land Plant

Problem

How is a fern plant adapted for life on land?

Materials

mature fern plant with spores
forceps
well slide
dropper
microscope
glycerol
magnifying lens
prepared slide of fern

Procedure

1. Draw your fern plant. Identify and label the following parts: frond, rhizome (underground stem), roots, spore cases.
2. Using forceps, carefully remove a brown spot from the underside of the leaf and place it in the well of your slide. Examine the spore cases under the low-power lens of your microscope.
3. If your spore cases are mature (rust-coloured), add a drop of glycerol. Using a magnifying lens, observe what happens.
4. Examine a prepared slide of fern tissue that shows the vascular cells. Draw what you see.

Analysis

1. What adaptations for life on land did you observe?
2. From your observations, what could cause a spore case to burst?

Extension

3. Use a reference book to find out about the life cycle of a fern. Report on its asexual and sexual reproduction during its life cycle.

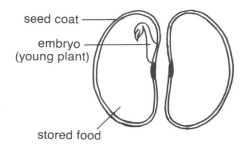

seed coat

embryo
(young plant)

stored food

Cross-section of a bean seed. Most seeds, like this one, have food stored in them for the young plant to use as a source of energy for growth.

Seed-Bearing Plants

A major adaptation to life on land is the seed. It contains the embryo of a new plant, sealed inside a protective coat. The embryo is thus kept from drying out and can remain inactive for many years—even hundreds of years. Furthermore, a seed allows a plant rooted in one spot to disperse its offspring to new areas. Consequently, seed-bearing plants are much more widespread and diverse than plants such as mosses and ferns, which do not produce seeds.

The cones of gymnosperms and the flowers of angiosperms are involved in the production of seeds, a stage in the sexual reproduction of these plants.

Probing

Read and prepare a report about some of the different adaptations that seeds have to help them travel away from the parent plant.

Seed-bearing plants are classified into two major groups. One group, the **gymnosperms**, bear their seeds exposed either on the surface of cone scales or on stalks among their leaves. The best known gymnosperms are in phylum Coniferophyta, which includes **conifers** (cone-producing trees) such as pines, spruce, firs, and cedars. Gymnosperms do not require a water environment for sexual reproduction. They produce pollen (containing the male gamete), which is blown about by the wind. Some of the pollen encounters the female part of a plant of the same species and may begin the growth of a new individual. In the second group, **angiosperms** (flowering plants), seeds develop inside a protective tissue or fruit. There is much more diversity among angiosperms (phylum Angiospermophyta) than among gymnosperms. Angiosperms include magnolias, lilies, roses, daisies, tulips, grasses, and every other familiar tree, shrub, and garden plant that produces flowers.

Grasses, the most widespread plants on the Earth, cover the prairies, savannahs, and grasslands of every continent except Antarctica. Cereal grain plants, such as wheat, oats, rice, corn, and barley, are all members of the grass family. One of the largest grasses is bamboo. The enormous amounts of pollen produced by the flowers of grasses may cause an allergic reaction known as "hay fever" in some people.

Angiosperms are the dominant type of plants on land. Apart from seed-bearing, their most important adaptation is the production of flowers, which provide them with an efficient method of sexual reproduction. Flowers contain the male and/or female gametes. The shape, colour, smell, or nectar of most flowers attract insects or other animals that transfer the pollen to other plants of the same species. In some angiosperms, the pollen is distributed by the wind, as it is in gymnosperms. Even at some distance from one another, individual flowering plants can thus breed with others of their species.

Despite the huge variety in the appearance of flowers, all flowering plants have the same basic parts. This fact is evidence of their close relationship to one another. Variations in the structure of flowers are used to classify angiosperms into different families, as you will see in the next Activity.

Water lily

Tiger lily

Wild geranium

Classifying Flowers

Problem

Which flowering plants can be grouped together?

Materials

four kinds of simple flowers
magnifying lens or dissecting
    microscope
dissecting needle

Table 5-11   *The Characteristics of Different Flowers*

FLOWER	NUMBER OF PETALS	TYPE OF SYMMETRY	TYPE OF OVARY
1.			
2.			
3.			
4.			

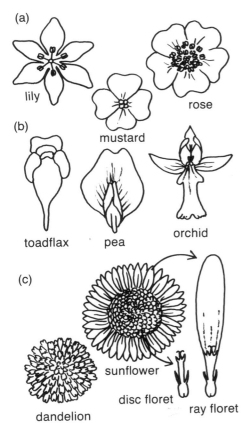

(a)

lily

rose

(b)

mustard

toadflax   pea

orchid

(c)

sunflower

disc floret

ray floret

dandelion

Typical flower shapes. All flowers in (a) have radial symmetry; the petals radiate like the spokes of a wheel from the centre. Flowers in (b) have bilateral symmetry. They can be divided down the middle into two identical halves. Composite flowers in (c) have many tiny florets arranged in a head.

Procedure

1. Draw a table like Table 5-11.
2. Carefully count the number of petals in each flower. Record the number in your table.
3. Compare the arrangement of the flower parts to the illustration of flower shapes. Is the symmetry more like (a), (b), or (c)? Record your answer.
4. Compare the position of the ovary in each flower to the illustration of ovary position. Record the ovary type in your table.

Analysis

1. Based on your comparison of the flower structures, which flowers would you place in the same group? Why?

Further Analysis

2. Why do you think some plants have large, brightly coloured flowers while others, such as grasses, have small and inconspicuous flowers?
3. Why do you think different flowers have different colours?
4. Construct a dichotomous key that identifies each of the four flowers you have looked at on the basis of their different characteristics.

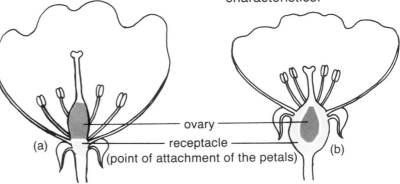

ovary
receptacle
(point of attachment of the petals)

(a)   (b)

Flower (a) has a superior ovary; that is, the base of the ovary is above the receptacle. Flower (b) has an inferior ovary: that is, the base of the ovary is below the receptacle.

Similarities and Differences Among Animals

All of the organisms classified as animals are **heterotrophs**—that is, they obtain energy for their life processes by consuming other organisms. All animals are multicellular and most have specialized tissues and organs. Animals live in water, on land, or inside other organisms. Most are able to move from place to place. All species of animals reproduce sexually, and some can also reproduce asexually. In this Topic, you'll survey some of the major groups into which animals are classified.

Table 5-12 *Major Animal Phyla*

PHYLUM	EXAMPLES
Porifera	sponges
Cnidaria	jellyfish, corals, anemones
Platyhelminthes	flukes, tapeworms, planarians
Nematoda	roundworms
Annelida	earthworms, leeches
Mollusca	snails, squid, clams
Echinodermata	starfish, sea urchins
Arthropoda	insects, spiders, crayfish
Chordata	fish, frogs, birds, mammals

Sponges

The members of the phylum Porifera are among the simplest of animals. **Sponges**, or **poriferans**, are shaped like a hollow bag full of tiny holes or pores, with a single large opening at one end of the body. The cells of sponges do not form true tissues, although some cells have specialized structures and functions. (See page 365 for information on tissues, organs, and systems.) For example, cells on the outside wall of the sponge have pores for drawing in water;

sea urchin
Phylum Echinodermata

ant
Phylum Arthropoda

snail
Phylum Mollusca

jellyfish
Phylum Cnidaria

kangaroo
Phylum Chordata

Sponges may appear more like plants than animals, but a close look at their structure reveals that they are simple animals.

cells facing into the body cavity have hair-like structures called flagella, which make waving motions to keep water circulating. Sponges have no systems or organs. They do not move from place to place and must rely on the continual flow of water to carry food and oxygen into their bodies. Sponges can reproduce asexually by budding, yet can also reproduce sexually. Most sponges have both male and female reproductive cells, but each are active at different times so the individual sponge cannot fertilize itself. The fertilized egg develops into a tiny larva that drifts through the sea for a brief period before settling onto a solid surface. There, it attaches itself and grows into an adult sponge.

Cnidarians

The phylum Cnidaria includes jellyfish, corals, sea anemones, and hydroids. There are two different types of body structure found among **cnidarians**. Some cnidarians are **polyps**—cylindrical animals with a ring of tentacles at one end. Hydra (shown on page 233) and sea anemones (shown below) are typical polyps. Others in this phylum are **medusae**—animals shaped like umbrellas with their tentacles trailing from the downward-facing edge. Jellyfish are typical medusae. Some cnidarians alternate between these two forms during their life cycle. Like the sponges, cnidarians have a single body opening, through which they take in food and expel waste particles. The opening is surrounded by tentacles.

Probing

Although cnidarians are small and simple, members of this phylum have built some of the largest structures on earth—coral reefs. The Great Barrier Reef, off the coast of Australia, is about 2000 km long and 145 km wide. In places, it rises 120 m from the ocean floor. In a reference book, find out how coral reefs are built up by colonies of coral polyps.

An anemone and a jellyfish. Both these cnidarians have stinging cells on their tentacles, used for capturing prey.

Worms

Biologists classify most kinds of worm-like creatures into three phyla. All worms are more complex in structure than sponges and cnidarians. They have organs, various simple systems, and definite front (anterior) and back (posterior) ends.

The simplest worms are called **flatworms** (phylum Platyhelminthes). Most, like the flukes and tapeworms, are parasites. They live and feed on or in the bodies of various animals, including humans. Others, such as planarians, live in salt water, fresh water, or damp soil. A planarian has a simple digestive system with a single opening and an excretory system that regulates the amount of water in its body. At its anterior end, the planarian has sensory lobes that respond to chemicals and touch. It also has eye spots that are sensitive to light. Together with specialized cells sensitive to stimuli throughout its body, these organs make up its simple nervous system. Planarians can reproduce asexually (by dividing in two parts) or sexually.

nervous system

digestive system

excretory system

eye spots

opening to digestive system

The planarian flatworm has several systems; three are shown here. It does not have a system of vessels to transport oxygen and nutrients (circulatory system), nor a system for gas exchange.

A flatworm.

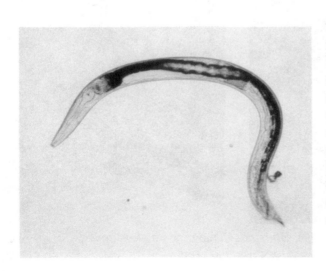

A second group of worms is called **roundworms** (phylum Nematoda). The most numerous kind of worm, nematodes are usually parasites on plants or animals, but some are scavengers. Their bodies are long and slender and taper to a point. Many species are microscopic. The digestive tract of a roundworm has two openings: a mouth at the anterior end and an anus at the posterior end. Roundworms also have a nervous system and an excretory system. Because of their small size, they need no systems to take in and distribute oxygen or nutrients (a gas-exchange system and a circulatory system). Materials simply pass from cell to cell to the various body parts.

A nematode.

Earthworm, an annelid.

The third group of worms, called **segmented worms** (phylum Annelida), includes the familiar earthworm as well as numerous fresh- and salt-water species such as leeches and sandworms. All **annelids**, as segmented worms are also known, have bodies that are divided into a series of similar segments. You can see these segments clearly on the exterior of an earthworm. The animal's interior is divided in the same way, with segments separated by thin walls. Annelids have a circulatory system, needed because of their larger size, to transport materials throughout the body.

Molluscs

They may appear very different, but a clam, an octopus, and a slug are all molluscs and have similar body structures.

Mollusca is the second largest phylum of animals. Most **molluscs** live in the sea. Slugs, snails, clams, octopods, and squid are all molluscs. Members of this phylum have soft bodies, but many of them produce hard outer shells for protection. The body of a mollusc is divided into three parts: a head that contains the mouth and sensory organs, the main body that contains the internal organs, and the mantle, a fold that covers the internal organs and secretes the shell. The shells of slugs and cuttlefish are small and lie under the mantle. Molluscs have a well-developed nervous system, with sense organs for smell, taste, touch, and sight, as well as circulatory, digestive, excretory, and gas-exchange systems.

Echinoderms

The phylum Echinodermata includes animals such as starfish, brittle stars, sea urchins, and sea cucumbers. All **echinoderms** live in salt water. Instead of having a well-defined head region, their bodies have radial symmetry — that is, they have a number of similar parts (usually five) radiating from the centre like the spokes of a wheel. Echinoderms have numerous projections called tube feet, which they use for locomotion. Most echinoderms are covered with hard spines. They have well-developed digestive and nervous systems, and a unique water-vascular system. This system circulates water and serves as a circulatory, excretory, and gas-exchange system all rolled into one.

Starfish and sea cucumbers are echinoderms.

Arthropods

The phylum Arthropoda contains the largest number and greatest diversity of animals of any phyla. **Arthropods** include such organisms as insects, crabs, lobsters, barnacles, millipedes, centipedes, spiders, scorpions, and ticks. All these animals have a number of structures in common. A hard shell helps protect them against predators and from drying out. Jointed limbs allow them to make complex movements. Unlike the worms, where all the segments are alike, arthropods tend to have bodies divided into distinct parts.

There are five major classes of arthropods, illustrated here: **crustaceans**, **arachnids**, **centipedes**, **millipedes**, and **insects**. In Activity 5-12, you'll have a chance to make your own classification of arthropods.

Probing

Overcoming desiccation (drying out) presents the same challenge for land animals as it does for plants. Find and list three ways that land animals are adapted to overcome desiccation.

Crustacea, such as crabs, lobsters, crayfish, and shrimps, have gills and two pairs of antennae.

Chilopoda (centipedes) are elongated and segmented. Each segment, except the first one and the last two, bears one pair of legs.

Diplopoda (millipedes) have rounder bodies than centipedes; each body segment bears two pairs of legs.

Arachnida, such as spiders, scorpions, ticks, and mites, have four pairs of legs and a body divided into two parts. They lack antennae, but have two claw-like fangs with a poison gland at the base of each one.

Insecta (insects) have a distinct head, thorax, and abdomen region. The thorax bears three pairs of legs. Most insects have two pairs of wings, although some may have one pair and others none.

Classifying Arthropods

1. Carefully examine the arthropods in the figure. Divide them into three major groups, based on differences and similarities in their structure.
2. Next, divide the largest group into subgroups.

Analysis

1. What structures do all the arthropods have in common?
2. Why is leg number useful for classifying arthropods?
3. What other characteristic could you use to distinguish between subgroups of arthropods?

Further Analysis

4. Name one characteristic that an insect shares with
   (a) a worm,
   (b) a mollusc,
   (c) a sponge.
5. Would you say that an insect is more similar to a worm or to a sponge? Why?

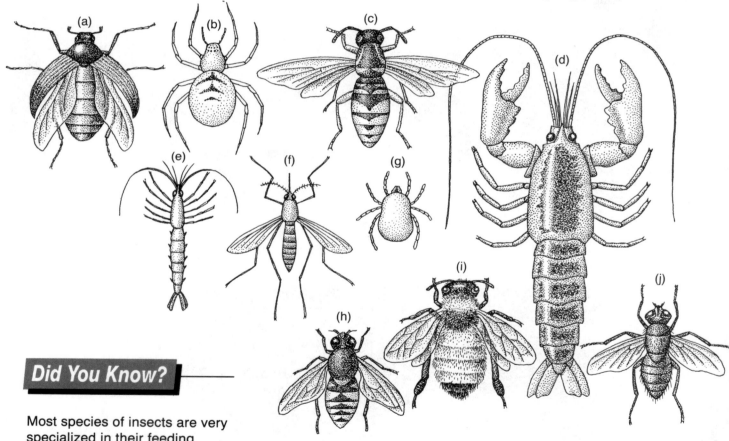

The ten arthropods shown here represent crustaceans, arachnids, and insects. To identify the individual organisms, turn to page 283.

Did You Know?

Most species of insects are very specialized in their feeding behaviour, allowing for the great diversity of this group in a wide range of environments. Even among one type of insect—the blood-feeding mosquitoes—there are 148 known species in North America alone.

Chordates

The phylum Chordata includes the organisms you probably first think of when you hear the word "animal." Organisms such as dogs, pigeons, turtles, frogs, and catfish are all members of this phylum, in the sub-phylum Vertebrata, the **vertebrates**. As well as these familiar vertebrates, **chordates** include small water-dwelling animals called lancelets and tunicates. All chordates have a structure called a notochord at some stage in their lives.

All vertebrates have a nerve cord that extends along their back and a protective bony spinal column. For this reason, they are sometimes known as "animals with backbones." All other animals are termed **invertebrates**, or "animals without backbones."

There are six main groups of vertebrates illustrated here. They are **jawless fishes**, **jawed fishes**, **amphibians**, **reptiles**, **birds**, and **mammals**. This Unit's final Activity will allow you to try classifying various vertebrates into the last four of these groups.

A colony of tunicates. Tunicates spend their adult life attached to a solid surface under water. Young tunicates (larvae) are free-swimming and have a notochord and other features of phylum Chordata.

Salmon are jawed fishes. Fishes with jaws can capture and eat other animals more efficiently than can those without jaws. About half of all living vertebrates are jawed fishes. They are divided into two main classes. One class, including sharks, skates, and rays, is made up of fishes that have a skeleton made of cartilage —a soft, light material. The other class, which is much larger, contains all fishes with bony skeletons.

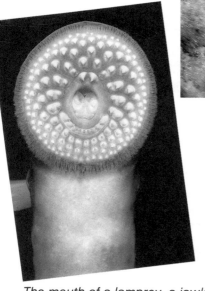

The mouth of a lamprey, a jawless fish, is adapted for sucking. The lamprey is a parasite, attaching itself to the body of a fish and scraping away the flesh. There are only two living species of jawless fishes, but fossils of many extinct jawless fishes have been found.

Amphibians, such as frogs, toads, newts, and salamanders, are adapted for life on land by structures such as legs and lungs. They are less well adapted than other land animals, however, because their skin and eggs can dry out rapidly. Most amphibians therefore live and deposit their eggs in water or damp places. Their larvae are aquatic and obtain dissolved oxygen using gills.

Reptiles, such as crocodiles, lizards, turtles, and snakes, have a dry, scaly skin and lay eggs that protect the embryo from drying out. Because of these adaptations, reptiles can live in many more land habitats than can amphibians.

Unlike fishes, amphibians, and reptiles, birds are endotherms. This ability to maintain a constant body temperature allows them to live in a wider range of environments. All birds have feathers, and most can fly. Some, such as the ostrich and penguin, are flightless.

Mammals are the most diverse class of vertebrates, including members that inhabit the sea (whales and seals) and the air (bats). The skin of mammals is covered with hair or fur rather than with scales or feathers. Except for a few species in Australia and New Guinea that lay eggs, mammals bear their young alive. All mammals nourish their young with milk produced from mammary glands.

Did You Know?

Although snakes have no legs, the skeletons of some species have small leg bones and shoulder and pelvic girdles. These structures are evidence that snakes are descended from reptiles that had legs. The loss of legs may be an adaptation for moving through small crevices or burrows.

Classifying and Comparing Vertebrates

Problem

How are some classes of vertebrates subdivided?

Materials

4 large envelopes (per group)
reference materials on vertebrates (loose leaf or individual page format)
paper clips

Procedure

1. Working in groups of four, collect pictures of different vertebrate animals in these four classes: amphibians, reptiles, birds, and mammals. Cut the pictures from magazines or copy from textbooks.
2. Have each person in your group choose one of the four vertebrate classes and label an envelope with the name of that class.
3. From the collection of pictures, select all the animals belonging to your chosen vertebrate class and place them in the envelope.
4. Swap envelopes with another student in your group. Examine the pictures in your new envelope and see if you agree that all belong in the same class. If not, discuss any difference of opinion with your partner. Re-evaluate the classification if necessary. List some of the characteristics of all the animals in your class.
5. On the basis of the characteristics of each animal, divide the pictures in your envelope into groups that you think represent different orders. You may have only one animal representing some orders. If you have several animals in the same order, fasten them together with a paper clip. Put the pictures back in the envelope and swap it with a different student in your group.
6. Examine the pictures and determine if you agree with the division into orders. If not, discuss any difference of opinion with your partner. When you are satisfied, list some of the characteristics of each order.
7. Check your classification with the classification given in a reference book. Note any errors and record the characteristics that define each order. Share your final classification with the others in your group.

Did You Know?

If you compare the forelimbs of members of different classes of vertebrates, you'll see that the basic structure is the same. The limbs are specialized for running, swimming, or flying, and are therefore different in size and shape. However, the same bones are present in all.

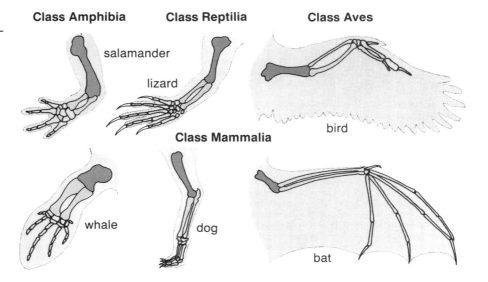

Class Amphibia — salamander
Class Reptilia — lizard
Class Aves — bird
Class Mammalia — whale, dog, bat

Working with Living and Once Living Things

Biology and Art

Celia Godkin loved both biology and art while in school and took biology in university. But she realized that a career in biology wasn't what she wanted. She found that scientific illustration was an ideal combination of art and science.

"As a scientific illustrator, I work from specimens in a laboratory or museum. My work must be absolutely correct in every detail in order to be used by scientists."

She works for a variety of clients; her work appears in textbooks and zoo exhibits. Her scientific background helps her produce illustrations that convey information clearly and accurately. (Celia's work appears throughout this book. In this Unit, for example, she produced the illustrations on pages 240, 241, 252, 253, 254, 260, and 269.)

Celia studied at the Ontario College of Art after she completed extensive biology studies at university. Another route to this career is to take studio courses at a community college or art school and biology courses at university. Celia feels that artistic skills and understanding of living things are of equal importance to anyone pursuing her chosen career.

Biology and Geology

Kevin Seymour has taken his life-long fascination with living things in another direction. He is a paleontologist who specializes in the fossil remains of vertebrates. Because he was interested in living things, he studied biology in high school and university. Then he discovered another branch of science—geology, the study of rocks and fossils in the Earth's crust. Fascinated, he decided to combine his two interests by studying paleontology—the branch of geology dealing with fossils.

In his search for specimens, he travels to such different places as Florida, British Columbia, and New Mexico. The type of rock around a fossil gives clues to how and when the fossil was formed and why it's in a particular location.

The work is sometimes frustrating. "Most often, fossils are incomplete or partially damaged. This means that you don't have enough information to completely answer your questions.

"My main interest is in trying to discover how species become adapted to their environment, and how some species do not adapt—and so become extinct. It's amazing what we can discover from old bones."

Checkpoint

1. Copy Table 5-13 below into your notebook and complete it.

Table 5-13 *Comparing Major Groups of Plants*

GROUP	SPERM SWIM TO EGG CELL FOR FERTILIZATION	ROOTS, STEMS, LEAVES	VASCULAR TISSUE	SEEDS	FLOWERS
Bryophyta	yes	no	no		
Filicinophyta					
Coniferophyta					
Angiospermophyta					

2. For each of the following, name two adaptations they have for living on dry land.
   (a) animals
   (b) plants
3. Match each word in the list with the description that best describes it.
   (a) autotroph
   (b) liverwort
   (c) vascular tissue
   (d) gymnosperm
   (e) angiosperm

   • an example of a bryophyte
   • a flowering plant
   • organism that can photosynthesize
   • plant with exposed seeds
   • a network of conducting vessels

4. Scientists classify plants, animals, and all other organisms according to their structure. In Row A below, *all* the imaginary organisms are classified as "Trips." In Row B, however, *none* of the organisms is classified as Trip. In Row C, *two* of the organisms are Trips—but which two?

   In identifying the two Row C organisms that are Trips, you should consider the characteristics that might be considered as "structures." These characteristics may include the number of "parts" (shapes) in each organism, *and/or* the colour of the shapes (dark or white), *and/or* the kind of shapes. Which two Row C organisms are trips?

5. Visiting an aquarium with your younger brother, you see a sponge, a sea anemone, and a jellyfish. How would you explain to him that the sea anemone and jellyfish are more similar to each other than either is to the sponge?

6. Each of the following groups lists a phylum and some organisms. Find which organisms are incorrectly classified; where there is an error, provide the correct classification and explain it.
   (a) Cnidaria—corals, jelly-fish, hydra, anemones
   (b) Mollusca—slugs, snails, worms, octopus
   (c) Platyhelminthes—flukes, tapeworms, planarians, leeches
   (d) Chordata—whales, sharks, snakes, penguins

The ten arthropods shown on page 278 are (a) beetle, (b) spider, (c) wasp, (d) lobster, (e) shrimp, (f) mosquito, (g) mite, (h) syrphid fly, (i) bee, (j) housefly.

Focus

- Organisms show a huge diversity of shapes, structures, and behaviour.
- Every organism has adaptations that allow it to survive and reproduce.
- The different structures found among different species are adaptations to different environments and ways of life.
- Diversity of appearance may be found within a species as well as among different species.
- Some of the characteristics of an organism, called traits, are inherited from its parents. These inherited traits are determined by genes.
- By the process of artificial selection, people control and develop the characteristics of a group of organisms by breeding only individuals with the desired traits.
- According to the theory of natural selection, organisms with traits best adapted to their environment are more likely to survive and reproduce. Natural selection may thus result in large changes in the traits of a species over time.
- Organisms are classified into groups on the basis of their similarities and differences. All organisms are grouped into one of five kingdoms.

- Organisms can be identified by the use of a dichotomous key.
- Plants are classified as bryophytes, gymnosperms, or angiosperms.
- Animals are classified into nine major phyla. Phylum Chordata includes all the vertebrates.

Backtrack

1. In your notebook, write the words in Column A. Beside each, write the correct definition from Column B.

COLUMN A	COLUMN B
(a) adaptation	having several forms within the same species
(b) metamorphosis	unit that controls the development of traits
(c) sexual dimorphism	trait that increases an organism's chance of survival
(d) polymorphism	change in form during an individual's life cycle
(e) asexual reproduction	presence of a different appearance in males and females of the same species
(f) sexual reproduction	reproduction involving two individuals
(g) gene	survival of individuals that are best adapted to their environment
(h) artificial selection	breeding of domestic animals or plants to develop particular traits
(i) natural selection	reproduction without sex

2. Describe one adaptation of a plant and one adaptation of an animal that lives in an area near you.

3. What is the advantage to a species of sexual reproduction over asexual reproduction?

4. Explain, with examples, the difference between artificial selection and natural selection.

5. "A rabbit that can run fast always has a better chance of survival than a slower rabbit." Explain why you agree or disagree with this statement.

6. (a) Name two traits you might expect to find in an organism that lives in a cold environment.
(b) How does each trait help the organism to survive?

7. The two animals shown at left are both grouped in the same class of vertebrates.
(a) Name two ways in which they are similar.
(b) Name two ways in which they are different.
(c) Name another vertebrate that is in the same class.
(d) Name a vertebrate that is in a different class.

Synthesizer

8. A particular species of lizard feeds on vegetation. One day, a pair of lizards is feeding on a tree branch when it breaks off and falls into a river below. Clinging to the branch, the lizards are washed out to sea. After several days, the branch and the lizards arrive at a distant rocky island. There are no trees and little vegetation on the island, but there is a lush growth of algae in the sea around it. Suggest what traits might change among the descendants of these lizards so that their chances of survival in this new environment are improved.

9. Examine the skull and the foot shown here. They both belong to the same animal. From the appearance of these parts, infer answers to the following questions, and explain your answers.
(a) Does the animal eat meat or vegetation?
(b) Is its habitat mainly terrestrial or aquatic?

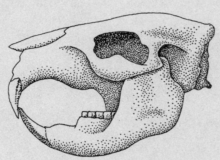

10. Butterflies, birds, and bats all have wings.
(a) Why are birds and bats classified in one phylum (Chordata) and butterflies in another (Arthropoda)?
(b) Why are birds classified in one class of chordates and bats in another class?

11. Why do you think there is a greater diversity of species in a tropical forest than in a desert?

12. Many species of animals that live on islands are found nowhere else in the world. Using the theory of natural selection, explain this observation.

13. The mole and the earthworm both live underground. Name two adaptations that suit each organism to this way of life. Suggest why a mole and an earthworm look so different from one another even though they share the same habitat.

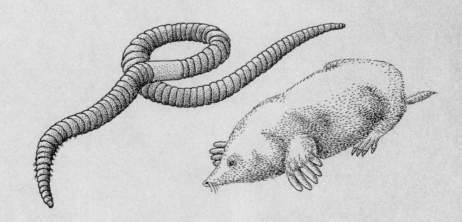

Environmental Quality

These photographs show some ways in which we make use of our environment. In each scene, how do you think human activity is affecting the environment? How does each scene make you feel?

For millions of years, humans have used natural materials from the environment—primarily for food, clothing, and shelter, but increasingly to satisfy "wants" as well as "needs." Discovering how to use the energy stored in fuels such as coal and oil, as well as the energy stored in water held back by dams, enabled people to make major changes in their environment.

Because of these technological achievements, we now have efficient transportation systems, comfortable home and work environments, and abundant food in our part of the world. However, for everything we use—materials and energy alike—we produce **waste** (anything considered to be useless). The more we use, the more waste results.

Everyone wants to live in an environment that has clean air, clean water, and healthy food. How can we be certain that each of these conditions is met? And how can we reduce or dispose of the waste we produce?

In this Unit, you will look at some current issues in environmental quality. You will choose an environmental topic of special interest to you and look at how science is a part of that issue. Having studied chemistry, heat transfer, and the diversity of organisms, you're now equipped with many of the science skills needed to study one problem in detail: What should a growing city do about its sewage treatment? As a final issue in the Unit, you and your classmates will decide!

People and the Environment

You may live in a community that is not very crowded. If you do, it is hard to imagine that a global ''population explosion'' has occurred in the past 100 to 150 years. Until the mid-1800s, the world's population was less than 1 billion. Until that time, the human population had increased very slowly. Within 80 years (by 1930), the world's population doubled, reaching 2 billion. By 1960, only 35 years later, the population doubled again, reaching 3 billion. What about now? And in the future? In Activity 6–1, you'll analyse trends in world population change.

Human Population Growth

PART A

Problem

How is the world population changing?

Table 6-1 *World Human Population, 1890–2010*

YEAR	POPULATION (billions)
1890	1.5
1900	1.6
1910	1.7
1920	1.8
1930	2.0
1940	2.3
1950	2.5
1960	3.0
1970	3.6
1980	4.4
1987	5.0
1990	—
2000	—
2010	—

Procedure

1. Draw a line graph of the data in Table 6-1, plotting time on the horizontal axis and population on the vertical axis. Include the years 1990, 2000, and 2010 on the horizontal axis.

2. From the graph, extrapolate the population for the years 1990, 2000, and 2010, assuming the present trend continues. Extend the line on your graph, using a dotted line.

Analysis

1. Do you think the predicted trend in population change that you have plotted until the year 2010 will occur? Explain a reason for your answer.
2. What factors might change the trend in population increase between now and the year 2010?

PART B

Problem

How is the distribution of the Canadian population changing?

Procedure

1. Select any five provinces from Table 6-2.
2. Plot the data in Table 6-2 for those provinces on one graph, using a different coloured line for each province. Choose a scale for the vertical axis that is appropriate to plot the population percentages.
3. Answer the questions on the following page.

Table 6-2 *Percentage (%) of Canadian Population Living in Urban Areas, 1851–1981*

PROVINCE	1851	1871	1891	1911	1931	1951	1961	1971	1981
British Columbia	–	9	43	51	62	69	73	76	78
Alberta	–	–	–	29	32	48	63	74	77
Saskatchewan	–	–	–	16	20	30	43	53	58
Manitoba	–	–	23	39	45	56	64	70	71
Ontario	14	21	35	50	63	73	77	82	82
Quebec	15	20	29	45	60	67	74	81	78
New Brunswick	14	18	20	27	35	43	47	57	51
Nova Scotia	8	8	19	37	47	55	54	57	55
Prince Edward Island	–	9	13	16	20	25	32	38	36
Newfoundland	–	–	–	–	–	43	51	57	59

Influences on Human Population Growth

Analysis

1. Describe similarities and/or differences in the shapes of the five lines on the graph.
2. What does the graph tell you about the trend in urban vs. rural living in Canada from 1851–1981?
3. Do you think this trend will continue in each of the provinces in your graph? Explain why or why not.
4. What is the advantage of expressing the populations in Table 6-2 as percentages instead of actual numbers of people?

What caused the human population explosion? Two main factors were responsible. One was the discovery that micro-organisms cause disease and the actions that were taken to reduce the number of such organisms. Until the 1800s, unsanitary conditions allowed the spread of many fatal diseases that today are rare; as well, lack of adequate medical treatment meant that many diseases that today cause few deaths, like pneumonia, were often fatal. Since the early 1800s, improved sanitation—keeping food and water free of disease-causing micro-organisms—and improvements in food production and medicine have greatly increased our chances of surviving. Therefore, humans live longer and produce more children, who themselves have a greater chance of surviving and producing children.

The second factor was the 18th- and early 19th-century historical development known today as the Industrial Revolution. In this period of **industrialization**, methods of producing materials and methods of transportation changed rapidly, due largely to the invention of new machines. For example, the invention of the steam engine enabled farmers to plough using machinery instead of following a horse on foot; steam-powered locomotives changed methods of transportation. Improved technologies meant that much work that had been done by hand was now done with machines, making people's lives easier.

Steam-run tractor, about 1900.

Cities and Technology

The change towards industrialization prompted people to move near the industries that were springing up, so that they could find work. This trend of movement to urban areas is called **urbanization**.

As a result of increased technology and urbanization, we can manufacture products that few people could even imagine 100 years ago — televisions, airplanes, space satellites, computers. To make more and more of these goods, however, we need more and more materials from the environment.

The two photographs show the impact of urbanization and technology on the environment. Taken at an intersection on the outskirts of a Canadian city, these photos show changes that occurred over a period of just 15 years! You have probably noticed changes like this yourself — farmland just outside a city one year is replaced by buildings the next year. Think about the two scenes. How much food production capacity was lost when this fertile farmland became an industrial and residential area? How much more energy is being used now because of the changes to the environment? How much more waste is being produced in the area?

A large, densely populated city highlights the impact that people can have on a relatively small area. A city full of people requires a large daily input of materials and energy and produces a large daily output of waste.

We are becoming increasingly aware that although technology has given us time-saving products and easier lives, it has also threatened environmental quality. The spill of a hazardous waste, the escape of a deadly gas from a factory, the greenhouse effect — these are just some of the problems we hear about. But how can we decide what to do about them? And how can science help?

An intersection on the outskirts of a large Canadian city.

The same intersection only 15 years later.

People in cities use large amounts of energy and materials. They also return large amounts of waste to the environment. The diagram shows the typical daily input and output for a North American city of 500 000 people.

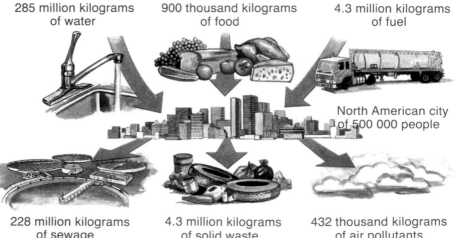

285 million kilograms of water

900 thousand kilograms of food

4.3 million kilograms of fuel

North American city of 500 000 people

228 million kilograms of sewage

4.3 million kilograms of solid waste

432 thousand kilograms of air pollutants

Analysing Issues

Although most people are concerned about the quality of the environment, they often disagree about the cause of a problem, how serious it is, and how to solve it. Most of this Unit will prepare you for a case study that simulates a real environmental issue. An **issue** is a matter about which people have different viewpoints (such as those listed on the next page). An issue is science-related when science can provide information relevant to the issue. Prepare for the case study by examining the issues illustrated here, then do Activity 6–2, which will give you first-hand experience analysing issues.

Issue A: Many types of industries use tall smoke stacks that release pollutants high into the atmosphere, so that they cause less damage to the surrounding environment. Although this benefits the communities near the stack, it does not benefit the distant regions where many of the pollutants fall to Earth. These industries provide jobs for local people and may produce products or energy to be used by people farther away. Should these industries be made to pay for the damage caused by their airborne pollution when it harms other regions or countries?

Issue B: Sour gas plants separate sulphur from gaseous fossil fuels. The sulphur is used to make products such as fertilizers, drugs, and industrial chemicals. The separation process is not 100% efficient. Several tonnes of sulphur are released every day from each smokestack as sulphur dioxide, a pollutant. New technology would result in less than 1% of the sulphur dioxide escaping into the atmosphere. However, it is very expensive to change the equipment in an older plant so that this new technology can be used. Installing the equipment could raise the costs of manufacturing. This, in turn, could mean higher prices for fuels, fertilizers, drugs, and metal products. Should all sour gas plants be required by law to reduce their sulphur dioxide emissions to less than 1%?

Issue C: The gas and oil industry provides fuel for human use as well as many jobs. Areas rich in these fossil fuels also provide habitats for wildlife. The drilling needed to obtain the gas and oil may disturb or destroy wildlife. Should oil and gas companies be allowed to drill in wilderness areas? Should they be restricted from drilling in areas that contain rare plants and animals?

Viewpoints to Consider

Each of the issues summarized here can be evaluated from a number of different viewpoints. People usually evaluate environmental issues from one or more of the following viewpoints or opinions:

- Ecological: concern for the protection of natural ecosystems.
- Economic: concern for financial gain and job creation.
- Educational: concern for acquiring and sharing knowledge.
- Egocentric: concern for self-interest.
- Ethical/Moral: concern that an action is morally right or wrong.
- Health-related: concern for physical and mental well-being.
- Recreational: concern that the environment be usable for leisure activities.
- Political: concern about how an action affects a government, a political party, or a politician.
- Scientific: concern that knowledge be obtained by objective observation and experiment.
- Technological: concern for practical problem-solving and application of scientific knowledge.

Issue D: National parks provide habitats for many kinds of wildlife, and many people enjoy these wilderness areas. Ski resorts disturb wildlife. But skiing is a form of recreation enjoyed by many people. Should developers be allowed to build ski resorts in national parks?

Issue E: Burning coal is now the least expensive way to produce electricity in countries with a large coal supply, such as Canada. It is also the most abundant fossil fuel. However, burning coal emits more carbon dioxide (one of the gases responsible for the greenhouse effect and possible global warming) than other fossil fuels. Without expensive control devices, burning coal also emits sulphur dioxide and other materials into the atmosphere. Should Canada convert its coal-fired plants to burn other, more expensive fuels?

Researching an Environmental Issue

Problem

How can scientific information be used to support or oppose a viewpoint on an environmental issue?

Materials

newspapers
magazines (e.g., *Environment News, Nature Canada, Probe Post, Discover, Science Digest, Popular Science, Equinox, Scientific American*)
references listed in the articles, if appropriate
government reports
opinions of specialists

Procedure

1. Select an issue from those described on pages 292 and 293, or choose another science-related issue that interests you.
2. Express your issue in the form of a written statement that can be supported or opposed.
3. Describe two (or more) viewpoints related to the issue.
4. Obtain at least two science-related articles from newspapers, magazines, or reports that pertain to your issue.
5. Summarize your articles.
6. Underline or list the scientific information in each article.
7. Make a list of the non-scientific information in the articles.
8. Prepare a table in your notebook using each of the viewpoints as headings.
9. Identify the main viewpoints in each article by referring to the viewpoints described on page 293.

Analysis

1. (a) Which viewpoints are presented in the articles? Explain.
   (b) Have you expressed a viewpoint (an opinion) by the way in which you worded your issue statement (Step 2)? Explain.
2. Are any inferences made in the articles that are not based on observations? Give examples to support your answer.
3. Describe any direct changes in the environment that are discussed in your articles.
4. Describe any indirect or unintended changes in the environment that are either mentioned or implied in the articles you selected.

Further Analysis

5. Continue to collect information about your issue throughout this Unit, using newspapers, magazines, reports, and other sources. File as much information as possible in a scrapbook or a folder to help you form your own viewpoint. Prepare a summary report on your issue after you have completed this Unit.

Environmental Quality and Pollution

You have seen that a great variety of issues on environmental quality exists today. Although we may not agree on the solutions to specific environmental issues, most people do agree on one matter. If asked what kind of environment they want to live in, most will say, "An unpolluted one." Just what is pollution? Put simply, anything people add to the environment that causes harm to living things is **pollution**. Pollution is caused in two major ways. Sometimes people produce too much of an otherwise harmless substance that occurs naturally in much smaller quantities in the environment. In other cases, they produce materials that don't occur naturally in the environment, some of which are highly **toxic** (poisonous).

Phosphates, for example, are substances that normally have a beneficial effect. Found naturally in the soil, **phosphates** are nutrients that enhance the growth of plants. But when fertilizers containing phosphates add even more phosphates to the soil, problems can arise. The excess phosphates are carried along by water runoff into lakes and rivers. Here, they stimulate the growth of algae and weeds. "Green slime" and underwater weeds clog boat motors and make swimming difficult. Other effects, which you will learn about later in this Unit, are even more serious. In such cases, the lakes have been polluted by too much of a normally desirable substance.

The overabundance of algae in this pond is due to phosphate pollution.

An example of a highly toxic material that is added to the environment is **dioxin**, a chemical found in certain pesticides and industrial wastes. Even trace (very small) quantities of dioxin are potentially very dangerous. Several years ago, an explosion in a chemical plant in Seveso, Italy, released a small quantity of dioxin into the air. It caused severe illness and skin disease in people living in the vicinity, and the town had to be evacuated. Later, babies were born with birth defects because the dioxin damaged the mother's or father's reproductive cells.

Pollution is not always made up of matter. For example, **noise pollution** and **thermal pollution** can also have negative effects on the environment. Excessive noise can cause hearing loss and other damage to various types of organisms. Thermal pollution often occurs where industries or power plants discharge hot waste into nearby lakes and streams. It can eliminate species unable to tolerate the increase in temperature.

In the following Topics, you'll investigate the types of waste we produce, and what happens to waste once it is produced. Does it cause pollution? If it does, what can be done about it? Issues can be resolved by looking at them from many different viewpoints and using scientific information to help.

What Can We Do with Our Output?

This Topic title is another way of asking "What can we do with our waste?" As you have seen, from all the materials that we "take in" or use to make other materials (our input), we produce a variety of products, most of which eventually become waste (our output). This Topic discusses the kinds of waste we produce and how we can control both its volume and its effect on the environment.

On average, each person in Canada discards more than 400 kg of household garbage each year. Just think how much space that much garbage would take up if you had to accumulate it for a year! This kind of waste is generally known as **solid waste**. The illustration on page 291 showed you that in a city of 500 000 people, about 4.3 million kilograms of solid waste are produced by all households and industry in one day. It is not surprising, then, that as more and more people move to cities, local governments must find ways to deal with the increasingly huge quantities of solid waste.

I THROW AWAY 400 kg? WHAT ABOUT THE WHOLE CITY?

Measuring Solid Waste — A Home Activity

Problem

How much of different types of waste do you produce?

Materials

a grocery bag for each type of waste
bathroom scale or spring scale*

Procedure

> CAUTION: If you touch any of the waste, wash your hands thoroughly. Be careful in handling glass waste.

1. Use a different bag for paper and paper products, glass, kitchen scraps, plastic, and miscellaneous waste. All members of your household should collect each of these types of waste in separate bags for a two-day period (the weekend is a good time).
2. Before you begin, discuss as a class how you will deal with factors that might influence your results. For example, if a household member has a large party and produces much more glass or kitchen waste as a result, you probably won't want to use that day in your two-day period. As well, if you go to a fast-food restaurant for a meal, you will need to include the waste from that meal.
3. At the end of the full two-day period, stand on a bathroom scale, holding each bag of waste in order to determine its mass (and subtract your mass from the total), or attach the bag to a spring scale in order to determine the total waste of each type produced by members of your household. Divide by the number of members in your household in order to determine the average mass of each type of waste per person, and divide by two to determine the mass of each type of waste per day.
4. Rank each type of waste by its bulk or volume. Record which type has the greatest volume, the least, etc.
5. Compile class results based on your findings.

Analysis

1. (a) Which type of waste produced the greatest mass?
   (b) Which type of waste was most bulky (had the greatest volume)?
2. How would you improve your analysis if you were to do it again?

* Spring scales and bathroom scales measure the weight of an object in newtons (N). However, these scales are often also calibrated in kilograms. If the scale you use shows only a newton scale, convert the weight of the solid waste to its mass in kilograms. (On Earth, a mass of 100 g has a weight of 1 N.)

Non-biodegradable materials can harm marine organisms. Millions of helium-filled balloons are released each year to commemorate special occasions. Many end up in the ocean. Some marine animals mistake them for jellyfish and choke to death when they try to eat them.

Biologists working along the coasts of North America have found marine turtles washed ashore with sheets of plastic blocking their digestive systems, and birds strangled by plastic six-pack holders. Pollution by these materials is more serious than many people have realized.

Waste Disposal

What happens to your garbage when it leaves your home? Solid waste was once deposited in open garbage dumps. Today it may be burned, or it may be taken to a **sanitary landfill site**. This is a supervised solid waste disposal site, where incoming waste is compacted to reduce its bulk and covered over with soil to prevent its spreading.

Some of the waste left at landfill sites is material that was once living (such as food scraps). Most of these materials can be used as food by organisms, such as bacteria and fungi. Materials that can be broken down (decomposed) and used by organisms are said to be **biodegradable**. However, many of the wastes taken to landfill sites are **non-biodegradable**: They cannot be decomposed by living things. Instead, they are eventually broken down by other chemical or physical processes. Non-biodegradable wastes tend to persist, or remain unchanged, in the environment longer than biodegradable materials. A metal takes several years to rust away completely. A styrofoam cup may take as long as 500 years to be completely broken down to molecules that can be cycled through an ecosystem. (Although plastics are made from oil, which is biodegradable, most plastics are not biodegradable.)

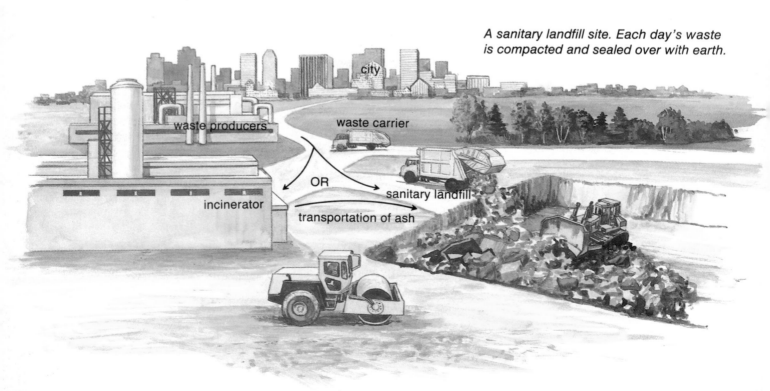

A sanitary landfill site. Each day's waste is compacted and sealed over with earth.

Did You Know?

Until fairly recently, most people did not realize that marshes were important breeding grounds for many kinds of fish, birds, and other animals. Many people thought these marshes were useless, so they were used as waste disposal sites, destroying this important habitat for many organisms.

Selecting a Sanitary Landfill Site

Sanitary landfill sites are intended to provide safe, long-term disposal of solid waste. Sites are chosen to avoid the possibility of any toxic wastes entering the water supply by seeping into ground water. Areas that are prone to flooding are not used as sanitary landfill sites.

The landfill site must be monitored even after it has been filled and completely covered over with soil, to make sure that no toxic wastes are leaching (leaking or seeping) into the ground water, and to ensure that no dangerous gases, such as methane produced by the decomposing wastes, are being released into the air.

Do only industries produce toxic wastes that can leach into ground water if the disposal site is inadequate? Unfortunately, no. For example, many household items and appliances contain metals that are not dangerous when used within the home. When they are discarded as waste, many are still not a problem, as long as only water washes over them. For example, the metal cadmium, which was used in old refrigerators and in many cans, is insoluble in water. But if the water is slightly acidic, cadmium and many other metals will dissolve in it and will leach into ground water, which then makes its way to rivers and lakes. Lead was once used commonly in paints, as was mercury in electrical equipment. These materials are also poisonous if their soluble forms are released into our waterways.

Open dumps such as this one are no longer permitted in most areas. The site selected for this garbage dump will not prevent leaching. Also, because the dump is open, gases can escape from it into the air.

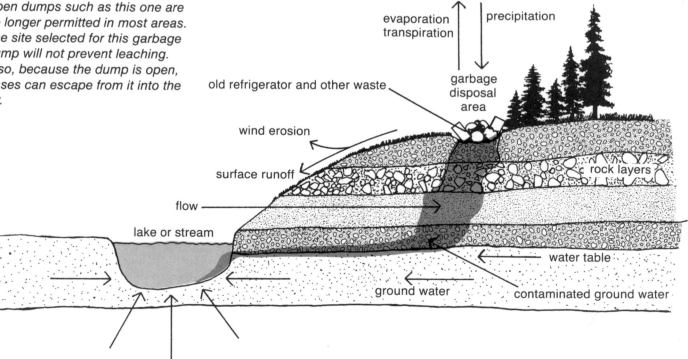

The illustration shows the danger to humans if large quantities of these toxic metals are released into waterways due to leaching from disposal sites or as a result of illegal dumping directly into lakes or streams. Recall from your Grade 8 studies that substances become more concentrated as they are consumed at each higher level of a food chain.

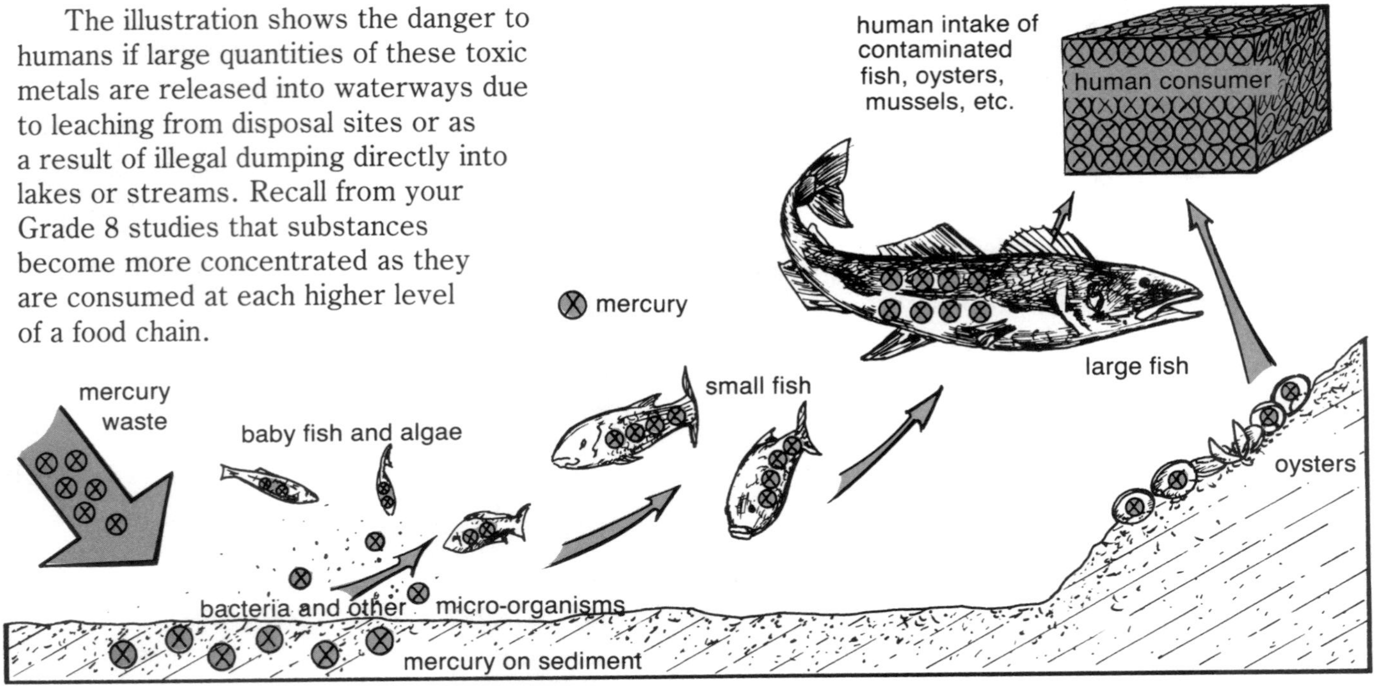

human intake of contaminated fish, oysters, mussels, etc.

human consumer

⊗ mercury

large fish

small fish

mercury waste

baby fish and algae

oysters

bacteria and other micro-organisms

mercury on sediment

This simplified diagram shows how toxic substances can be concentrated in a food chain. At each level, greater amounts accumulate in each individual.

The need to monitor a landfill site for many years after the site has been filled is one of the problems involved in managing the disposal of solid waste. Another problem is the steadily increasing volume of solid waste that must be disposed of. Landfill sites quickly become filled. It is increasingly difficult for local governments to find new sites for solid waste disposal because many people do not want these sites near their community.

One way to approach the solid waste problem is to somehow reduce the volume of solid waste that has to be managed, in other words, to simply produce less waste.

Incineration

Many large Canadian cities burn some or all of their garbage in an incinerator. The incinerator burns the solid waste at a very high temperature. The volume of the ash that is left is only about 5% of the original material. The ash can easily be disposed of at a landfill site. However, although burning reduces the volume of the waste, the gases produced contribute to air pollution.

An incinerator receiving domestic waste.

What Can Each Person Do?

People responsible for waste management have the enormous task of trying to ensure that our waste doesn't harm the environment. They offer this advice: reduce, reuse, recycle, and recover. These are known as the "Four Rs of waste management." Waste managers urge each person to consider *reducing* the amount of garbage produced. Then, if you can't reduce, reuse; if you can't reuse, recycle. What exactly do waste managers mean by each of these terms?

How much of this packaging is really needed?

Reduce simply means to decrease the amount of waste produced. The clearest example of this is reducing the amount of packaging, since a lot of packaging becomes waste. For example, if you buy cereal in bulk, you don't have to find a new use for the cardboard box that surrounds the "bag" that actually holds packaged cereal, nor do you have it to throw away. It is most important to reduce the kind of packaging that persists in the environment for a long time and emits dangerous gases when burned, as so many plastics do. Reducing is the best and cheapest way to decrease the amount of waste produced (both in terms of energy and material consumption), but it also requires the most noticeable changes in lifestyle. For example, it is less convenient to transport and store large bags of cereal than it is smaller cereal boxes so some people may not be willing to make this change.

Reuse means to use things either for the same purpose a second time or for a new purpose. The reuse of returnable soft drink bottles is one example. Reuse is a good approach to cutting down waste because it requires only enough energy to clean and refill the bottles, and no new materials are needed. In poor countries, and in poorer regions of our own country, people often come up with very inventive ways to reuse materials, out of need.

In places where returnable bottles are used, each one can be refilled, on average, 20 times. Their glass can then be recycled into new bottles.

Collecting aluminum cans for recycling.

Aluminum cans in a recycling centre.

Recycle means to collect waste of a certain type, in order to break it down and rebuild it into other products. Recycling requires energy to break down and rebuild materials, as well as the input of some new materials. Recycling is often more convenient for consumers than reusing and entails the least noticeable changes in lifestyle. It also means that stores and manufacturers do not have to alter their packaging and operating procedures very much. For example, some store owners object to the time and effort involved in sorting bottles for reuse. At the very least, recycling lets us know where our wastes are, and it helps ensure they are being managed in a way that does not risk possible leaching of dangerous materials into our waterways.

Recover means to reclaim either waste material or energy, in order to put it to another use. Recovery is effective only when large amounts of waste material or energy are produced. For example, in Japan and several European countries, when waste is burned in a large incinerator, the resulting thermal energy is recovered by transforming it into electrical energy. Several large companies in Canada have similar recovery programs. Our government, however, is emphasizing reuse programs instead of recovery programs.

Probing

A promising technique for reducing the volume of waste is the use of huge machines to shred the material at a landfill site before burying it. Design a controlled experiment to demonstrate how shredding a waste such as paper can increase the amount by which it can be compacted.

Automobile "graveyards" such as this were common throughout much of North America until recently. Recovery projects by steel manufacturers are rapidly eliminating such blights on the landscape.

1. Define each of the following terms and write a sentence that uses it.
   (a) urbanization
   (b) environmental issue
   (c) pollution
   (d) biodegradable waste

2. Explain how each of the following has contributed to environmental problems:
   (a) the human population explosion
   (b) urbanization
   (c) industrialization

3. Explain how a metal such as cadmium or lead, which was originally part of an appliance, can end up in a river some distance from where the metal was discarded in a garbage dump.

4. How are substances concentrated in a food chain?

5. Why is leaching a process that people in charge of waste management must know about?

6. Which is worse for the environment—tossing an apple core or a plastic candy wrapper into the woods? Give a reason for your answer.

7. Give three examples of how a household could reduce the amount of solid waste it produces.

8. (a) Why is the waste accumulated daily at a sanitary landfill site compacted by heavy machinery?
   (b) Why are landfill sites surrounded by high fences during the time they are being filled?

9. (a) If you have the choice of buying peanut butter in a glass or a plastic container, which would you choose? Explain why.
   (b) You have just finished a large jar of peanut butter. Describe three ways you could reuse the container for other purposes.

10. (a) If you were given a choice, would you ask for paper or plastic bags to carry your groceries from the store?
    (b) Give a reason for your answer to (a).
    (c) Give a reason why someone else might choose the other alternative in (a).
    (d) What alternative to both plastic and paper bags could be considered?

11. (a) Suggest ways a person packing a lunch could reduce the amount of packaging material used and still have fresh, nutritious food to eat.
    (b) What food items would you avoid in a cafeteria if you want to minimize the amount of packaging to be thrown away?

12. Design a reusable package for any product you use whose packaging is currently disposed of as waste.

13. List at least three ways in which these photographs show increased use of energy or materials between 1906 and the present.

▲ Same city today.

◄ Edmonton, Alberta, in 1906.

The Air We Breathe and the Water We Use

"I am a passenger on the space ship Earth."

Many people have begun to think seriously about this statement expressed by the architect and philosopher R. Buckminster Fuller. Being on Earth is very similar to being on a space ship in that Earth, like a space ship, carries its own air and other supplies, and its inhabitants depend on the cycling of materials within the "ship." You have seen that solid waste produced in one location may affect other areas if its disposal is not carefully controlled. Likewise, the substances put into our air may eventually affect our water, and substances put in water at one location may eventually affect the environment both there and in other areas. In this Topic, you'll investigate our major sources of air and water pollution.

A view of Earth from space.

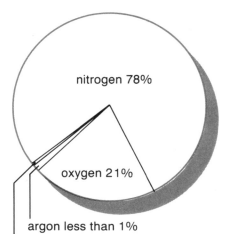

nitrogen 78%

oxygen 21%

argon less than 1%

Remaining fraction is mainly carbon dioxide (0.03%) and tiny amounts of hydrogen and neon.

Composition of the air.

The Air We Breathe

Have you ever been in a very crowded room and suddenly felt an urgent need for "fresh air"? If so, you know what it feels like to have the air you breathe altered in some way. The depletion of oxygen and the resulting build-up of carbon dioxide in the room has made you acutely aware of the need for a normal air supply.

The pie graph shows the usual composition of the air we are accustomed to breathing. As you can see, it is made up largely of nitrogen and oxygen, with very small amounts of other substances.

A change in the composition of the air can have drastic effects, as people who were living in London, England, in 1952, know. During a five-day period that year, the smog in London was so heavy that 4000 people died from causes that could be directly linked to the smog. This disaster—the worst case known in which deaths could be directly related to intense smog—shook British citizens and horrified people throughout the world. What substances in smog are damaging to our health and the environment? Read about fossil fuels in the next section in order to find out.

Fossil Fuels

Forests like the one shown here became buried under sediments and have become the coal and oil we use today.

Recall from previous studies that coal, oil, and natural gas are fossil fuels. Each was formed from organisms that lived millions of years ago. When organisms die, they are usually decomposed by micro-organisms. This process supplies the micro-organisms with energy that had been stored as chemical energy in the organisms' bodies, and it also supplies the soil with nutrients. The nutrients are then cycled through ecosystems. Not so with fossil fuels.

Some of the organisms that died many millions of years ago did not decompose. Instead they were buried by sediments, and the energy stored as chemical energy in their bodies remained "locked up." They became coal, oil, or natural gas, depending on the conditions under which they died and were buried. For the past two centuries or so, we have made extensive use of these long-buried organisms to fuel our homes, industries, and, more recently, our automobiles.

Did You Know?

No one used the word "smog" until this century. In fact, it originated when a word was needed to describe the combination of coal smoke and fog that began to hover over major industrial cities late in the 1800s. Eventually the word smog came to be used to describe any dirty-looking gases in the air.

Unfortunately we did not know that these fossils fuels pollute air. Because these fuels were once living organisms, they contain the same elements that make up living things—oxygen, carbon, and hydrogen, with lesser amounts of nitrogen, sulphur, and other substances.

When fossil fuels are burned, large amounts of carbon dioxide and water are produced. Other products of their combustion are hydrocarbons (compounds of hydrogen and carbon, such as methane), carbon monoxide, and oxides of sulphur and nitrogen. Where do these substances go? Into our air—as pollutants.

Activity 6-4

Sources of Air Pollution

Use the bar graphs and the information in this section to analyse air pollution from the use of fossil fuels in Canada.

1. What is the major source of each of the pollutants?
2. All sources considered (gases and particulates), what is the greatest source of air pollution?
3. What natural process causes a large percentage of carbon monoxide, hydrocarbon, and particulate pollution?
4. In light of the information given here, what are some recommendations you could make to your local government to help reduce air pollution in a heavily populated area?

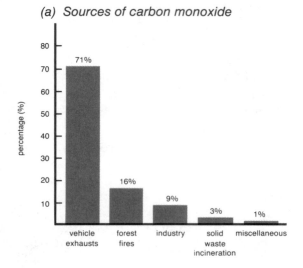

(a) Sources of carbon monoxide

Air Pollutants

What are some of the major air pollutants and where do they come from? In Activity 6-4, you will examine bar graphs describing the main sources of pollutants, listed here:

• **Carbon monoxide**—an odourless, invisible, poisonous gas produced when the amount of oxygen is limited so combustion of fossil fuels is incomplete.

carbon   +   oxygen   →   carbon monoxide
(from a fuel)   (in air)

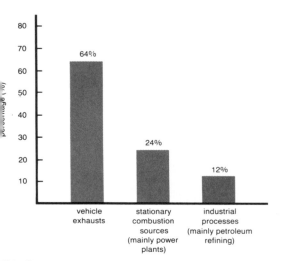

(b) Sources of nitrogen oxides

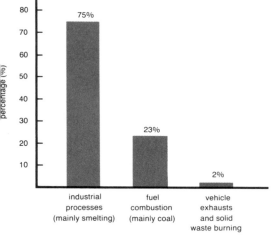

(c) Sources of sulphur oxides (mainly sulphur dioxide)

- **Nitrogen oxides**—gases formed when fuels are burned at high temperatures in engines and furnaces. Nitrogen from a fuel can combine with oxygen, forming nitrogen monoxide. Nitrogen monoxide (NO) then reacts with oxygen to form another pollutant, nitrogen dioxide (NO_2). This yellowish brown gas is often visible following morning rush hour in large cities. Nitrogen oxides contribute to the formation of acid rain.

$$\text{nitrogen} \quad + \quad \text{oxygen} \quad \rightarrow \quad \text{nitrogen monoxide}$$
(from a fuel)   (in air)

$$\text{nitrogen monoxide} \quad + \quad \text{oxygen} \quad \rightarrow \quad \text{nitrogen dioxide}$$
   (in air)   (in air)

- **Sulphur dioxide** and other sulphur oxides—gases formed when oxygen and sulphur combine. Coal contains more sulphur than oil and much more than natural gas. Sulphur dioxide is a clear, colourless gas with a pungent, overwhelming odour detectable only when the gas is in high concentrations. It is highly corrosive and, along with nitrogen dioxide, is one of the major components of acid rain.

$$\text{sulphur} \quad + \quad \text{oxygen} \quad \rightarrow \quad \text{sulphur dioxide}$$
(from a fuel)   (in air)

- **Hydrocarbons**—gases such as benzene, ethylene, and methane, which are compounds of hydrogen and carbon.

- **Particulates**—tiny particles of solids and liquids released into the air mostly when fuels and other materials are incompletely burned. They often remain suspended in the atmosphere, travelling great distances on air currents. Particulates include soot and ash, lead, zinc, arsenic, dust, pollen, and fine mists of liquid aerosols.

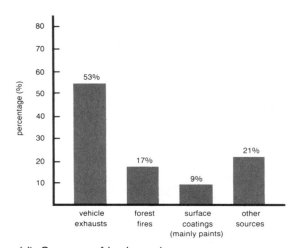

(d) Sources of hydrocarbons

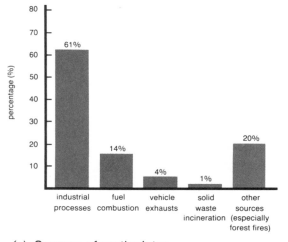

(e) Sources of particulates

The carbon monoxide molecule (CO) is similar to the oxygen molecule (O_2). Carbon monoxide is poisonous when inhaled because our red blood cells "mistake" the CO molecule for an O_2 molecule and carry it instead of oxygen to various parts of the body. Therefore, the cells of the brain and other organs are supplied with carbon monoxide instead of the oxygen they need, and death can result.

Winds carry pollutants over long distances — hundreds or even thousands of kilometres.

Acid Rain

The sulphur and nitrogen oxides produced by an ore smelter, a coal-fired electrical generating station, or a paper mill like the one shown here may have a widespread effect on the environment. The sulphur and nitrogen oxides combine with water vapour in the air, producing sulphuric and nitric acid.

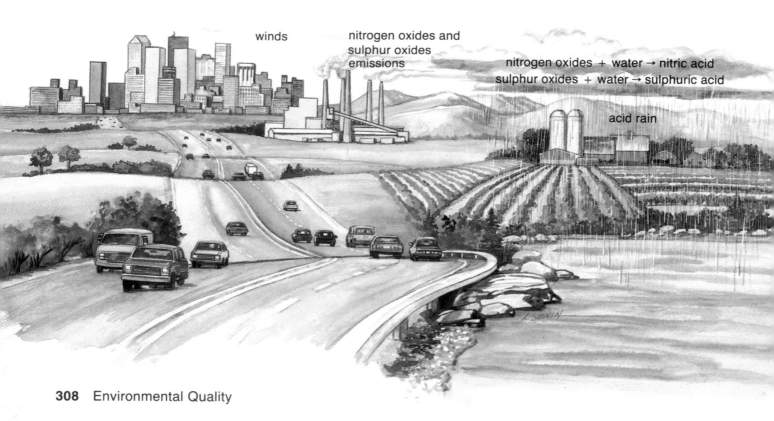

winds

nitrogen oxides and sulphur oxides emissions

nitrogen oxides + water → nitric acid
sulphur oxides + water → sulphuric acid

acid rain

A neutral solution has a pH of 7.0. Substances with a higher pH are basic, those with a lower pH value are increasingly acidic. A decrease of one unit indicates the acidity has been multiplied by a factor of 10. Thus, a solution of pH 4 is 10 times more acidic than a solution of pH 5 and 100 times more acidic than a solution of pH 6.

These air pollutants then fall to the ground as acid rain. Rain usually has a pH of about 5.6. (See the illustration of the pH scale on page 43.) **Acid rain** has a pH lower than 5.6. The lower the pH, the more acidic the solution. Acid rain may feel or taste no different from ordinary rain. But its effects on the environment are far different and are very serious.

Acid rain causes chemical changes in soil and water that, over many years, can reduce soil fertility, retard tree growth, and kill animal and plant life in rivers and lakes. In Ontario alone, 4000 lakes have already been made uninhabitable by all species of sport fish due to acid rain. By the year 2000, if drastic measures are not taken, 48 000 lakes could be affected as severely.

Acid rain is strong enough to corrode exposed metal surfaces and eat away stone statues and limestone buildings. It can also result in the leaching of toxic chemicals, such as mercury, from soil into our waterways.

Why is acid rain more destructive in some areas than in others? In some areas, a type of rock called limestone (calcium carbonate) makes up much of the land and also lies below many lakes. Limestone neutralizes acids. In other areas, however, such as on the Canadian Shield in much of Ontario, the rock is largely granite. Unlike limestone, granite does not neutralize the effects of acid rain. As a result, lakes in the region have become increasingly acidic.

These trees are many kilometres away from a plant that releases sulphur from its smokestacks, yet they show the effects of acid rain.

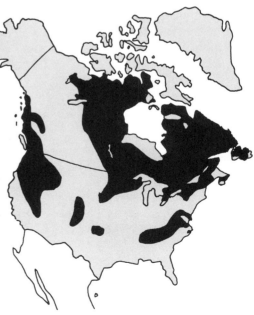

Regions of North America most sensitive to acid rain are shown in black.

Acid Shock

Mature fish are not generally killed in large numbers by the acidity in lakes affected by acid rain. However, in lakes that are becoming more acidic, it is common to see no young fish, only mature adults. This is because female fish seem to have trouble releasing their eggs in acidic water, and eggs and newly hatched young die at a higher rate in acidic lakes.

Much of Canada is covered by snow in winter. In the spring, when all of this snow melts and enters rivers and lakes, the waterways may become 100 times more acidic. Although the period of extreme acidity, sometimes called **acid shock**, may last just a few days, it can have disastrous effects on hatchling fish and, ultimately, on all organisms.

Salamanders lay their eggs in water. They are among the first species to disappear from areas where acidity is a problem.

Did You Know?

Most Canadians know that acid rain is a problem in eastern Canada. Is it a problem in western Canada? Monitoring devices have shown that rainfall along the west coast is becoming more acidic. It seems that residents of the West can no longer consider themselves safe from acid rain.

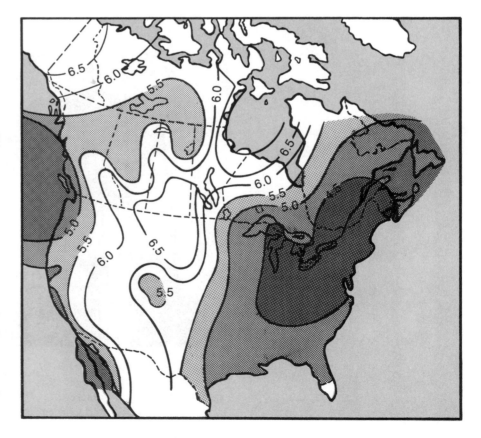

Regions receiving acid rain. The numbers represent average pH levels for various regions of North America. In the state of New Hampshire, in the northeastern United States, a pH between 4.0 and 4.2 was common in a ten-year period of monitoring (1964–1974).

Controlling Air Pollution

To decrease the damage caused by emissions of sulphur dioxide, some smelting operations have built tall smokestacks so that the winds can carry the pollutants farther from their source of production and allow them to be mixed with larger amounts of air. However, dilution is not the best solution to the acid rain problem. Some industries and electrical generating plants now use scrubbers to remove most of the sulphur dioxide gas from other gases before they leave the smokestacks. Many plants still haven't introduced these scrubbers because they are expensive. Other solutions are: using fuels with less sulphur, such as natural gas; switching to alternative forms of energy, such as solar power, wind power, or tidal power. But these changes, too, cost money and are not practical in all places.

Even stricter controls must be placed on automobile use and exhaust emissions. The catalytic converters required in new cars in the 1970s were a big step forward. In these devices, carbon monoxide is converted to carbon dioxide before it is released into the air through the car's exhaust system. The shift from leaded to non-leaded gasoline was also important; it reduced the amount of lead released into our air, soil, and waterways.

Many devices are now available to reduce particulate emissions from industries. One such device is the electrostatic precipitator shown here. The level of particulates in the air in North America has gone *down* in the last 20 years because of measures such as these.

In 1976, the Canadian Network for Sampling Precipitation (CANSAP) was established specifically to monitor contaminants of all kinds. This network of stations regularly samples environmental quality. All of the data are submitted to a central station where any researcher studying a problem has access to the information. The network has discovered patterns and effects of acid rain and other types of air pollution; it has documented them so that appropriate action can be taken.

In an electrostatic precipitator, dirty air flows between negatively charged wires and metal collecting plates. The particulates in the flowing air become negatively charged and are attracted to the positively charged plates. The plates hold the accumulated particulates until they are released through hoppers into dump trucks for disposal while clean air flows out through the smokestack.

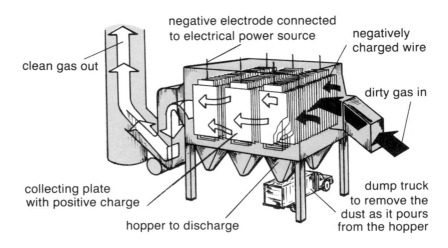

negative electrode connected to electrical power source

negatively charged wire

clean gas out

dirty gas in

collecting plate with positive charge

hopper to discharge

dump truck to remove the dust as it pours from the hopper

The Water We Use

You now know that pollutants can reach our lakes and rivers indirectly from the solid and gaseous waste we produce. Pollutants are carried through ground water to waterways and through the air from the burning of fossil fuels. Pollutants are also added to the environment directly, through our waste water. Table 6-3 and the illustration summarize the major types and sources of water pollutants.

Table 6-3 *Major Types of Water Pollutants and Their Sources*

TYPE OF POLLUTANT	EXAMPLE	MAJOR SOURCES
Disease-causing agents	bacteria, viruses, protozoa, parasitic worms	• poorly treated domestic sewage, animal wastes (agricultural)
Carbon compounds	fecal waste (solid waste produced by animals)	• domestic sewage, animal feed lots
	oil	• machine wastes, spills, pipeline breaks
	chlorine compounds	• water purification, paper-making plants
	pesticides	• agriculture, forestry
	detergents, plastic	• homes and industries
Water-soluble acids and salts	sulphuric and nitric acids	• mining, industrial wastes
	salt	• irrigation, mining, de-icing of roads
Water-soluble poisons	lead	• leaded gasoline, poisons, pesticides, smelting of lead
	mercury	• industrial waste, fungicides
	various herbicides and pesticides	• agriculture, forestry, mosquito control
Plant nutrients	phosphates, nitrates	• domestic sewage, industrial waste, natural runoff, agricultural runoff, food-processing industries
Sediment and suspended undissolved matter	soil, silt	• runoff from agriculture, forestry, construction
Radioactive substances	uranium, strontium	• natural sources (rocks, soils), uranium, mining, nuclear-weapons testing
Heat (from water used as a coolant)		• electrical power and industrial plants

sewage via pipes to
sewage treatment plant

Cropland, animal feedlots, rural homes, suburban developments, industries, and cities are all potential sources of water pollution.

Controlling Water Pollution

The average Canadian produces about 238 L of waste water per day for personal uses, such as showers, laundry, and flushing toilets. Have you ever wondered what happens to the waste water that runs down the pipes from your bathroom, kitchen, and laundry?

The liquid and solid waste carried away from homes in underground pipes is known as **sewage**. In many rural and suburban areas, waste water and sewage from houses run into a specially designed large underground container called a **septic tank.**

Since a septic tank system requires a large area beside each house, such a system is not feasible in a city where many people live close together. In urban areas, sewage is taken by pipes to sewage treatment plants where some of the impurities are removed. The sewage treatment plant receives and treats domestic waste water (from homes, schools, etc.), industrial waste water (from factories and other industries), and water from street drains. The partially purified waste water, called **effluent**, then runs into a river, lake, or ocean. During heavy rainstorms, overflow waste water is sometimes carried directly to a lake without any treatment.

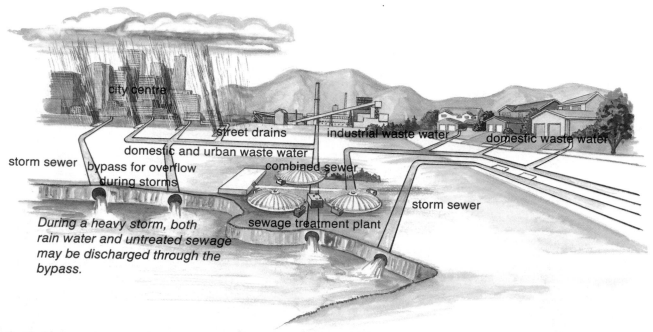

city centre

street drains

industrial waste water

domestic waste water

domestic and urban waste water

storm sewer

bypass for overflow
during storms

combined sewer

storm sewer

sewage treatment plant

During a heavy storm, both rain water and untreated sewage may be discharged through the bypass.

Types of sewer systems commonly used in cities.

Sewage treatment may involve up to three stages : primary, secondary, and tertiary. The levels differ according to how extensively, and by what methods, the water is treated. The type of sewage treatment used by a community depends on the demands of the community's members, on the cost, and on local government regulations.

Primary sewage treatment is done by physical processes: first, filtering or sieving followed by settling. Waste water flows through a series of metal screens that can trap solids ranging from particles as small as a grain of sand to larger objects, such as diamond rings or tree branches. The water then enters into a tank, where waste material is allowed to settle out as **sludge**. After some more treatment and drying, the sludge can be returned to the environment as fertilizer or as landfill material (so long as it is not contaminated with a pollutant). The water remaining above the sludge is treated with chlorine, which kills harmful micro-organisms. The water may be returned to the environment as effluent.

If it is not returned to the environment, the effluent passes on to a secondary treatment facility. **Secondary sewage treatment** is biological. The waste water is held in tanks where bacteria and other micro-organisms decompose much of the remaining biodegradable waste (material that failed to settle out into the sludge during the primary treatment stage). Oxygen is needed for this process, so air is bubbled through the tanks. Once again, the waste water that remains can either be returned to the environment as effluent, or passed on for still more processing. Most of the biodegradable material in waste is removed by primary or secondary treatment.

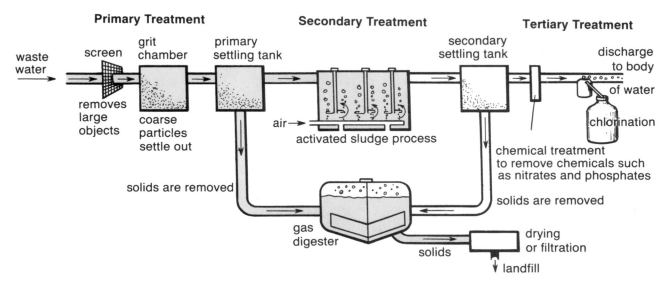

Stages of sewage treatment.

Tertiary treatment is designed to remove dissolved nitrates, phosphates, and remaining suspended undissolved solids from the effluent. This is mostly by a series of chemical processes. As you have learned, nitrates and phosphates are plant nutrients that often flow into the waste water as a result of fertilizer use, domestic sewage, and industrial plants. If they are not removed from waste water and are allowed to enter rivers or lakes, they can cause overabundant growth of algae and water plants. When this rich plant growth dies, it provides food for large populations of bacteria and other micro-organisms that decompose it. Because the decomposition process uses the oxygen that is dissolved in water, too much decomposition can make the oxygen levels drop very low. This means that other organisms, such as fish and aquatic insects that require oxygen for survival, will die of suffocation.

List at least five steps that could be taken to control solid waste production, emission of materials into the air, and treatment of water itself that would reduce water pollution in your area. Give reasons why you think each of these steps might be an issue in your community.

Table 6-4 *Some Substances Found in Effluent from Primary, Secondary, and Tertiary Sewage Treatment*

SUBSTANCE	PERCENTAGE (%) REMAINING AFTER TREATMENT		
	PRIMARY	SECONDARY	TERTIARY
suspended solids	about 40	about 10	less than 10
biodegradable waste	about 70	about 10	less than 10
phosphates	about 90	about 70	less than 10
nitrates	about 80	about 50	less than 10

Science and Technology in Society

Foods that Pollute

Both of the foods in the photograph have been the cause of serious water pollution! The key is not *what* they are, but rather *how* they are made ready for us to eat. Let's examine the facts to get to the "roots" of this story.

Carrots and potatoes both grow under the ground. Before you can eat them, the first step is to wash off any soil. People often also peel the skins off both vegetables before eating them. When you prepare these vegetables at home, you may wash the soil down the kitchen sink. The peels and tops may be composted or tossed into a garbage container. Already the vegetables have become potential pollutants.

A nutritious vegetable and a popular snack—what could possibly harm the environment?

Imagine not one, but millions of carrots or potatoes being prepared for processing. Large amounts of water must be used to wash them free of soil. In a large processing plant, water is used to wash away the peels as well. In fact, water was once used to peel the carrots! The carrots were dipped into a strong solution of sodium hydroxide (lye) to soften and loosen their skins. High-pressure jets of water then blasted the skins free of the carrots and rinsed them away. This type of carrot-processing plant used 6.9 million litres of water per day—as much as the waste water from 2000 people. All this waste water had to be cleansed of solid wastes before the effluent could be safely returned to the environment. The large amounts of very dirty water could overload the waste-water treatment plant.

A Way to Reduce Carrot Pollution

Basically, the problem was how to remove the skins from the carrots without using so much water. People in the food-processing industry found that rubber provided a useful tool. The solution is to dip the carrots in lye (as in the old process) and then send them through revolving rubber discs. These discs rub off the softened peel, leaving the useful portion of the vegetable. The result—up to 75% less water is used by the carrot-processing industry, and 85% less solid waste ends up in the water sent to the waste-water treatment plant.

Better than Expected

This technological solution to a particular pollution problem has had benefits beyond those expected. The dry-peeling process actually improves the quality of the processed carrots because the carrots retain nutrients that would otherwise be rinsed away in the water. The dry peeling is more efficient because it leaves 10% more of the carrot for processing. And the dry peel is easily collected for high-quality animal feed once the lye is removed.

Potato Pollution

One potato processing plant had a daily potato intake of 175 000 kg. It produced a daily waste of 40 500 kg of potato ends and peels, 11 250 kg of dirty or damaged chips, and 550 kg of used cooking oil. All these waste products used to be washed away with the plant's waste water, often overloading the city's treatment plant.

Co-operation Saves the Day

For the potato chip producer, the way to reduce its pollution was to join forces with the city government and the local farmers. The city accepted the waste water from the producer but charged the company a special fee based upon the amount of water and how much treatment it required. This money was spent on improvements to the treatment plant so that it could better handle the waste. The chip producer then installed equipment to recover the bulk of the waste (peels, chips, and oil) from its waste water. The dried waste (peels and chips), amounting to several hundred tonnes a year, is now delivered to local farmers as livestock feed. The recovered oil is used by other industries.

Any large-scale procedure, even washing vegetables like these sweet potatoes, is a potential source of pollution.

Think About It

A meat packing plant uses 660 million litres of water per year, a paper towel manufacturer uses 425 million litres, and a typical dairy 88 million litres per year. Most small cities have several of these and similar industries. For one of the industries above, or for a local industry of your choice, find out (a) what happens to the waste water from the industry, and (b) what steps have been taken or are being considered to improve the quality of the waste water and/or to reduce its volume.

Water Quality

Canadians are fortunate to have abundant surface water in lakes and rivers. This water is available for drinking and for use in personal cleanliness, manufacturing processes, and recreation — swimming, boating, and fishing. But all is not well with our water supply. Many of our lakes and rivers have become acidic due to acid rain. Others contain pollutants from industry or from untreated domestic sewage. What happens when a river or lake is used for several conflicting purposes? The following case study of Rapid Steps, a city with a water problem, will give you some idea of the difficulties of dealing with issues of environmental quality. Read "The Issue in Rapid Steps" to understand the concerns of different people involved in the issue.

The Issue in Rapid Steps

The city of Rapid Steps has grown up along the banks of the Rapid River. Farms and cattle ranches surround the city. Both visitors to the city of Rapid Steps and its residents enjoy canoeing and fishing on the river and hiking along its banks. The location has attracted many new businesses and industries. Some companies have moved their main offices and factories to Rapid Steps. This has brought people from many parts of the country to live and work here, and the population of Rapid Steps has grown rapidly in the last 25 years, reaching 300 000. The city's population is expected to double in the next 25 years.

Rapid Steps and surroundings.

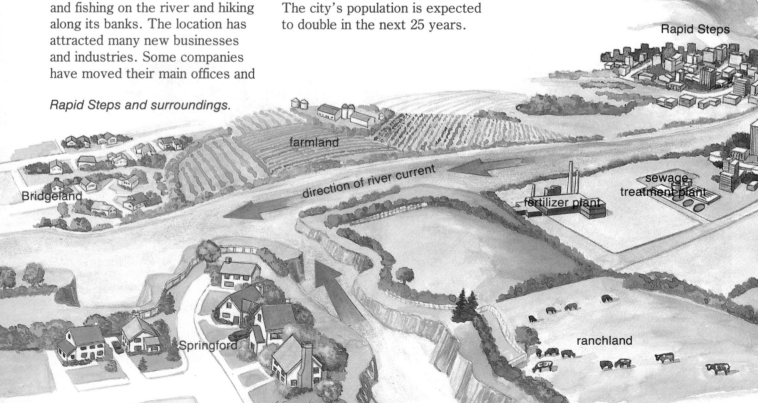

Rapid Steps

farmland

direction of river current

Bridgeland

fertilizer plant

sewage treatment plant

Springford

ranchland

The Concerns of People Downstream

Downstream from Rapid Steps are the small towns of Bridgeland and Springford. The people in these communities have been having problems with the drinking water that comes from Rapid River. It is cloudy and some days it "tastes bad." The people blame the city of Rapid Steps for allowing too much waste (especially phosphates) to be dumped into the river. They have sent a petition, signed by 400 people, to the city council of Rapid Steps. They have also talked to newspaper reporters and have written letters to the editors of local newspapers explaining their concerns.

The Concerns of Politicians

An election will be held in Rapid Steps next year, and the city council is worried about the negative publicity that their city's waste water treatment has received. The council has no direct evidence that Rapid Steps is responsible for all the problems reported downstream from the city. They do know, however, that the city's secondary sewage treatment plant removes less than half the phosphates from the effluent before it enters the river from the plant. If the city is responsible for the problems with the river's water, then the council members must find ways to improve the quality of the water. To do so is expensive. The cost of buying and installing phosphate removal equipment to upgrade the city's sewage treatment plant to the tertiary level would be $40 million. The cost of the chemicals needed for the process would be another $3 million per year. Maintaining the additional equipment and monitoring the process would drive the costs even higher.

To pay for these improvements, the city council would have to raise taxes. But people are already finding it difficult to live in the city because of high costs. Many complain because the taxes have been raised annually for the past five years. Tax money is needed to pay for so many basic items: new sewer pipes, new water pipes, new gas lines, road improvements, school maintenance, and city maintenance in general.

Money from taxpayers pays for new sewer pipes, among other things.

The Concerns of Industry

The fertilizer plant near Rapid Steps removes the solid wastes from their plant wastes and then releases the effluent into the river. The plant managers know that the effluent contains phosphates. If the city acquires phosphate removal equipment, the fertilizer plant could send its waste water to the city's sewage treatment plant and pay the city to remove the phosphates. If the city insists that the fertilizer plant clean up its own waste water, then the extra expense might be too great for the company to remain profitable. The plant might even be forced to close, which would mean the loss of several hundred jobs.

Some Other Viewpoints

The detergent plant in Rapid Steps specializes in producing a phosphate-free detergent. The manufacturers of the detergent argue that if the people in the city use only their phosphate-free detergent, there will be less phosphate entering the river (and the detergent manufacturers will make more money). So the detergent manufacturers are telling the city council that there is another solution besides adding very expensive tertiary sewage treatment equipment.

An agriculturalist at the city college also thinks there are ways other than an expensive addition to the city's sewage treatment plant to deal with the phosphate levels in the Rapid River. The farmers and ranchers along the river use large amounts of fertilizer containing phosphates. Some of the fertilizer is dissolved in rainwater and runs off fields and into the river. If farming practices were changed, this source of phosphates could be reduced. For example, holding basins could be built to prevent the runoff from entering the river. The collected runoff could then be reused as irrigation water. However, building holding basins will also be expensive for the farmers and ranchers if they have to pay the costs.

A local resident who is active in promoting programs to ensure the environment is kept as natural as possible is encouraging all efforts to improve the condition of Rapid River. He thinks each contributor to the pollution should be responsible for carrying out a program to reduce its own phosphate contamination of the river. But he thinks everyone should pay, because everyone ultimately will benefit from the clean-up.

Who should pay for cleaning up the river—everyone, or only those who use it for recreation?

Holding basins can be built to help prevent runoff.

runoff from feedlots or cropland

holding basin (water can be used again for irrigation)

river

Preparing for the Hearing

Now you know the major concerns surrounding the environmental issue in Rapid Steps. When you conduct your own hearing on the issue at the conclusion of your study, you will examine your evidence to determine whether the source of pollution is the fertilizer plant, the city residents, the runoff from fields, or all three of these. You will decide whether or not the city should upgrade its sewage treatment plant. And finally, if action must be taken, you will decide who should pay — the fertilizer plant, the city residents, the farmers and ranchers, or everyone — and how much they should pay.

Who Will Speak at the Hearing?

The following people have asked to speak at the hearing. Divide into groups; each group will take the role of one of these people in Activity 6–9:

- a member of city council,
- a taxpayer from Rapid Steps,
- the manager of the detergent factory,
- the manager of the fertilizer plant,
- a worker at the fertilizer plant,
- a representative of the people of the towns of Bridgeland and Springford,
- a guide who makes a living taking people fishing on Rapid River,
- a local teacher who likes to take students canoeing on Rapid River,
- a member of an environmental group,
- an agricultural researcher,
- a farmer/rancher,
- a sewage treatment technologist,
- a freshwater biologist.

Your previous studies in this Unit have outlined the different viewpoints on issues that could be represented at your hearing. However, you are still lacking some critical information that you need to conduct the hearing — the scientific information. What information does the freshwater biologist need to have about water quality in order for a decision to be reached? By investigating both the physical characteristics of water and the different types of organisms that live in water, the subject of this Topic and Topic Five, you will be conducting the same kinds of tests that a scientist must do in order to assess water quality.

Suspended Solids

The people living downstream from Rapid Steps had three major complaints about the water in the river. First, they complained that the water was cloudy. Usually water is cloudy or **turbid** if there are undissolved solids suspended in it. Solids may enter a river or lake by erosion from the land or by means of sewage or industrial waste. (Excessive algal growth can also make water appear cloudy.)

As well as making the water look unpleasant to people, turbid water can also cause problems for the organisms living in the water. Suspended solids block sunlight, preventing it from penetrating into the water. This reduces the ability of plants and algae to carry out photosynthesis, which returns oxygen to the water. When there is insufficient oxygen in water, the organisms living there die.

Rapid City's secondary treatment sewage plant should be removing most of the suspended solids from waste going into it unless the city is now producing more waste than the tanks can fully process. In order to determine why the water was cloudy downstream from Rapid City, the biologist took water samples from several sites, as is being done in this photograph.

The sites the biologist selected for this and other measurements were as follows: *Site 1*, just upstream from Rapid City; *Site 2*, directly downstream from the sewage treatment plant; *Site 3*, directly downstream from the fertilizer plant; *Site 4*, just upstream from Bridgeland; and *Site 5*, 60 km downstream from Rapid City. The next Activity shows the biologist's results.

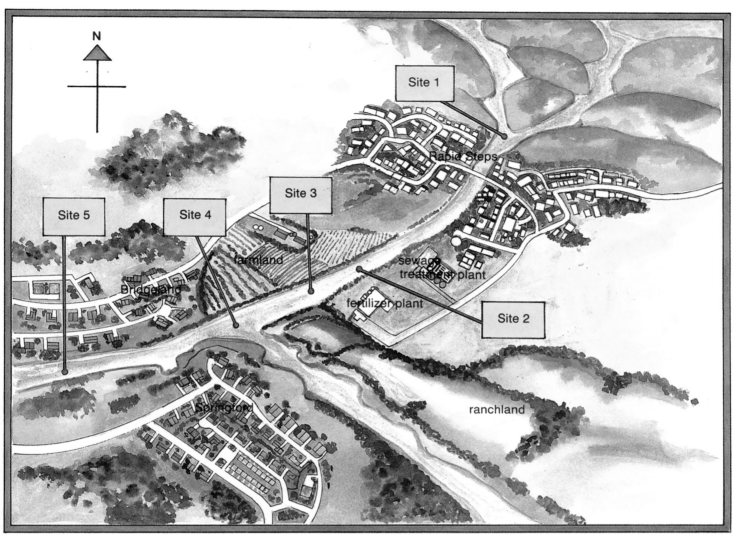

Sampling sites.

Biologist taking a water sample.

Measuring Undissolved Solids

You have used the process of filtration in your previous studies to separate mixtures into two parts, the **residue** and the **filtrate**. This is one of the tests done by biologists to determine the amount of suspended solids in a water sample. Read the description of the biologist's experiment, then copy her results for each site into a table in your notebook for later use. Your column headings should be: Undissolved Solids, Dissolved Phosphates, Dissolved Oxygen. (Part B of each of the Activities that follow will give you the remaining data you need to complete your table.)

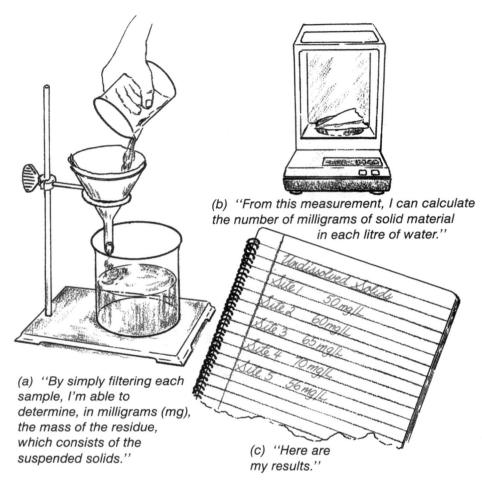

(b) *"From this measurement, I can calculate the number of milligrams of solid material in each litre of water."*

Undissolved Solids	
Site 1	50 mg/L
Site 2	60 mg/L
Site 3	65 mg/L
Site 4	70 mg/L
Site 5	56 mg/L

Analysis

1. How does filtering separate the residue from the filtrate?
2. What is (are) the source(s) of suspended solids in the river that flows to Bridgeland and Springford?

(a) *"By simply filtering each sample, I'm able to determine, in milligrams (mg), the mass of the residue, which consists of the suspended solids."*

(c) *"Here are my results."*

Chemistry is used to determine which solutes are present in water.

Measuring Phosphates in a Sample

The second complaint of people living downstream from Rapid Steps was that there was increased algal growth in the river. As you know, algae use phosphates as nutrients. Abundant algal growth, therefore, indicates the presence of large amounts of phosphates.

If you suspect that phosphates are dissolved in water, you can add magnesium sulphate to the water sample, and then make the solution slightly basic. The magnesium in this compound will combine with the phosphates to form a solid insoluble magnesium phosphate. As crystals of magnesium phosphate form, they cause the previously clear solution to become cloudy. Try this in the next Activity.

Phosphate Identification

PART A

Problem

How can you determine if phosphates are present in various water samples?

Materials

safety glasses
safety gloves
safety apron
water samples to be tested:
  distilled water
  tap water
  water sample from a local
    stream, river, pond, or lake
  water containing fertilizer
  water containing detergent
dilute ammonium hydroxide
  solution
magnesium sulphate solution
test tubes
test tube rack
medicine dropper
graduated cylinder (10 mL)
watch glass
hand lens

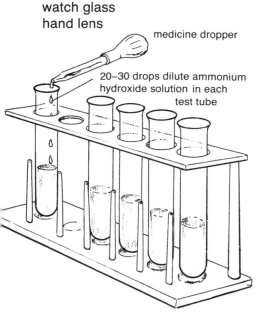

medicine dropper

20–30 drops dilute ammonium hydroxide solution in each test tube

Procedure

1. Predict which samples contain phosphates.
2. Pour about 15 mL of each of the water samples into separate, labelled test tubes in a test tube rack.
3. Carefully add about 20 to 30 drops of dilute ammonium hydroxide to each sample. (This ensures that the solution is basic, so that magnesium phosphate can form.)
4. Add about 2 mL of magnesium sulphate solution to each test tube.
5. Let the test tubes stand for several minutes. Record your observations in a table.
6. If crystals form in any of the test tubes, pour a small amount of the material onto a watch glass and examine the crystals using a hand lens.

> **CAUTION: Ammonium hydroxide can burn your skin. Handle it carefully. Be sure to wear glasses, safety gloves, and a safety apron.**

Analysis

1. Which sample(s) contained crystals after magnesium sulphate was added?
2. Describe the crystals that formed.

3. Which sample(s) contained phosphates? Give a reason for your answer.
4. What was the purpose of the sample of distilled water in this experiment?

Further Analysis

5. Describe two ways in which phosphates get into surface water.
6. Why might testing the phosphate concentration of water be part of the operation of a sewage treatment plant?

PART B

A normal concentration of phosphates in waterways around Rapid City is about 10.0 μg/L of water. The results of the biologist's tests for phosphates at each of the five sites shown on page 323 are as follows: *Site 1*, 10.0 μg/L; *Site 2*, 22.0 μg/L; *Site 3*, 25.0 μg/L; *Site 4*, 28.0 μg/L; *Site 5*, 18.0 μg/L. Record each of these amounts in the table in your notebook in which you recorded the amount of undissolved solids at each site.

Analysis

1. At which sites do the largest changes in concentration of phosphates occur?
2. What is (are) the possible source(s) of phosphates in the river that flows to Bridgeland and Springford?

Measuring Dissolved Oxygen

The third complaint of people downstream from Rapid City was that there were fewer fish in the river than in the past. Measuring oxygen — one of the more important dissolved substances in the water — can give the biologist one kind of important evidence about why there are fewer fish. As you know, most organisms cannot survive without oxygen. The amount of oxygen dissolved in the water of a lake or river depends on several **abiotic factors**, including the temperature of the water and whether the water is calm or turbulent (moving in a rapid, irregular flow). (Abiotic factors are the non-living parts of the environment.)

Biotic factors (the living parts of the environment) also determine the amount of oxygen in the water. Organisms such as fish, snails, insects, bacteria, and fungi consume oxygen dissolved in the water. Where there is a large amount of biodegradable material, the decomposers multiply and use up most of the available oxygen. The lack of oxygen then kills the fish and other organisms that need it.

The concentration of dissolved oxygen in water is measured in milligrams per litre (mg/L) or parts per million (ppm). Most organisms living in water need oxygen levels above five milligrams per litre (5.0 mg/L) of water. This means that in every litre of water there must be more than five milligrams of oxygen. The amount of oxygen in the water can be determined by an indicator that changes colour in the presence of oxygen. The amount of colour change depends on the amount of dissolved oxygen in the water.

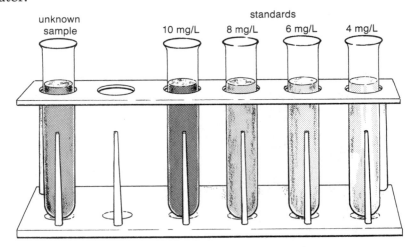

Measuring dissolved oxygen. The colour of the indicator after it has been placed in the water sample is compared to a set of standard colours. In this example, the colour of the unknown sample appears to be midway between the colours of the 8 mg/L and the 10 mg/L standards. The oxygen level of the sample is therefore about 9 mg/L.

Oxygenated Water

PART A

Problem

How do temperature and turbulence affect the amount of oxygen dissolved in water?

Materials

tap water
100 mL beaker
hot plate
beaker tongs or oven mitts
dissolved-oxygen measuring kit
sealed jar of boiled, cooled
    water

Procedure

1. Using the dissolved-oxygen kit, determine the amount of dissolved oxygen in a sample of tap water at room temperature.

> CAUTION: Avoid getting burned by the boiling water or steam.

2. Place about 75 mL of water in a 100 mL beaker, bring it to boiling using a hot plate. Determine the amount of dissolved oxygen in it immediately afterwards.
3. Unseal the jar of boiled, cooled water and pour a sample into a 100 mL beaker. Measure the amount of dissolved oxygen in milligrams per litre.
4. Empty some of the water from the jar until the jar is half full, reseal it, and shake it vigorously for about 30 s.
5. Remove the lid of the jar for 10 s, then reseal the jar and shake it again.
6. Repeat Step 5 twice more, then measure the amount of dissolved oxygen in the water in the jar.
7. Record the class results in a table, and calculate average values for the amounts of dissolved oxygen in each sample of water tested.

Analysis

1. What was the average amount of dissolved oxygen in:
   (a) water at room temperature?
   (b) boiled water?
   (c) boiled, cooled water?
   (d) shaken water?
2. What is the relationship between water temperature and the amount of dissolved oxygen?
3. Account for differences in the class results for the amount of oxygen in the turbulent water.

Further Analysis

4. (a) Name two circumstances that might raise the temperature of the water in a river.
   (b) What might be the effect of warmer water on the organisms living in a river? Explain.
5. What natural features of rivers help to oxygenate the water?
6. A student measured 7.2 mg/L dissolved oxygen in a stream at sea level. A stream of comparable size with the same rate of flow and amount of turbulence at a 4000 m altitude had a dissolved oxygen content of 4.0 mg/L. Give a possible reason for the difference.

PART B

The results of the biologist's tests for dissolved oxygen at each of the five sites shown on page 323 are as follows: *Site 1*, 10.0 mg/L; *Site 2*, 4.0 mg/L; *Site 3*, 3.5 mg/L; *Site 4*, 2.2 mg/L; *Site 5*, 8.0 mg/L. Record each of these amounts in your table, along with suspended solids and phosphates.

Analysis

1. At which site(s) does the oxygen concentration show a decrease?
2. What possible source(s) of pollution could have caused this decrease?

1. Use the clues to solve the mystery word and give a definition of this term once you've solved the word puzzle.

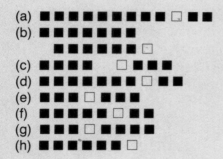

Clues

(a) tiny particles of solids and some liquids released into the air

(b) the burning of coal is our greatest source of this air pollutant (two words)

(c) precipitation with a pH below 5.6

(d) substances found in fertilizers

(e) solid material that remains on a filter

(f) material that passes through a filter

(g) the most common substance in air

(h) water treatment that involves filtering or sieving, followed by settling

2. List two sources for each of the following water pollutants:
(a) compounds that are broken down by micro-organisms,
(b) acid rain,
(c) nitrates,
(d) phosphates,
(e) suspended solids.

3. Which level of sewage treatment (primary, secondary, or tertiary) removes most of the phosphates in the water?

4. Suppose no phosphates or nitrates were removed from our cities' wastes. Changes in the number of different species of organisms and level of oxygen would probably take place. Copy the list below in your notebook; beside each item indicate whether it would increase or decrease. In a third column give a reason for each of your decisions.
(a) water plants and algae
(b) bacteria
(c) dissolved oxygen
(d) insects
(e) fish

5. Explain why aquariums have aerators.

6. (a) Explain two ways in which human activities can cause water to be turbid.
(b) What effect does increased turbidity have on photosynthesis carried out by aquatic algae and plants?

7. An industrial plant uses water from a lake to cool its machines. The heated water is returned to the lake. What effect will this have on:
(a) the amount of dissolved oxygen in the lake?
(b) the survival of fish?

8. How might the industrial plant in Question 7 solve the problem of releasing heated water into the lake?

9. Which viewpoint does each of the comments in the illustration reflect? For example, (a) reflects an economic viewpoint. Categorize viewpoints (b) through (g).

(a) "How much money will it cost to clean up the water?"

(b) "We can remove phosphates and pesticides from sewage and runoff."

(c) "Are the people interested in paying more taxes for cleaning up the water?"

(d) "The water quality must be improved."

(e) "It isn't a pretty sight."

(f) "What is in the water that killed the fish?"

(g) "It's wrong to kill fish!"

Testing Water Quality by What Lives in It

You have learned how to measure water quality directly by examining some of the substances it contains. Water quality can also be assessed indirectly by studying the organisms that live in the water. Water that is heavily polluted will generally have fewer kinds of organisms than similar areas of "clean" water.

Some organisms are used as **biological indicators** of water quality. For example, certain species of bacteria, invertebrates (animals without backbones), and fish can live only in water that is clean and highly oxygenated. Other species that require less oxygen, however, can live in polluted waters. For example, only certain kinds of worms and midge larvae (sometimes called blood worms) can live in highly polluted streams and ponds. Identifying organisms in a sample of water is therefore one way to assess the concentration of dissolved oxygen.

Most species of organisms cannot live in polluted water, but some species of worms and insect larvae (young) thrive in polluted conditions.

AH, THIS IS THE LIFE!

In this Topic, you'll test the quality of water in your own area by analysing the organisms that live in it. In so doing, you will conduct the other major test that the biologist would do before presenting the scientific results to the hearing in Activity 6–9. Table 6-5 shows the oxygen requirements of some common freshwater invertebrates. Become familiar with these requirements and the habitats of common freshwater invertebrates before proceeding with Activity 6–8.

Table 6-5 *Oxygen Requirements of Some Freshwater Invertebrates*

OXYGEN CONCENTRATION (mg/L)	ORGANISMS PRESENT
8 and above (excellent level for many species of fish and other desirable organisms)	• large variety of invertebrates (insect larvae [young] of many kinds, beetles, worms, etc.)
6 (good level)	• a few mayfly larvae • some stonefly larvae • some beetles • many midge larvae • many worms, including leeches
4 (critical level—some species have difficulty living in waters with this little oxygen)	• freshwater shrimp • many midge larvae • many worms, including leeches
2 (low level)	• many midge larvae • some worms, including leeches
less than 2 (very low level—no species of game fish exist)	• some midge larvae • some worms

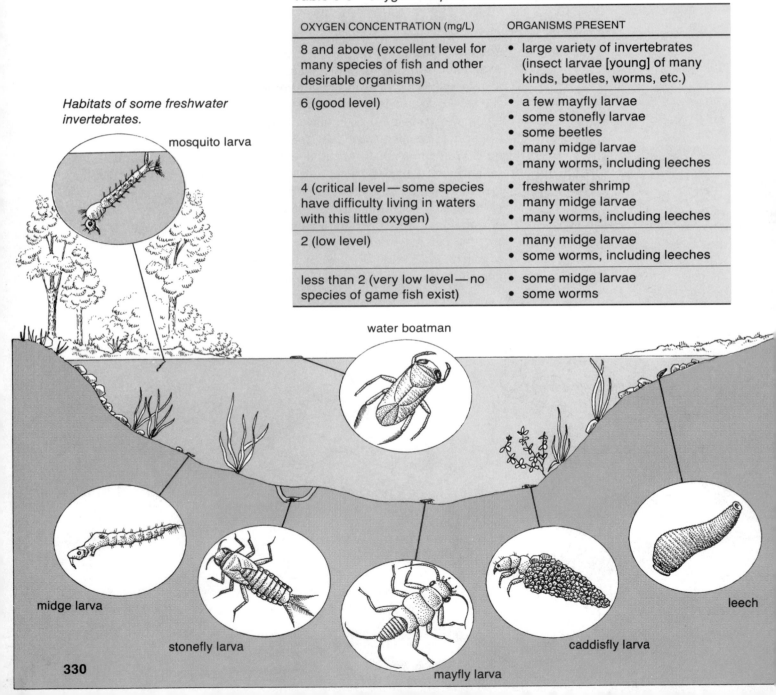

Habitats of some freshwater invertebrates.

mosquito larva

water boatman

midge larva

stonefly larva

mayfly larva

caddisfly larva

leech

Invertebrates as Indicators of Water Quality

PART A

Problem

Using invertebrates as indicators, what is the quality of water in a river or stream?

Materials*

net with a metal rim
appropriate clothes for collecting (rubber boots, etc.)
shovel
dissolved-oxygen kit
container with a lid (such as an ice-cream bucket or margarine tub)
forceps
petri dish
hand lens or dissecting microscope
70% ethanol solution (optional)
classification key with illustrations of freshwater invertebrates

* If your class is unable to collect samples, examine samples provided by your teacher.

CAUTION: Some streams are very dangerous; only collect samples from a stream that your teacher has designated, and only with the rest of the class.

Procedure

1. Use the dissolved-oxygen kit to measure the amount of dissolved oxygen in a sample of stream water. (Do this as soon as you collect the water. The amount of oxygen in your sample may change over time and may therefore not be the same as the stream's oxygen level if left too long.)

2. Use the net and shovel as shown in the illustration to collect a sample of organisms from the river or stream bottom. (If you are unable to examine the samples within 24 h, preserve them in the 70% ethanol solution.)

Collection Procedure:

(a) Hold the net in the water with the open end facing upstream.

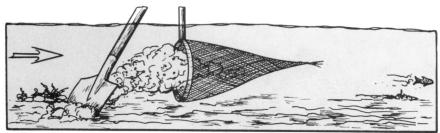

(b) Disturb the stream bottom (substrate) by twisting the shovel in the substrate in front of the net.

(c) Place the organisms and gravel that have entered the net in a bucket of water. (If you are not going to examine the sample within 24 h, remove the organisms from the bucket and place them in a container of 70% ethanol.)

3. Copy Table 6-6 into your notebook. Leave some space to add any invertebrates you might find that are not listed in the table.
4. Carefully place a few organisms from your collection into a petri dish.
5. Use a hand lens or dissecting microscope to examine each organism.

Table 6-6

NAME OF ORGANISM	NUMBER
Mayfly larvae	
Stonefly larvae	
Caddisfly larvae	
Beetles	
Midge larvae	
Leeches	
Other worms	
Other	

6. Using a classification key and the illustrations shown here, identify as many different organisms as you can and record them in your table.
7. Count the numbers of each type of organism in your sample, or decide whether they are "abundant," "rare," or "moderately abundant." Record this category in your table.
8. Return any live organisms to the bucket.

Some organisms commonly found in fresh water (not to scale).

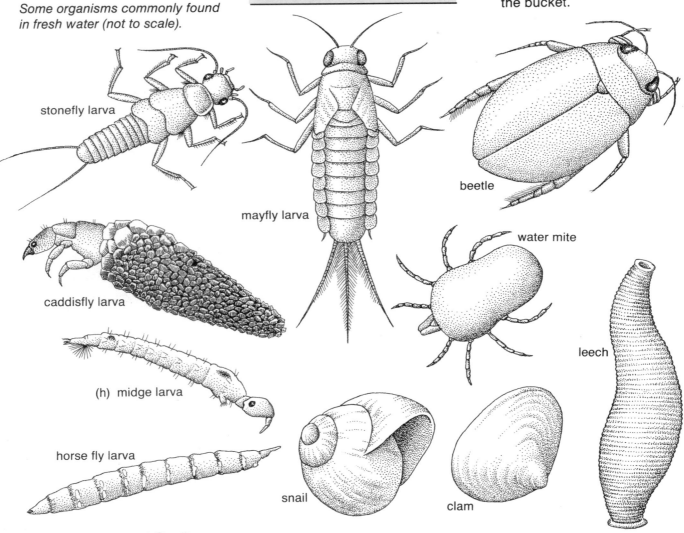

stonefly larva

mayfly larva

beetle

caddisfly larva

water mite

(h) midge larva

horse fly larva

leech

snail

clam

Analysis

1. What was the amount of dissolved oxygen you measured in the sample?
2. Refer to Table 6-5. According to the organisms you counted and identified, what should the dissolved oxygen concentration in the water be?
3. Are your answers to Questions 1 and 2 the same? If not, try to explain why.
4. Give an assessment of the water quality of your sample, based on the numbers and kinds of organisms present.
5. What are some possible sources of error in this experiment?

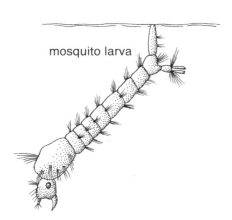

mosquito larva

freshwater shrimp

PART B

The data in Table 6-7 were obtained from the five sites shown on page 323.

Analysis

1. Use the data in Table 6-7 to make a bar graph. Plot the kind of organism on the horizontal axis and its abundance on the vertical axis, using different coloured pencils for each site.
2. Explain why there are fewer insects of all kinds, except midges, and more worms immediately downstream from the sewage treatment plant.
3. Explain the possible cause for changes in the numbers of both insects and worms found at the other sites.
4. Does this analysis agree with the data the biologist obtained from testing for concentration of dissolved oxygen (Activity 6–7, Part B)? If not, explain how it differs.
5. Does this river have a water quality problem? Give reasons for your answer.

Extension

6. Research and prepare a report on the life cycle of one of the organisms you identified.
7. How do various freshwater invertebrates get the oxygen they need? Find out what structures are used by at least two organisms common in your sample.

Table 6-7 *Biological Data Obtained from Rapid River near the City of Rapid Steps*

NAME OF ORGANISM	NUMBER OF ORGANISMS IN A STANDARD SAMPLE				
	SITE 1	SITE 2	SITE 3	SITE 4	SITE 5
Mayfly larvae	254	1	1	0	180
Stonefly larvae	120	0	0	0	75
Caddisfly larvae	43	13	5	0	27
Beetles	16	1	0	0	8
Midge larvae	135	247	250	200	187
Leeches	18	98	80	50	20
Other worms	22	135	130	50	35

The Public Hearing on the Issue

The city council of Rapid Steps is ready to hold its public hearing on the issue of whether the city should improve its sewage treatment plant. The scientists and technicians have carried out their tests, and the other participants in the hearing have prepared their speeches. Before you conduct Activity 6–9, review "The Issue in Rapid Steps" on page 318.

Here is an example of what each speaker's viewpoint might be:

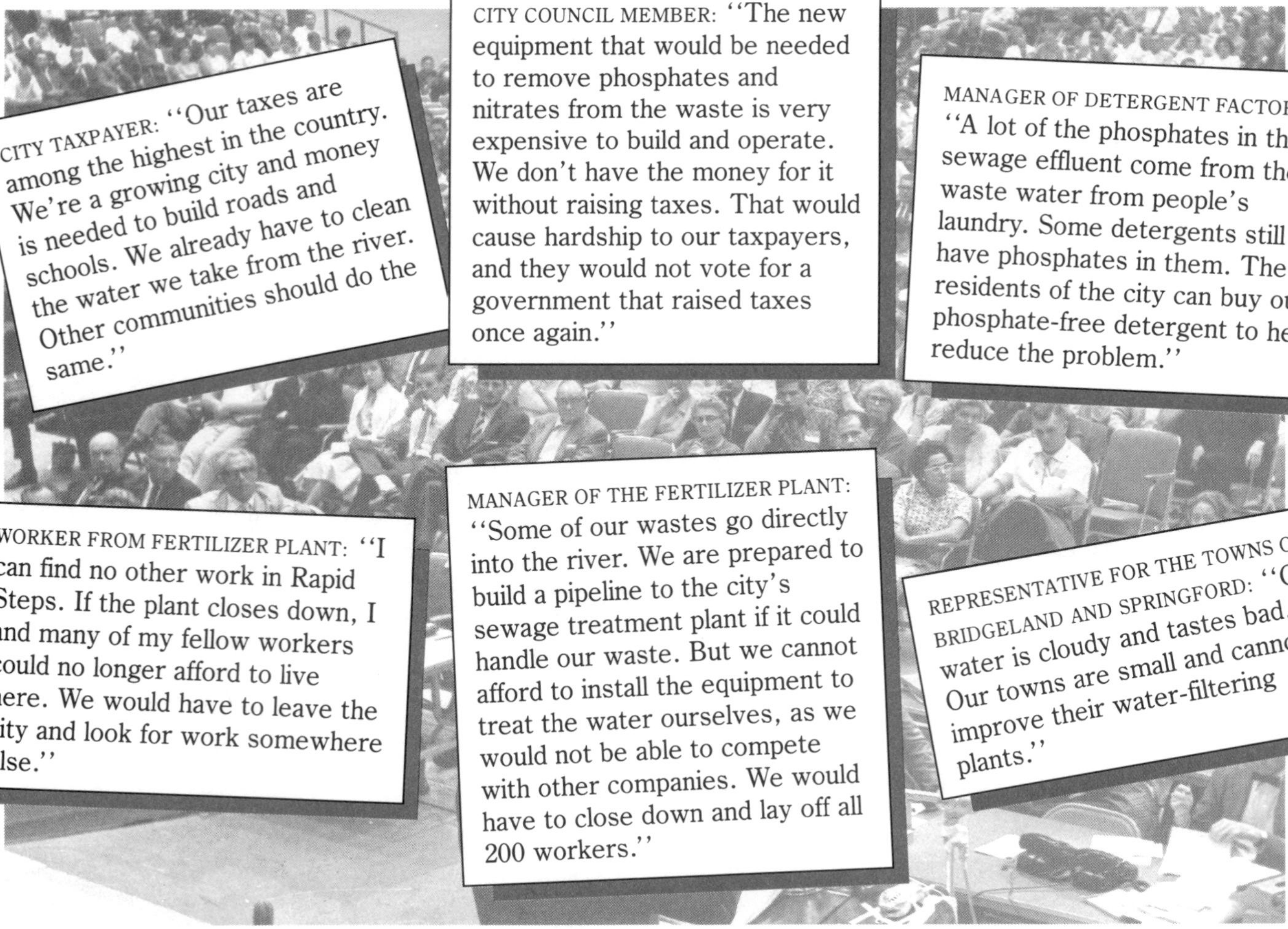

CITY TAXPAYER: "Our taxes are among the highest in the country. We're a growing city and money is needed to build roads and schools. We already have to clean the water we take from the river. Other communities should do the same."

CITY COUNCIL MEMBER: "The new equipment that would be needed to remove phosphates and nitrates from the waste is very expensive to build and operate. We don't have the money for it without raising taxes. That would cause hardship to our taxpayers, and they would not vote for a government that raised taxes once again."

MANAGER OF DETERGENT FACTORY: "A lot of the phosphates in the sewage effluent come from the waste water from people's laundry. Some detergents still have phosphates in them. The residents of the city can buy our phosphate-free detergent to help reduce the problem."

WORKER FROM FERTILIZER PLANT: "I can find no other work in Rapid Steps. If the plant closes down, I and many of my fellow workers could no longer afford to live here. We would have to leave the city and look for work somewhere else."

MANAGER OF THE FERTILIZER PLANT: "Some of our wastes go directly into the river. We are prepared to build a pipeline to the city's sewage treatment plant if it could handle our waste. But we cannot afford to install the equipment to treat the water ourselves, as we would not be able to compete with other companies. We would have to close down and lay off all 200 workers."

REPRESENTATIVE FOR THE TOWNS OF BRIDGELAND AND SPRINGFORD: "Our water is cloudy and tastes bad. Our towns are small and cannot improve their water-filtering plants."

SEWAGE TREATMENT TECHNOLOGIST: "We know that we can remove phosphates from sewage effluent by adding a chemical such as aluminum sulphate or magnesium sulphate to the sewage. However, the process is expensive. The equipment and its installation will cost about $40 million and the chemicals will cost about $3 million per year."

AGRICULTURAL RESEARCHER: "Phosphates and nitrates are found both in animal wastes and in fertilizers. These chemicals can get into the river in rainwater runoff from the feedlots and fields. This pollution can be reduced by such things as using smaller amounts of fertilizer, leaving vegetation between the fields and river to trap runoff, and diverting the runoff into holding basins."

LOCAL TEACHER: "Every year, I take a group of students canoeing on Rapid River. We saw plenty of fish in the river ten years ago — all the way from the centre of the city to downstream from the sewage treatment plant. Now all we see is water weeds and slimy algae."

MEMBER OF LOCAL ENVIRONMENTAL GROUP: "The government — and we as citizens — must take responsible action to keep the river safe for all the other species that live there, as well as for ourselves. This is too important an issue to ignore — we must take whatever steps are necessary to eliminate phosphate pollution."

FISHING GUIDE: "We used to be very busy in the summer. People came here not only from all over the country but also from other countries. Now, the number of fishing parties I take out has fallen because there are fewer fish. We tried for three years to restock the river with fish, but the number of fish is still much smaller than it was ten years ago."

FRESHWATER BIOLOGIST: "Measurements show that the amount of algae downstream from Rapid Steps has increased to almost four times what it was 25 years ago. Phosphate input from sewage effluent alone has increased from 190 t (tonnes) per year to 600 t per year. These figures do not include phosphate added directly by industry or washed from farmers' fields. I'd now like to present the results of analyses of water quality at five different sites along the river. . . ."

FARMER/RANCHER: "We need to use all of that fertilizer on our fields to make them productive for food crops, so that we've got feed for our livestock in the winter. This is the way to make sure you people in the city can continue to have enough to eat. It costs money to change farming practices — first to explain to everyone what to do and how to do it, and then actually to do it. It would be much easier for the city to monitor and remove its phosphate input."

The Public Hearing — A Role-Play

Problem

Should the city of Rapid Steps upgrade its sewage treatment plant?

Procedure

1. Divide the class into 13 groups. One group will represent the chairperson at the hearing, and each of the other groups will represent one of the participants. If you feel there should be other participants, add them.
2. Decide whether your group is for, against, or neutral about spending money to upgrade the sewage treatment plant of Rapid Steps. Together develop your presentation to the council. You may wish to use the example viewpoint listed and build on it, or you may decide on a new point of view.
3. Select one person from your group to make the speech.
4. Set up your classroom for the hearing.

5. The chairperson will make an opening statement about the purpose of the hearing and will then call on each participant in the order shown in Table 6-8. Each speech must be less than 2 min.
6. As you listen to the proceedings, decide whether each participant supports, opposes, or is neutral about upgrading the sewage treatment plant.

7. Identify the points of view of each of the participants, using the viewpoint descriptions from Topic One. A participant may have more than one viewpoint.
8. Record your observations and conclusions in a summary sheet like the one shown below.
9. If you need additional facts to make a decision, ask appropriate questions of one or more participants.
10. Use your table to help you decide what course of action the city should take.

Table 6-8 *Order of Proceedings*

A. Opening remarks by the chairperson.
B. Presentations by participants, in the order their viewpoints were listed on pages 334 to 335.
C. Cross-examination or questions by members of the audience. Be prepared to answer questions about the sub-issues if your group feels it is appropriate, in your role, to express a viewpoint on these two matters:
 (a) What source(s) is (are) responsible for the phosphate pollution?
 (b) Who should pay and how much, if a decision to upgrade the sewage treatment is reached?
D. Closing remarks by the chairperson.

Suggested arrangement of the classroom for the hearing.

Analysis

1. What was your decision?
2. What was the decision made by most students in the class? (Take a class vote.)
3. Was there any attempt at compromise at the hearing? Explain your answer.
4. Now examine the two sub-issues if they have not yet been addressed.
   (a) What source(s) is (are) responsible for the phosphate pollution?
   (b) Who should pay for the sewage treatment upgrading, if it is decided upon, and how much should each group pay?

Further Analysis

5. In a situation such as this, does an individual have much influence? Explain why or why not.
6. Write a paragraph explaining the role that science played in this community decision.
7. List at least three consequences of your decision in a table with headings such as in Table 6-9. Note the duration of the results of each of your decisions. Is the consequence short term (S—less than a century) or long term (L—at least a century)?
8. Assess the *magnitude* of each consquence (how important is its effect?) and the *probability* of occurrence (how likely is it?). Express each of these two assessments as high (H), moderate (M), low (L), or none (N).
9. Discuss your decision based on the analysis you have just done in Questions 7 and 8 above.
10. (a) Did any speaker stress the need to reduce, reuse, recycle, or recover?
    (b) Give ways that at least two of these actions on waste management could be used to help solve this issue.

Table 6-9 *Decision Analysis*

CONSEQUENCE	DURATION (S or L)	MAGNITUDE (H, M, L, N)	PROBABILITY (H, M, L, N)
1.			
2.			
3.			
4.			

Working with the Environment

Preventing Pollution

Paul Wilson has been a field services technologist for over 12 years. He is a government investigator who helps protect the environment and the public by tracking down polluters.

"In one day I may deal with oily waste in a stream, black smears on someone's front lawn, and buried tanks of unknown chemicals at a building site." Each case must be investigated; this includes interviewing the people who reported the pollution.

Paul also works closely with industries, helping them assess their waste. If legal action must be taken to stop an industry from polluting the environment, evidence from his investigation will be used in court. Writing and record keeping are thus important skills for his job. Paul obtained a community college degree in chemistry and worked for different industries before becoming a field services technologist. His experience with industry helps him understand the problems faced by companies in managing their wastes.

Paul enjoys his work, but he has an unusual ambition. "Although my job is interesting and useful, I hope one day it's no longer necessary. On that day people will have stopped polluting the environment."

Protecting People

You may have seen pictures of workers in bulky protective suits cleaning up hazardous wastes. Have you ever wondered who looks after *their* safety as they make the environment safer for us?

Catherine Cull is an industrial hygienist—the person responsible for safety procedures at a clean-up site. She specializes in the area of waste management.

One of the challenges in her job is to keep workers safe in conditions that are often very dangerous. "The air at a site may be contaminated and special breathing apparatus will be needed. I have to be able to identify the hazards and know the latest techniques available."

Her job involves reading and researching, but she enjoys the "people part" as well. It is important that the people who are going to use the equipment understand it, so she ensures that all the workers who are going to use it are properly instructed.

"I entered this field by first obtaining a four-year science degree in university. I then found a job which trained me as a hygienist. Colleges now offer hygienist programs, but the best jobs will go to those students who take these after solid university training in chemistry and biology."

1. In a water sample taken from a stream or river, what does the presence of large numbers of mayfly larvae signify?

2. (a) Which invertebrate— mayflies, stoneflies, midge larvae, or leeches—would you most expect to find in water with a dissolved oxygen concentration of 3.0 mg/L?
   (b) What does this oxygen level and the presence of this type of invertebrate tell you about the water quality?

3. In which of the following locations would you expect worms to be more numerous than most kinds of insects? Explain your answers.
   (a) downstream from a sewage treatment plant
   (b) upstream from a large industrialized city
   (c) in a river with a high dissolved oxygen concentration
   (d) in a river with a low phosphate concentration

4. A logging company plans to build a sawmill upstream from several small towns and many farms that line the banks of a major river.

   (a) How could this plan become an issue?
   (b) Select any five of the viewpoints listed on page 293 and state how each could be used to evaluate this plan.

5. Alice Gunning had run a successful trout farm for the past five years. One summer, her fish began to die. Her neighbours suggested that a new cattle ranch upstream was responsible. To determine whether this was really the source of the problem, she hired a freshwater biologist to test the water quality of the river. The biologist measured dissolved oxygen and phosphate concentrations at four sites in the river. The test sites are marked A, B, C, and D on the map. The test results are shown in Table 6-10.

   (a) Suggest which types of organisms might have been found at each site.
   (b) Infer from the data what might have killed Alice Gunning's fish.
   (c) Suggest a reasonable solution to the problem.

Table 6-10

TEST SITE	DISSOLVED OXYGEN (mg/L)	PHOSPHATE (μg/L)
A	10.0	4.0
B	1.5	50.0
C	2.0	9.0
D	4.0	7.0

Map of river showing test sites and locations of Alice Gunning's trout farm and the new cattle ranch.

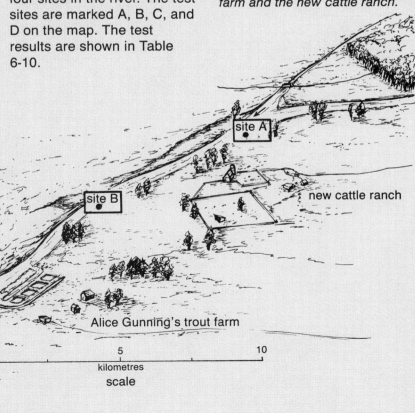

site A

new cattle ranch

site B

site C

flow

Alice Gunning's trout farm

site D

0   5   10

kilometres

scale

Focus

- Increasing human populations, urbanization, and technological developments have led to changes in the environment.
- People often disagree about the causes of environmental problems, their seriousness, and the solutions to the problems.
- The huge quantities of solid waste discarded by our society pose a problem; some of these, such as most plastics, are non-biodegradable.
- Today solid wastes are burned in appropriate incinerators or compacted in sanitary landfill sites.
- Reducing, reusing, recycling, and/or recovering certain wastes can help to decrease the total quantity of solid waste to be managed.
- Pollution is the addition of anything to the soil, air, or water that can be harmful to living organisms.
- Air becomes polluted when harmful chemicals are released into the atmosphere. Many of these chemicals come from vehicle exhausts and from industrial plants and electrical generating stations that burn fossil fuels.
- Some air pollutants are formed when other pollutants react with substances that occur naturally in the atmosphere. One example is acid rain.

- Water pollution results from inadequately treated sewage from homes or industries, runoff from farms and other sources, heated water that enters lakes and rivers after being used in industrial cooling processes, and acid rain.
- Many of the suspended and dissolved materials in water can be identified, measured, and removed by various chemical and physical processes.
- Organisms that live in water require oxygen. Water quality can be assessed by studying the quantity and type of living organisms found in lakes and streams.
- Science and technology can further our understanding of environmental problems and can help society to find ways to deal with these problems.

Backtrack

1. Provide the term that best fits the definition.
   (a) water that enters streams or lakes from the land
   (b) a matter about which people have different viewpoints
   (c) anything people add to the environment that causes harm to living things
   (d) non-living parts of the environment
   (e) to extract and reuse
   (f) to leak or seep into ground water
   (g) a supervised waste disposal site
   (h) materials that can be broken down and used by other organisms
   (i) the viewpoint expressed in the statement, "We must consider that taxes were already raised last year, and next year is an election year."
   (j) the partially purified waste water that enters a river
   (k) the process in which solid waste material is allowed to settle out
   (l) a gas needed in the production of energy by living organisms
   (m) liquid and solid waste carried away from homes in pipes
   (n) aquatic organisms that photosynthesize
   (o) anything considered to be useless
   (p) cloudy water containing undissolved solids

2. Which viewpoint listed on page 293 is being expressed in each of the following statements?
   (a) "It is time that government insist that less sulphur dioxide is released into the air."
   (b) "Eggs should be sold only in cardboard cartons, not plastic ones."
   (c) "It takes too much time to store cans and then take them to a recycling depot."
   (d) "It doesn't matter to me if these chemicals are pollutants; I don't want ants on my patio."
   (e) "I prefer to buy frozen orange juice, which I mix up at home, rather than buying a new jug every time."

3. List several ways in which an individual has a *direct* effect on the amount of solid waste in his or her local landfill site. What are several *indirect* ways she or he could have an effect on the volume of solid waste?

4. Explain how water quality can be "measured" by finding out what organisms live in the water.

5. During a flood one spring, a truck carrying fertilizer becomes stuck in a ditch. In order to get it out, the driver has to dump its load. The next summer there is more algae in the pond where this ditch empties than there ever has been before. The following year, there are almost no fish in the pond, although it had previously been a favourite fishing hole. Explain what has happened.

6. (a) Describe two methods of treating solid waste.
   (b) Describe the primary, secondary, and tertiary levels of sewage treatment.

7. The illustration shows the major components of acid rain. Describe what is happening at each number. Assume the lake at No. 5 is largely on granite rock, not limestone.

Synthesizer

8. Why should the manufacturers of chemical products such as pesticides or other hazardous substances suggest methods for safe disposal on the label, along with instructions for safe use?

9. It is easy to lose sight of the benefits of technology when studying the environmental problems that result from a particular technology. For example, automobile exhausts emit nitrogen oxides, which contribute to acid rain, and carbon dioxide, which contributes to the greenhouse effect. List three positive effects that the automobile has had on people's lives in this century, and list alternative ways in which each benefit might have been achieved.

10. Explain the relationship between the average Canadian lifestyle and the amount of waste produced.

11. Give at least one consequence of each of the following alternative actions to reduce acid rain.
    (a) By law, industry must reduce sulphur dioxide emissions.
    (b) A limit is put on the amount of fuel each household can burn.
    (c) Gasoline for cars and trucks is rationed.
    (d) All industries and generating stations must build very high smokestacks.
    (e) Any electrical generating station using coal must switch to natural gas within 20 years.

12. Why might the wide availability of biodegradable products discourage people from reducing and reusing?

Units of Measurement and Scientific Notation

At the time of the French Revolution about 200 years ago, a group of scientists met in Paris to design a standard system of measurement. They chose as their basic unit of length one ten-millionth of the distance from the North Pole to the equator, measured along the meridian through Paris. This distance was called a **metre**, and two lines a metre apart were engraved on a metal bar to serve as a standard. Later, when more accurate measurements of the Earth were possible, scientists were forced to abandon this definition. The metre is now defined very accurately using the distance travelled by light in a certain length of time. In 1960, the metric system was modernized to form the International System of Units. This system is usually referred to as the SI, from the French name — *Le Système international d'unités*. Now, more than 90% of the world's population in more than 100 countries uses the SI. Table A shows the seven base units, from which all others are derived. Table B shows all the SI units that are used in this book.

Table A *The SI Base Units*

QUANTITY	UNIT	SYMBOL
length	metre	m
mass	kilogram*	kg
time	second	s
electric current	ampere	A
temperature	kelvin	K
amount of substance	mole	mol
luminous intensity	candela	cd

* The kilogram is the only base unit that contains a prefix. The gram proved to be too small for practical purposes.

Table B *Some SI Quantities, Units, and Symbols*

QUANTITY	UNIT	SYMBOL
length	kilometre	km
	metre	m
	centimetre	cm
	millimetre	mm
mass	kilogram	kg
	gram	g
	tonne	t
area	square metre	m^2
	square centimetre	cm^2
	hectare	ha
volume	cubic metre	m^3
	cubic centimetre	cm^3
	litre	L
	millilitre	mL
time	second	s
temperature	degree Celsius	°C
force	newton	N
energy	kilojoule	kJ
	joule	J
pressure	kilopascal	kPa
	pascal	Pa

Note: one tonne (t) = one megagram (Mg); one hectare = one square hectometre (hm^2).

Larger and smaller units are closely related to these. For example, the prefix *"mega"* means "multiplied by one million." Therefore, one megametre is equal to one million metres. The prefix *"kilo"* means "multiplied by one thousand," so one kilometre is equal to one thousand metres. Similarly, each unit can be divided into smaller units. For example, the prefix *"milli"* means "divided by one thousand," so one millimetre is equal to one one-thousandth of a metre.

As you can see in Table C on the next page, all the units are related to each other in this way by multiples of ten. Thus, to convert from one unit to another, you can simply move the decimal point. Move it to the right when the new unit is smaller and to the left when the new unit is larger.

Table C *Metric Prefixes*

PREFIX	SYMBOL	FACTOR BY WHICH THE BASE UNIT IS MULTIPLIED		EXAMPLE
exa	E	10^{18}	= 1 000 000 000 000 000 000	
peta	P	10^{15}	= 1 000 000 000 000 000	
tera	T	10^{12}	= 1 000 000 000 000	
giga	G	10^{9}	= 1 000 000 000	
mega	M	10^{6}	= 1 000 000	10^6 m = 1Mg
kilo	k	10^{3}	= 1 000	10^3 m = 1kg
hecto	h	10^{2}	= 100	
deca	da	10^{1}	= 10	
		10^{0}	= 1	m
deci	d	10^{-1}	= 0.1	
centi	c	10^{-2}	= 0.01	10^{-2} m = 1 cm
milli	m	10^{-3}	= 0.001	10^{-3} m = 1 mm
micro	μ	10^{-6}	= 0.000 001	10^{-6} m = 1 μm
nano	n	10^{-9}	= 0.000 000 001	
pico	p	10^{-12}	= 0.000 000 000 001	
femto	f	10^{-15}	= 0.000 000 000 000 001	
atto	a	10^{-18}	= 0.000 000 000 000 000 001	

Some of the prefixes of metric units, such as exa-, pico-, atto-, etc., are not commonly used, and many people aren't familiar with them. Thus, scientists often use more common units and **scientific notation** (also called ''standard form'') to describe quantities measured. In this notation, one non-zero digit is placed before the decimal point, and the number is multiplied by 10 to the appropriate power. The best way to see the usefulness of scientific notation is to study examples of large and small numbers (Table D).

Table D *Examples of Scientific Notation*

QUANTITY MEASURED	APPROXIMATE MEASUREMENT	SCIENTIFIC NOTATION
Mass of the Earth	6 000 000 000 000 000 000 000 000 kg	6×10^{24} kg
Distance to the Andromeda Galaxy	19 000 000 000 000 000 000 000 m	1.9×10^{22} m
Distance from the Earth to the Moon	380 000 000 m	3.8×10^{8} m
Thickness of a spider web strand	0.000 005 m	5×10^{-6} m
Mass of a proton	0.000 000 000 000 000 000 000 0017 g	1.7×10^{-24} g

Moving Decimals

1. Convert these measurements in your notebook.
   (a) 7.85 m = ■ cm
   (b) 2.14 kg = ■ g
   (c) 5743 mm = ■ m
   (d) 850 m = ■ km
   (e) 1500 mg = ■ g

2. Estimate your increase in height in the last three years in millimetres, centimetres, and decimetres.

3. Change these measurements to scientific notation.
   (a) 713 000 000 000 g
   (b) 8 640 000 000 000 s
   (c) 0.000 08 m
   (d) 0.000 000 000 000 845 kg

4. Change these measurements from standard notation.
   (a) 3.7×10^4 L
   (b) 7.0×10^{11} s
   (c) 3.1×10^{-4} g
   (d) 2.9×10^{-10} m

5. The diameter of the Earth is 13 000 000 m. Record this number using scientific notation.

6. The distance from Earth to our nearest star is 4×10^{16} m. Change this scientific notation to its appropriate measurement in metres.

7. A human red blood cell is about 0.000 008 m in diameter. Express this in scientific notation.

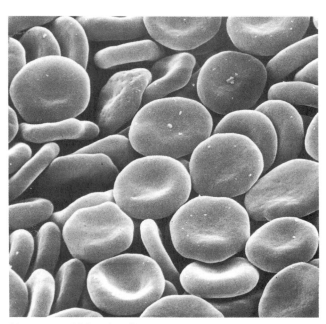

Human red blood cells.

The Periodic Table of the Elements

Chemists organize all the elements in a system called the **periodic table**. In the simplified periodic table on the next page, you can see that elements are listed in horizontal rows, in order of the number of protons in the nucleus of the elements. They are also arranged in vertical columns so that similar elements appear directly above or below one another. The elements in any one column are said to be a **family** of elements. In the hydrogen family, for example, you will find the closely related elements lithium, sodium, and potassium. These elements behave similarly in chemical reactions. Answer these questions by referring to the periodic table on the next page.

Skillbuilding Practice 2-1

Using the Periodic Table

1. Find chlorine (element number 17) in the periodic table. Name three elements that are in the same family as chlorine.
2. What element has atoms with the smallest nuclei?
3. What element has atoms containing 20 protons?
4. List as many metals as you can think of. Find each of them in the periodic table. Describe their general position in the table.
5. Helium belongs to the family of elements known as the "inert gases." (They are "inert" because they stay apart and do not readily form compounds.) Name five other elements in this family.

Did You Know?

The word "electricity" is related to "electron." Substances that are good conductors of electricity, such as most metals, allow electrons to pass easily from atom to atom. In electrical insulators, electrons are not so free to move.

The Periodic Table of the Elements

Key:

symbol	H
number of protons	1
name	hydrogen

H 1 hydrogen																	He 2 helium
Li 3 lithium	Be 4 beryllium											B 5 boron	C 6 carbon	N 7 nitrogen	O 8 oxygen	F 09 fluorine	Ne 10 neon
Na 11 sodium	Mg 12 magnesium											Al 13 aluminum	Si 14 silicon	P 15 phosphorus	S 16 sulphur	Cl 17 chlorine	Ar 18 argon
K 19 potassium	Ca 20 calcium	Sc 21 scandium	Ti 22 titanium	V 23 vanadium	Cr 24 chromium	Mn 25 manganese	Fe 26 iron	Co 27 cobalt	Ni 28 nickel	Cu 29 copper	Zn 30 zinc	Ga 31 gallium	Ge 32 germanium	As 33 arsenic	Se 34 selenium	Br 35 bromine	Kr 36 krypton
Rb 37 rubidium	Sr 38 strontium	Y 39 yttrium	Zr 40 zirconium	Nb 41 niobium	Mo 42 molybdenum	Tc 43 technetium	Ru 44 ruthenium	Rh 45 rhodium	Pd 46 palladium	Ag 47 silver	Cd 48 cadmium	In 49 indium	Sn 50 tin	Sb 51 antimony	Te 52 tellurium	I 53 iodine	Xe 54 xenon
Cs 55 cesium	Ba 56 barium	Lu 71 lutetium	Hf 72 hafnium	Ta 73 tantalum	W 74 tungsten	Re 75 rhenium	Os 76 osmium	Ir 77 iridium	Pt 78 platinum	Au 79 gold	Hg 80 mercury	Tl 81 thallium	Pb 82 lead	Bi 83 bismuth	Po 84 polonium	At 85 astatine	Rn 86 radon
Fr 87 francium	Ra 88 radium	Lr 103 lawrencium															

Lanthanide Series:

La 57 lanthanum	Ce 58 cerium	Pr 59 praseodymium	Nd 60 neodymium	Pm 61 promethium	Sm 62 samarium	Eu 63 europium	Gd 64 gadolinium	Tb 65 terbium	Dy 66 dysprosium	Ho 67 holmium	Er 68 erbium	Tm 69 thulium	Yb 70 ytterbium

Actinide Series:

Ac 89 actinium	Th 90 thorium	Pa 91 protactinium	U 92 uranium	Np 93 neptunium	Pu 94 plutonium	Am 95 americium	Cm 96 curium	Bk 97 berkelium	Cf 98 californium	Es 99 einsteinium	Fm 100 fermium	Md 101 mendelevium	No 102 nobelium

New Ideas on Measuring

Measuring Volume Accurately

For an accurate reading of the volume of a liquid in a graduated cylinder, you must have your eye at the level of the **meniscus**, or curved surface of the liquid. Read the level at the lowest point of the meniscus.

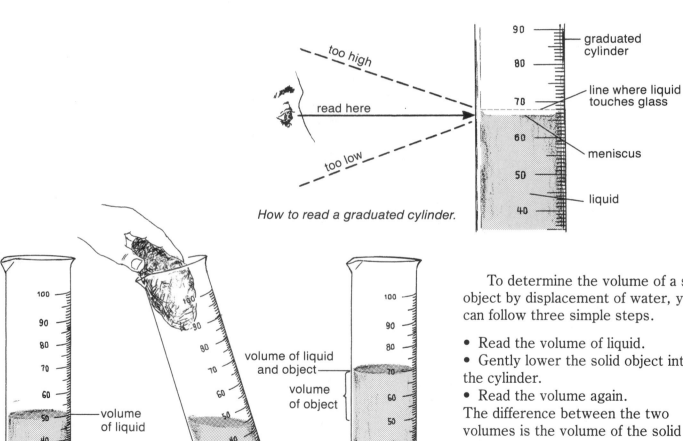

too high

read here

too low

How to read a graduated cylinder.

graduated cylinder

line where liquid touches glass

meniscus

liquid

volume of liquid

volume of liquid and object

volume of object

To determine the volume of a solid object by displacement of water, you can follow three simple steps.

- Read the volume of liquid.
- Gently lower the solid object into the cylinder.
- Read the volume again.

The difference between the two volumes is the volume of the solid object.

Determining volume by displacement.

For some experiments, you must measure an exact volume of liquid. To do this, pour into the graduated cylinder an amount just less than the volume you need. Use a dropper to add liquid until the lowest point of the meniscus is at exactly the right mark. Remember to read the volume with your eye at the level of the meniscus.

Measuring an exact volume of liquid.

Measuring an exact mass of solid.

Measuring an Exact Amount of Mass

If you need to measure an exact mass of a solid substance, you can follow three simple steps, using a balance.

- Balance the beams with the empty container on the balance.
- Move the weights until you have added the desired mass to the right hand side.
- Slowly add the solid substance until the beams are again exactly balanced.

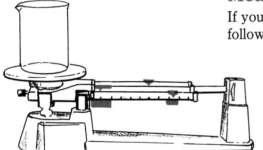

(a) A student has put an empty beaker on the balance and adjusted the weights. It is perfectly balanced.

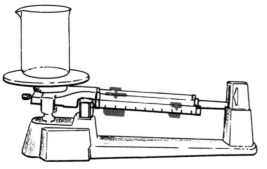

(b) The weights have been moved so the right side now has 50 g more mass than the left side has.

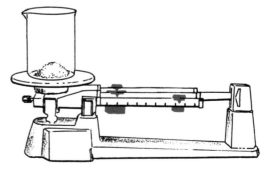

(c) Sugar has been added until the device is again balanced. The beaker contains exactly 50 g of sugar.

Density

Why is aluminum preferred as a construction material for airplanes? Why are car manufacturers replacing steel in cars with plastics or fibreglass wherever possible? In both these situations, the preferred material is one with a low *density*—that is, one with relatively little mass for a given volume of the material.

Density is defined as the mass of one unit of volume of a substance. In mathematical terms:

$$density = \frac{mass}{volume}$$

In the SI, the base unit of mass is the kilogram (kg) and the base unit of volume is the cubic metre (m^3), so density is stated in kilograms per cubic metre (kg/m^3).

Density.

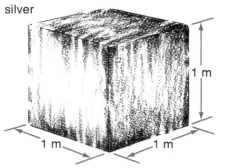

silver

1 m 1 m 1 m

The mass of 1 m^3 of silver is 10 000 kg. Therefore the density of silver is 10 000 kg/m^3.

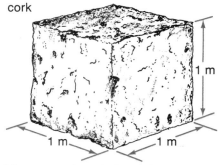

cork

1 m 1 m 1 m

The mass of 1 m^3 of cork is 222 kg. Therefore the density of cork is 222 kg/m^3.

Because a cubic metre is relatively large, it is sometimes more convenient to use other, smaller units to measure density. Density of a liquid is often expressed as grams per millilitre (g/mL) or grams per litre (g/L). Density of a solid is often expressed as grams per cubic centimetre (g/cm^3). The relationship of these units is shown in Table E.

Table E   *Units of Density*

	kg/m³	g/cm³ or g/mL	g/L
gold	19 300	19.3	
table salt	2 160	2.2	
water		1.0	1000
air		.0013	1.3

m³

cm³

mL

L

Because density is a characteristic property of pure substances, you can use it in identification tests. Suppose you find a large clear colourless crystal and you want to determine whether it is a diamond or a piece of quartz. It is beautiful, so you don't want to do anything to damage it. In a reference book, you notice that diamond and quartz have different densities (Table F). To determine the density of your crystal, all you need to do is:

- determine its volume
- determine its mass
- calculate the ratio $\dfrac{\text{mass}}{\text{volume}}$

The easiest way to determine the volume is by displacement of water. You gently lower the crystal into a graduated cylinder containing exactly 50 mL of water. The new volume reading is 58 mL.

volume of water and crystal	= 58 mL
volume of water	= 50 mL
volume of crystal	= 8 mL (or 8 cm³)

Table F

SUBSTANCE	DENSITY (kg/m³)
diamond	3510
quartz	2650

volume of water volume of water and object

Determining volume by displacement of water.

On a balance, you find that the mass is 21.6 g. Its density must be:

$$\frac{21.6\,\text{g}}{8\,\text{cm}^3} = 2.7\ \text{g/cm}^3 = 2700\ \text{kg/m}^3$$

You compare your results with the reference values, and sadly conclude that what you have found must be quartz.

(Experimental results involving measurement are almost never exactly accurate. Your results here, considering the error possible in using standard laboratory equipment, are quite respectable.)

1. (a) Calculate the density in kilograms per cubic metre of an unknown substance that has a volume of 8 m³ and a mass of 10.4 kg.
(b) Convert the answer so the units are in g/cm³.
(c) Refer to Table 2-5 on page 80 and indicate what substance this might be.

2. A graduated cylinder containing 40 mL of water has a mass of 60 g. As you carefully place an object into the graduated cylinder, the total volume increases to 70 mL and the total mass becomes 95 g. Determine the object's
(a) mass,
(b) volume,
(c) density.

3. A block of metal has the dimensions 10 cm × 5 cm × 3 cm. Its mass is 1.7 kg.
(a) Calculate its density in grams per cubic centimetre and in kilograms per cubic metre.
(b) Refer to Table 2-5 on page 80 and indicate what substance this might be.

Accuracy

Numbers are used frequently in science. Some are the result of counting, but most are the result of measuring. Any number that involves measurement always contains some error; no measurement is perfect. For example, suppose you are trying to measure 250 mL of sugar. No matter how many times you try, you will always have a different number of sugar granules in the cup. Or suppose you are asked to measure the length of a room using a stick a metre long with no subdivisions marked on it. Your measurement will contain an error because of the poor quality of the measuring instrument. With an unmarked stick such as this, your measurement would probably be accurate to only about one-fifth of a metre. But even with a higher quality metre stick, your measurement might be accurate to only about 0.1 m. There is a certain amount of **instrument error** with all measuring instruments, no matter what their quality.

If you were asked to measure the length of a tree's shadow, the error would be even greater, especially on a windy day. If the tree were swaying in the wind, you would need to decide what point to consider as the end of the shadow, before you could measure it. In a measurement like this, as well as instrument error, there is also a large amount of **experimental error**, caused by the difficulty of making the measurement.

Because most measurements contain both instrument error and experimental error, it is generally understood that the last digit in a measurement is uncertain or has been estimated. For example, if the length of a sidewalk is given as 7.4 m, it means that the length is between 7.3 m and 7.5 m. If the measurement were actually made more accurately, another digit would be included; for example the length could be stated as 7.38 m. This would mean that the length is between 7.37 m and 7.39 m. Careful techniques can reduce errors in experiments, but sources of small errors still exist.

Skillbuilding Practice 3-2

1. Identify as many sources of possible experimental error as you can in the following situations.
   (a) In an experiment, 50 mL of liquid was found to have a mass of 30 g.
   (b) The students found that it took 5 min to heat 100 mL of water from 20°C to 80°C.
2. Describe how much uncertainty there is in each of the following statements.
   (a) The winner's time was 10.67 s.
   (b) We took 45 min for the entire trip.
   (c) Drive 7 km past the gas station, then turn right.
   (d) They bought a 3 ha piece of land.

3. Some of the numbers in the following statements are believable, but others should raise questions in your mind. Which statements are you doubtful about? Explain your answers.
   (a) George is 1.6053 m tall.
   (b) The distance from the Earth to the Moon is 382 156.25 km.
   (c) My bank balance is $1479.85.
   (d) It takes 365.25 days for the Earth to revolve once around the Sun.
   (e) The soup recipe calls for 2.04 kg of potaoes.

Graphing

Interpreting Graphs

When numerical data are plotted on a graph, they become much easier to interpret. Because trends and patterns are obvious, it is also easier to make inferences based on the data. Try the following exercises for practice in interpreting graphs.

Skillbuilding Practice 4 – 1

Acid Rain

As snow and rain fall through the air, they may combine with substances such as sulphur dioxide and nitrous oxide, which are present in the atmosphere. The result is known as "acid rain." Acidity is measured in pH units. The lower the number, the more acidic the solution. For example, distilled water has a pH of 7.0, tomatoes have a pH of 4.2, and vinegar has a pH of 2.2.

In northern areas, acid snow accumulates during the winter. When the snow melts, a large quantity of acid is suddenly added to the lakes and streams. This results in "spring pH depression," as shown in the graph. The acidic water eventually mixes with the rest of the water in the stream and is brought into contact with sediment, which neutralizes some of the acid. The water usually returns to normal pH values by June or July.

1. When is the lake water most acidic?
2. Rainbow trout cannot live in water with a pH of 5.5 or less. Is this water ever acidic enough to harm rainbow trout?
3. Do you think the lake water would be more acidic in the spring if there were a heavy snowfall or a light snowfall the previous winter? Explain your answer.

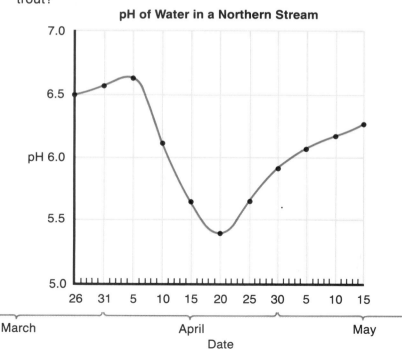

pH of Water in a Northern Stream

Airborne Lead

Lead is one of the substances that pollutes the air. Scientists measure the amount of lead in the air on a regular basis and then study the data obtained. In one experiment, the amount of airborne lead was measured near expressways and at several other locations. The graph shows the results of this study.

1. In what year was there the most pollution caused by airborne lead?
2. Is the amount of lead in the air increasing or decreasing?
3. Near expressways, is the amount of airborne lead more than or less than the average of all sites?
4. a) Make a hypothesis to explain the presence of lead in the air.
   b) Infer why the amount of lead pollution has changed in this period of time.

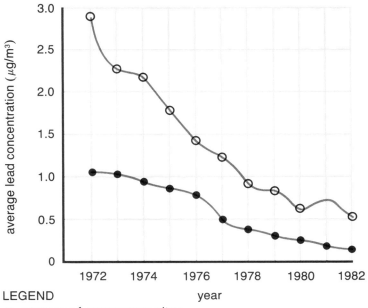

Concentration of Airborne Lead, 1972 to 1982

LEGEND
○ average of expressway sites
● average of all sites

The Growth of Bacteria Cells

All populations have a tremendous capacity for growth when they have sufficient food and space, an ideal environment, and few predators. For example, consider the growth of bacteria in a test tube of nutrient broth. The graph illustrates an experiment in which 1000 bacterial cells were added to a test tube of broth. Each cell divided into two cells every 30 min.

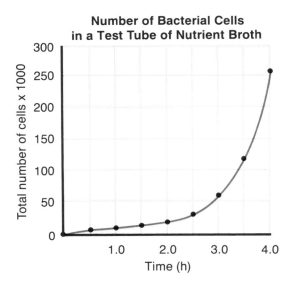

Number of Bacterial Cells in a Test Tube of Nutrient Broth

1. How long did it take for the number of bacterial cells to increase to 100 000?
2. (a) In the time interval from 0 h to 2.0 h, approximately how many new cells were produced?
   (b) In the time interval from 2.0 h to 4.0 h, approximately how many new cells were produced?
   (c) Does the rate at which new cells are produced increase or decrease with time? Explain your answer.

Graphing **355**

Steps in Making a Graph

1. Collect the data to be graphed. This will consist of pairs of numbers, called **variables**. Arrange the pairs of numbers in two columns in a table.

2. Prepare the axes. Draw a vertical line and a horizontal line that meet at the **origin**. Decide which is the **manipulated variable** (the condition that is set and systematically regulated by the experimenter), and label the horizontal axis (or x-axis) with this quantity and its units. Label the vertical axis (or y-axis) with the **responding variable** (the other condition being measured) and its units.

3. Mark the scales on both axes. You should keep in mind that all data should fit on one piece of graph paper and that the graph should be large enough to fill most of the page. It is not necessary for the scales to start at 0 if this is not suitable for the data.

4. Plot the points on the graph. For each point, make a dot surrounded by a small circle. (If you are plotting more than one set of results on one graph, use a different shape to surround the second set of points, or use a different colour for the points belonging to each set of data.)

5. Draw a smooth line through the points. Before you draw the line, look for a pattern. Do the data lie in a straight line? Do the data form a curve? If there is a pattern, show it by drawing a ''best fit'' line, with approximately equal numbers of points on each side, as shown in the examples. This process of drawing a ''best fit'' line is called **interpolation**. It is based on the assumption that if you did more experiments, all the results would fit the same pattern. When you have only a limited amount of data, some ordinary experimental error is to be expected. If just one or two pieces of data do not seem to fit the pattern of the rest, you may decide to ignore the pieces that don't ''fit.'' In such a case, you may need to repeat the experiment.

6. Complete the graph. Add a title. If there is more than one line on the graph, include a legend to identify the sets of points.

Two examples of ''best fit'' lines.

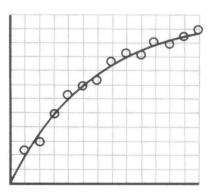

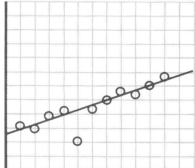

If all but one of your points fit a pattern, you might decide to check your calculations or repeat your measurements. If that is impossible, you could suggest what might have gone wrong.

Using Graphs for Predicting

If a graph shows a regular pattern, you can use it to make predictions. For example, using this graph, in which mass is plotted against volume, you could predict the mass of 5 cm³ of silver. To do this, you would **extrapolate** the graph, or extend it beyond the measured points, assuming the observed trend would continue.

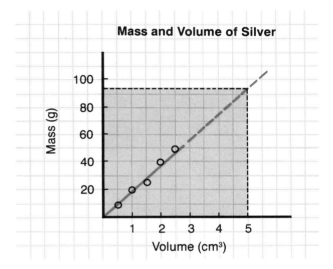

By extrapolating the graph, you can predict that 5 cm³ of silver will have a mass of 96 g.

Some common sense is needed in extrapolation. Sometimes a pattern extends only over a certain distance. For example, if you extrapolated the graph of bacterial growth on page 355, you would predict that after 24 h there would be 70 000 000 000 000 000 cells. (In scientific notation, that's 7×10^{16} cells.) These would have a mass of about 10 kg. In a day and a half, the cells' mass would be nearly the mass of the Earth. Obviously, something happens to break the pattern!

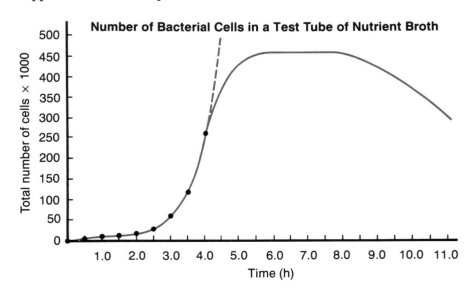

When food is used up and waste products accumulate, the population of bacteria levels off and then drops dramatically.

Using Graphs

1. The height of a pendulum was measured every 0.1 s as it swung back and forth. The results are shown in Table G.
   (a) Plot the results on a graph.
   (b) Attach another piece of graph paper, extend your graph by extrapolation, and predict the height of the pendulum at 1.5 s and 2.0 s.

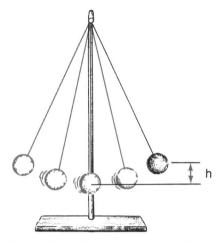

The motion of a pendulum can be described by measuring its height at different times.

Table G *Height of a Pendulum*

TIME (s)	HEIGHT (cm)
0.0	10.0
0.1	6.9
0.2	4.2
0.3	1.9
0.4	0.5
0.5	0.0
0.6	0.5
0.7	1.9
0.8	4.2
0.9	6.9
1.0	10.0

2. A family of five kept track of its home electricity bills over a period of two years. The cost of the electricity is shown in Table H. Plot the values for both years on a single graph, and use the graph to infer answers to the following questions.

   (a) In what month do you think the family takes a vacation?

   (b) Compare the cost of electricity during the two winters. What could have caused this difference?
   (c) What do you notice about the electricity bills in the spring and fall? Suggest an explanation for this.
   (d) Do you think the house is air-conditioned? Explain your answer.

Table H

MONTH	COST OF ELECTRICITY ($)	
	YEAR ONE	YEAR TWO
January – February	172	126
March – April	198	118
May – June	65	92
July – August	48	23
September – October	114	112
November – December	111	118

Using the Microscrope

The microscope is so commonplace a piece of equipment in science classrooms that it's easy to forget it is a complex, valuable, and delicate instrument. Examine the diagram showing the parts that make up the optical system, the light system, and the mechanical system of a typical compound-light microscope. When you are familiar with the parts, read the following hints. Before attempting any investigations, ensure that you know how to use your microscope effectively.

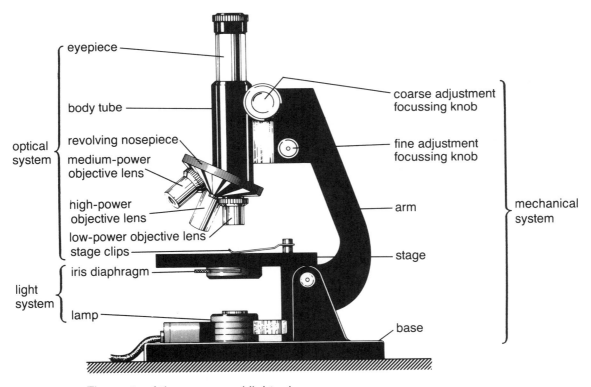

The parts of the compound-light microscope.

Using a Microscope

1. Carry your microscope with both hands. Use one hand to hold it vertically by the metal arm, and use the other hand for extra support under the base.
2. Place the microscope on the surface of a cleared lab bench or a clean desk.
3. Keep the microscope in an upright position at all times. Tell your teacher if your microscope is dirty or if the parts do not move freely. Do not try to force any parts of the microscope to move. If you are examining a liquid, use only a small drop. Keep the stage clean and dry.
4. Start by focussing with the low-power objective lens. (To focus means to make something sharp or clear.) Observe the microscope from the side as you use the coarse adjustment knob. Lower the low-power lens as close as possible to the stage without touching the stage.
5. When you focus, look through the eyepiece and slowly turn the coarse adjustment knob so that the lens moves **upward** from the stage. Never focus downward. The fine adjustment knob may then be used to sharpen your view of the object.
6. When you have finished using the microscope, remove the glass slide, place both stage clips so that they point forward, and ensure that the low-power lens clicks into place below the eyepiece.
7. Cover your microscope when it is not in use.

Cell Structure and Function

Before microscopes were invented, only about 400 years ago, people didn't know that living things were made of cells. During the 17th and 18th centuries, scientists discovered many kinds of cells, especially the cells of micro-organisms, with the early microscopes. Gradually, scientists came to realize that all living things were made up of cells. By the middle of the 19th century, the **cell theory** was developed. It can be summarized in three statements.

1. Cells are the basis of structure for all living things.
2. Cells are the basis of function for all living things.
3. All cells are formed from pre-existing cells.

Cells vary greatly in size. One of the smallest cells is the pneumonia bacterium (0.0001 mm in diameter). The ostrich egg is probably the largest single cell (75 mm in diameter). The longest cells in the human body are nerve cells, which can be up to 1000 mm long. A human egg cell measures about 0.2 mm across. Because they have so many different functions, cells also have many different shapes.

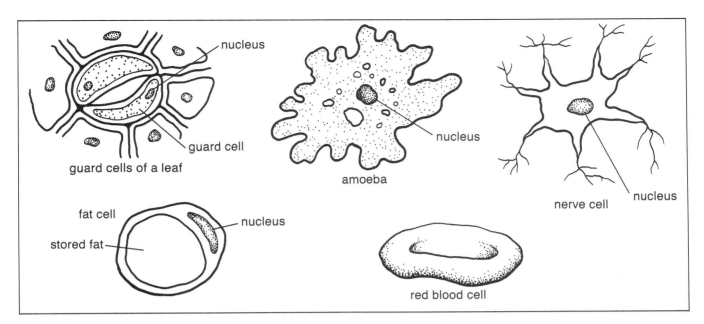

Some different kinds of cells (not drawn to scale).

All cells, no matter what their size or shape, have certain characteristics in common. They have three main parts that are responsible for much of the basic structure and function of the cell. These three parts are the *cell membrane*, the *cytoplasm*, and the *nucleus*.

- The **cell membrane** regulates the passage of certain substances into or out of the cell. It also separates one cell from other cells. Under a microscope, the cell membrane looks like a very thin line.
- The **cytoplasm** contains the cell structures, or organelles, that are critical for a cell to function. It is a jelly-like material where many processes are occurring all at once, rather like a factory. The cytoplasm receives materials from the cell membrane and expels waste materials back out through the cell membrane. Within it are many smaller structures in which important processes occur. Some of these smaller structures can be seen under a microscope.
- The **nucleus** is the "control centre" of the cell. It organizes and directs the functions of the cell and is essential for the production of new cells. Within the nucleus are the chromosomes containing the genes that are responsible for all inherited characteristics. Under a microscope, the nucleus looks like a dark blob.

Plant cells have an additional structure outside the cell membrane, called the **cell wall**, which animal cells do not have. When you look at plant cells, you can see this structure easily. Usually the cell membrane and the cell wall are so close together you can't see the difference between them using a light microscope. Whereas the cell membrane is a living part of the cell, the cell wall is non-living (made up of material produced by the cell).

When you look at cells under a microscope, you may be able to identify a few other organelles.

- Within the nucleus, you may see a darker area called the **nucleolus**. This organelle is involved with the transfer of information from the nucleus to the cytoplasm.
- **Vacuoles** are organelles surrounded by a membrane and filled with different substances. In plant cells, vacuoles may be large and easy to see. In most animal cells, they are small. (However, in certain animal cells, large vacuoles are used to store fat. There is a drawing of a ''fat cell'' on page 361.) Under a light microscope, the other organelles in animal cells look like granules.
- In plant cells only, you may be able to see **chloroplasts**. These organelles contain the green pigment called chlorophyll, which is necessary for photosynthesis.

The diagrams show typical plant and animal cells, as you might see them under a microscope.

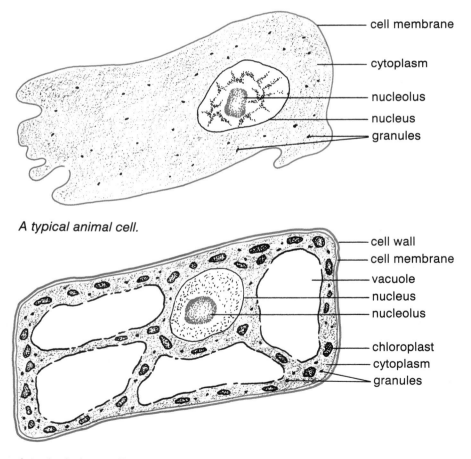

A typical animal cell.

A typical plant cell.

Looking at Plant and Animal Cells

PART A

Problem

What can you see in a living plant cell?

Materials

forceps
microscope slide
cover slip
compound microscope
paper towel or tissue
iodine solution
dropper
water
small section of white onion
(with outer dry skin removed)

Procedure

1. Put a small drop of water on a slide.
2. Using the forceps, strip a small, thin section of skin (membrane) from your onion. It will come off easily if the inner curved surface of the onion is folded, as shown here.
3. Continue using the forceps to place the membrane on the drop of water. Be careful to have a single layer. Don't let the membrane fold over.
4. Place another small drop of water on top of the onion membrane.
5. Place the cover slip on the slide.
6. View your prepared wet mount with the compound microscope, using the low-power objective lens.
7. Draw a diagram of at least three cells and label any parts you can see.

> **CAUTION: Iodine is a corrosive liquid that will stain the skin, clothing, desk tops, and floors. Take care to avoid spilling. Quickly wipe up any accidental spills, and rinse with water.**

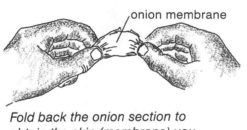

onion membrane

Fold back the onion section to obtain the skin (membrane) you will need.

Carefully place the cover slip on the slide.

Place a drop of iodine at one edge of the cover slip. The iodine will spread under the cover slip.

8. Remove the slide from the microscope and carefully draw off the water by placing a piece of paper towel or tissue at one edge of the cover slip as shown.
9. Using the dropper, place a small drop of iodine solution at one edge of the cover slip. Allow it to spread underneath the cover slip.

onion membrane
drop of water
forceps

Using the forceps, place the skin carefully on the slide.

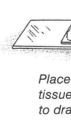

Place a piece of paper towel or tissue at the edge of the cover slip to draw off the water.

10. View the slide again using first the low-power objective lens, then the medium-power lens, and finally, the high-power lens.
11. Draw a diagram of one cell as it appears under the high-power lens. Label the nucleus, cell wall, and cytoplasm.

Analysis

1. What structures could you see without the iodine?
2. What structures could you see more easily after adding iodine?

Extension

3. Using forceps, peel a piece of thin skin from a section of either a tomato or a green pepper. Then carefully use a scalpel or single-edged razor blade to scrape the inner surface of the skin to remove the flesh. The skin should be clear. Place the skin carefully on a slide so that the outer surface is uppermost. Add two drops of iodine solution, then place the cover slip on top. Observe your slide under all magnifications, starting with low power. Draw a diagram of two cells that are beside each other, and label all visible parts.

CAUTION: The scalpel (or single-edged razor blade) is extremely sharp. Take care.

Scrape the inner surface of the tomato skin until it is clear (transparent).

4. Examine a prepared slide of a water plant such as elodea or spirogyra under low power. Find an area in which the cells appear quite distinct, then move the slide until that area is in the centre of the field. Examine it under medium power. Draw a large, labelled diagram of one cell.

PART B

Problem

What can you see in an animal cell?

Materials

compound microscope
prepared microscope slide(s) of one or several of the following: human skin cells, human smooth muscle cells, frog blood cells

Procedure

1. Place the prepared slide on your microscope stage. Observe the slide first with low power, then with medium power, and finally, with high power.

Analysis

1. Draw a diagram of two or three cells as they appear under high power.
2. Label any structures you recognize.

Further Analysis

3. What is the general shape of the cells?
4. Can you see all structures clearly within the cell? If not, explain why you think some are difficult to observe.
5. Describe the ways in which these cells look different from the cells you observed in Part A.

Sharing the Work

Some kinds of organisms consist of only one cell. That single cell must therefore carry on all the functions of life, such as finding and digesting food, sensing and responding to stimuli in the environment, and eliminating wastes. Most familiar organisms, however, are **multicellular** (many-celled).

Each cell making up an organism contains, within its nucleus, the instructions for all life functions; however, as most cells develop, they become specialized to perform only a few functions. A group of such cells working together is called a **tissue**. For example, in your own body, cells work together forming muscle tissue, nerve tissue, skin tissue, and many others.

Specialized tissues that are grouped together to perform a common function form an **organ**. Some organs in the human body are the heart, brain, lungs, stomach, liver, and kidneys. Organs, in turn, act together in **systems** to perform major life functions, such as receiving and responding to stimuli, digesting food, obtaining oxygen, circulating blood, and eliminating waste products. Some of the major systems of the human body are shown in the illustration.

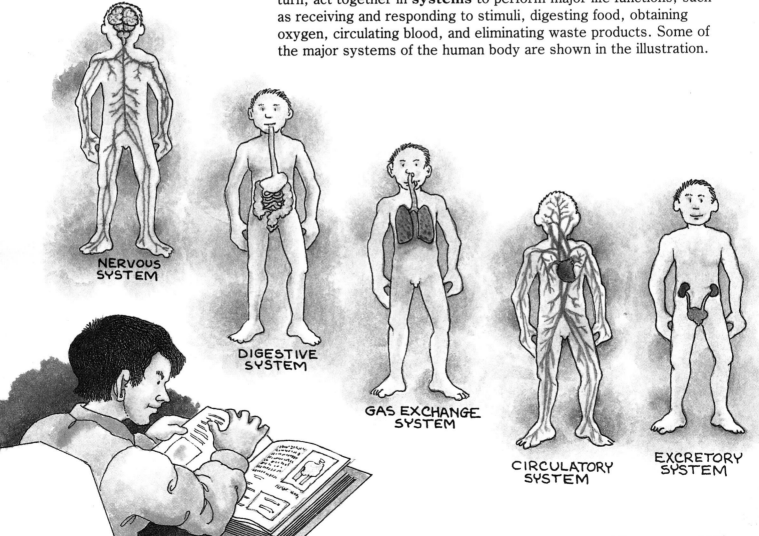

NERVOUS SYSTEM

DIGESTIVE SYSTEM

GAS EXCHANGE SYSTEM

CIRCULATORY SYSTEM

EXCRETORY SYSTEM

Glossary

How to Use the Glossary

The numbers in brackets after each definition tell you where to find the glossary word in the text. The first number is the Unit or Skillbuilder number; the number after the dash is the page number in the text. For example, at the end of the definition of acid, the number (1-40) tells you that this term is first used in the text in Unit One on page 40. A Pronunciation Key is included to help you pronounce difficult words.

a = **a**cid, m**a**sk
ae = r**ai**n, s**a**me
ah = c**a**r, f**a**ther
aw = h**o**t, l**a**wn
e = m**e**t, sh**e**lter
ee = cl**ea**n, sh**ee**t
ih = b**i**te, m**y**
i = s**i**mple, th**i**s
oh = h**o**me, l**oa**n
oo = sh**oo**t, b**oo**m
u = s**u**n, w**o**nder
uh = t**a**ken, f**o**cus
uhr = t**ur**n, st**i**r, w**o**rry, ins**er**t

abiotic factors   the non-living parts of the environment. (6-326)

acid   a type of compound that, when dissolved in water, produces a solution with a pH lower than 7. (1-40)

acid rain   precipitation with a pH below 5.6. (6-308)

acid shock   a period of extreme acidity in waterways caused by the runoff of acidic snow. (6-310)

active solar heating   the use of solar energy to heat structures indirectly; for example, by using a material to absorb the solar energy, releasing it later to heat the structure. (3-158)

adaptation   a structure or behaviour that increases an organism's chance of surviving or reproducing in a particular environment. (5-223)

aerodynamic   relating to the effect of air on objects moving through it, or concerning the flow of air around objects. (2-66)

allele [a-LEEL]   a form of a gene. (5-237)

alloy   mixture formed by two or more metals; for example, brass is an alloy of copper and zinc. (1-51)

ammeter   a device used to measure electric current. (4-175)

ampere   the unit used to measure electric current; the symbol is A. (4-175)

amphibians   vertebrates, such as frogs or salamanders, that live part of their lives in water and part on land (most lay their eggs in water); in the class Amphibia. (5-279)

aneroid   without liquid; as in an aneroid barometer. (2-97)

angiosperms [AN-jee-oh-spuhrms]   seed-bearing plants that produce flowers; in the phylum Angiospermophyta. (5-269)

Animalia [an-i-MAL-ee-uh]   one of the five kingdoms of living things; multicellular organisms that obtain their food by ingestion; most can move from place to place. (5-254)

annelids [AN-e-lids]   animals whose bodies are divided into a series of similar segments, such as earthworms, leeches, and many sea worms; in the phylum Annelida. (5-275)

antacid   a substance containing a mild base, used by some people to neutralize stomach acid. (1-44)

arachnids [a-RAK-nids]   animals with eight jointed legs and a hard shell, such as spiders, scorpions, and ticks; in the class Arachnida of the phylum Arthropoda. (5-276)

armature   the rotating electromagnet in an electric motor. (4-190)

arthropods [AHR-thruh-pawds]   animals with jointed legs and a hard shell, such as insects, crabs, lobsters, barnacles, millipedes, centipedes, spiders, scorpions, and ticks; in the phylum Arthropoda. (5-276)

artificial selection   the deliberate change in species of animals, plants, and other organisms, by breeding together only those individuals that have the traits desired. (5-239)

asexual reproduction   producing offspring from a single individual; for example, by a cutting or by budding. (5-233)

atomic theory   the theory that all matter is composed of atoms with a positive nucleus surrounded by negative electrons; atoms may combine to form molecules. (1-12)

atom   the smallest particle into which an element may be subdivided, and retain the properties of the element. (1-12)

autotroph [AHTOH-trohf]   an organism able to make its own food through the process of photosynthesis. (5-266)

balance   an instrument used to measure the mass of an object, by comparing its mass with that of a known mass. (SB3-349)

ballast   material used to weigh down or steady a vehicle, such as a ship, submarine, or balloon. (2-90)

bar graph   a graph showing a series of bars, each of which represents a particular object or quantity. Bar graphs are useful for making comparisons. (6-306)

barometer   an instrument for measuring air pressure. (2-97)

base   a type of compound that, when dissolved in water, produces a solution with a pH higher than 7. (1-40)

battery   two or more electric cells, joined and used as a source of electric current (in common usage, a single cell is often called a battery). (4-178)

biodegradable   able to be broken down and used as food by organisms. (6-298)

biohazardous infectious material   according to WHMIS usage, organisms such as disease-causing viruses, fungi, and protozoa, and the toxins produced by such organisms. (Intro-xiii)

biological indicators   organisms whose presence or absence can be used to indicate how polluted an environment is. As well, in some cases, the amount of a pollutant in the organism can be measured to determine if the pollutant has reached a level considered to be dangerous. (6-326)

biotic factors   the living parts of the environment. (6-326)

birds   vertebrates that have feathers, lay eggs with a shell, and are endotherms; in the class Aves. (5-279)

buoyancy   the tendency to float or rise in water or air. (2-72)

buoyant force   the upward force of a fluid. (2-76)

brush (electric)   the contact in an electric motor or generator, made of graphite (a form of carbon) or metal. (4-190)

bryophyte [BRIH-oh-fiht]   a plant that has no true roots or vascular tissue; includes liverworts and mosses; in the phylum Bryophyta. (5-267)

carbohydrate   a type of compound produced by living organisms; it consists of the elements carbon, hydrogen, and oxygen. (1-11)

carbon monoxide   an odourless, colourless gas produced by the incomplete combustion of fossil fuels. (6-306)

cell (electrical)   a device that converts chemical energy into electrical energy. (4-178)

cell (biological)   the basic unit of which all living things are made. (SB5-360)

cell membrane   the thin structure surrounding a cell; regulates the passage of substances into or out of the cell. (SB5-360)

cell theory   a theory stating that cells are the basic structural and functional units of all living things and that they develop from existing cells. (SB5-360)

cell wall   a rigid structure of non-living material surrounding the cell membrane of plant cells, but not animal cells. (SB5-360)

centipedes   arthropods with many segments having one pair of legs per segment; in the class Chilopoda. (5-276)

chemical   a substance; any form of matter. (1-2)

chemical change   a change in which one or more new substances are produced with properties different from those of the starting substance(s). (1-22)

chemical energy   potential energy stored in chemical compounds. (1-37)

chemical formula   the chemical symbols that represent a compound; for example, H_2O represents water. (1-14)

chemical reaction   any chemical change; for example, burning and rusting. (1-27)

chemical symbol   symbol for an element, consisting of either a single capital letter or a capital letter followed by a small letter; for example, H is the chemical symbol for the element hydrogen. (1-14)

chemical test   a distinctive chemical reaction that allows you to positively identify an unknown substance; for example, oxygen causes a glowing splint to burst into flame. (1-32)

chemistry   the science concerned with properties of matter and changes in matter. (1-1)

chemist   one who studies chemistry or uses it in practical ways. (1-1)

chlorophyll   the green pigment that converts the energy of sunlight into the chemical energy of food. (SB5-362)

chloroplast   the organelle that contains chlorophyll, in the cells of green plants and some protists. (SB5-362)

chordates   animals that have a hollow cord extending along their back at some stage of their lives; in the phylum Chordata. (5-279)

circuit (electric)   an uninterrupted conducting path for an electric current. (4-200)

circuit breaker   a device that protects an electric circuit from overheating by opening (breaking) the circuit. (4-213)

closed circuit   a circuit in which there is an uninterrupted conducting path for electric current. (4-200)

cnidarians [nih-DA-ree-uhns]   animals with stinging cells, such as jellyfish, corals, sea anemones, and hydra; in the phylum Cnidaria. (5-273)

combustible   easily set on fire. (Intro-xii)

combustion   a chemical reaction in which oxygen is one of the reactants and which occurs rapidly enough to produce heat and light. (1-35)

commutator   see **split-ring commutator**.

compound   a pure substance made up of two or more elements combined in a definite proportion. (1-10)

compound-light microscope   a microscope that uses a series of lenses to increase magnification. (SB5-359)

compressible   able to be squeezed into a smaller volume. (2-90)

concentration   the amount of solute dissolved in a specific amount of solvent at a certain temperature. (2-82)

conduction (electricity)   passage of electric current through a material. (4-172)

conduction (heat)   transfer of energy through a material by direct collision of particles. (3-124)

conductor (electricity)   a material that allows electric current to pass through it easily. (4-200)

conductor (heat)   a material that allows heat to transfer through it readily. (3-124)

confined fluid   a fluid in a completely enclosed container. (2-88)

conifers   seed-bearing plants that produce cones; for example, pine and cedar trees; in the phylum Coniferophyta. (5-269)

control (electric)   the part of an electric circuit by which the circuit is opened or closed; usually a switch. (4-200)

control (experimental)   the variable that does not change in a scientific test; it is compared with the variables that do change. (3-111)

convection   the process of transferring heat by the circulating motion of particles. (3-132)

convection current   the circulating path of a fluid caused by the application of heat. (3-128)

corrosion   a type of chemical reaction in which a metal is "eaten away" as it reacts with other substances in the environment. (1-51)

corrosive material   a material, such as sodium hydroxide or hydrochloric acid, that "eats away" another material. (Intro-xii)

crustaceans [krus-TAE-shuhns]   arthropods with gills and two pairs of antennae; for example, crabs, lobsters, and shrimps; in the class Crustacea. (5-276)

current electricity   the continuous movement of charged particles along a path. (4-172)

cytoplasm [SIHT-oh-plaz-uhm]   the jelly-like material within a cell that contains cell organelles; it receives materials from the cell membrane and carries waste materials back out through the membrane. (SB5-360)

density the mass of one unit of volume of a substance. (2-80) (SB3-350)

dichotomous key a list of pairs of alternative characteristics used for classification; in biology, the characteristics used to identify organisms are usually structural traits. (5-259)

dioxin a highly toxic chemical found in certain pesticides and industrial wastes. (6-295)

displace to push out of the way. (2-72)

drag a force that acts to slow a body that is moving through a fluid. (2-66)

echinoderms [ee-KIH-noh-duhrms] animals with spiny skin and radial symmetry; for example, sea urchins and starfish; in the phylum Echinodermata. (5-276)

ectotherm an animal whose body temperature changes when the temperature of its surrounding environment changes; all animals living today except birds and mammals are ectotherms. (5-258)

effluent partially purified waste water that runs off into a river, lake, or ocean. (6-313)

elastic energy a form of potential energy stored in compressed or stretched objects. (1-37)

electric charge positive or negative particles that exert an electric force. (4-169)

electric circuit See circuit.

electric current measure of the passage of electric charges along a conductor. (4-172)

electric discharge the removal of an electric charge from an object. (4-169)

electric forces the attraction or repulsion between objects or particles that have an electric charge. (4-169)

electrical energy the energy of electrically charged particles. (1-37)

electrode part of an electric cell that allows electrons to enter or leave; it is usually made of a metal or of carbon. (4-181)

electrolysis a chemical reaction caused by the passage of electricity through a substance. (1-38)

electrolyte a solution that conducts electricity. (4-181)

electromagnet a coil of wire, usually with an iron core, that becomes magnetized when a current exists in it. (4-188)

electromagnetic spectrum the entire range or spectrum of electromagnetic radiation, from low-energy radio waves through waves of infrared, visible, and ultraviolet light, to high-energy X rays and gamma rays. (3-136)

electromagnetic wave a form of radiant energy. (3-138)

electron a subatomic particle carrying a negative electric charge that is outside the nucleus of the atom. (1-16)

element a pure substance that cannot be broken down into any simpler substances by chemical means. (1-9)

endotherm an animal that maintains a constant body temperature, even if the temperature of its surroundings change; birds and mammals are endotherms. (5-258)

endothermic reaction a chemical reaction that absorbs energy from its surroundings. (1-38)

energy the ability to do work. (1-37)

evolve change over time. (5-246)

exothermic reaction a chemical reaction in which energy is released to the surroundings. (1-37)

experimental error errors or inaccuracies in measurement. (SB3-352)

family (of elements) a vertical grouping of elements in the periodic table. (SB2-346)

fat a type of compound produced by living organisms, consisting of the elements carbon, hydrogen, and oxygen. (1-11)

faucet a valve that can be turned on or off to control the flow of liquid. (3-109)

ferns spore-producing plants with vascular tissues, which require moisture for sexual reproduction; in the phylum Filicinophyta. (5-268)

filtrate the material that passes through a filter. (6-324)

flatworms simple worms, usually with a one-ended digestive system and simple nervous system; for example, tapeworms and planarians; in the phylum Platyhelminthes. (5-274)

flow rate speed at which a fluid flows under certain specified conditions. (2-65)

fluid any material that flows; gases and liquids. (2-60)

force a push or pull. (2-75)

fossil fuel fuels from once-living organisms: natural gas, oil, and coal. (6-305)

frond the leaf of a fern. (5-268)

Fungi [FUNG-ih] one of the five kingdoms of living things. It includes yeasts, moulds, and mushrooms. (5-253)

fuse a thin piece of conducting material that will melt rapidly if electric current in it exceeds a certain amount. (4-213)

galvanometer a device used to detect weak electric currents. (4-175)

gamete [GA-meet or ga-MEET] a special reproductive cell; a sperm cell or an egg cell. (5-233)

gene a unit of heredity within the nucleus of a cell, containing instructions that control the development of traits. (5-237)

generator a device that produces electrical energy from mechanical energy, usually from the kinetic energy of steam or running water. (4-186)

graph a diagram of numerical data that shows changes or comparisons. A graph is used to organize numerical information to make it easier to understand. (SB4-354)

gravitational energy the potential energy possessed by an object because of its position. (1-37)

greenhouse effect a warming of the temperature of the air, caused by the trapping of heat as in a greenhouse, where heat is trapped inside a glass structure; this effect also occurs in the Earth's atmosphere, where carbon dioxide molecules trap heat. (3-146)

grounded (electricity) connected so that a charge is conducted to or from the Earth. (4-213)

gymnosperms [JIM-noh-spuhrms] seed-bearing plants whose seeds are exposed on the surface of leaves or scales; the most common gymnosperms are conifers. (5-269)

heat energy transferred from a hotter substance to a cooler one. (3-123)

heat capacity the amount of heat transferred (absorbed or emitted) when the temperature of a substance changes by 1.0°C. (3-152)

heat conductor see conductor (heat).

heat insulator see insulator (heat).

heat rays (also **heat waves**) infrared radiation. (3-136)

heterotroph [HET-uhr-oh-trohf] an organism that obtains energy for its life processes by consuming other organisms. (5-272)

hierarchical [hihuhr-AHRK-i-kahl] a system of classification in which groups at each level are subdivided to produce smaller groups at a level below; for example, phyla are subdivided into classes. (5-254)

horizontal axis the number line of a graph that runs from left to right. The horizontal axis shows the manipulated variable. It is often called the x-axis. (SB4-356)

"hot" wires (also **live wires**) wires that are not grounded, bringing electrical energy into a home or building. (4-213)

hydraulic system system that works because of the movement of a liquid or the force of a liquid in a closed system. (2-102)

hydrocarbons compounds made up only of hydrogen and carbon; produced as gases when fossil fuels are burned. (6-307)

hydroelectric concerning electricity produced by using the force of running water as an energy source. (4-186)

hydrometer a device used to measure the density of liquids. (2-82)

hypothesis a set of ideas or models that provides an explanation of why something always occurs in the natural world. (SB4-355)

incompressible not able to be squeezed into a smaller volume. (2-90)

indicator a substance used to detect the presence of another substance. For example, iodine is an indicator of the presence of starch. (1-40)

industrialization changes in methods of production and transportation that occurred because of the widespread use of machines, such as the steam engine. (6-290)

inference an explanation for the possible cause of an event, based on a set of observations. (SB4-354)

infrared radiation a type of radiant energy; also called heat waves or heat rays. (3-136)

insects arthropods having distinct body parts: a head, thorax, and abdomen; in the class Insecta. (5-276)

instrument error inaccuracies in measurement that occur because of the limited accuracy of the instruments used. (SB3-352)

insulation (electric) see insulator (electric). (4-173)

insulation (heat) the prevention of a large amount of heat transfer. (3-144)

insulator (electric) a material that does not allow electrons to pass easily from one particle of the material to the next. (4-206)

insulator (heat) a material that is used to reduce the amount of heat transfer. (3-144)

interpolation in graphing, the process of drawing a "best-fit" line to show a pattern. (SB4-356)

invertebrate [in-VUHRT-ibret] an animal that lacks a bony spinal column or "backbone." (5-279)

issue topic about which there are two or more viewpoints. (6-292)

jawed fish vertebrates that live in water, breathe using gills, and have a covering of scales; most fish, both those with cartilaginous skeletons (such as sharks) and those with bony skeletons (such as perch and trout) are in this group. (5-279)

jawless fish vertebrates that lack jaws; the group is rare today, including only hagfish and parasitic lamprey. (5-279)

joule the SI unit for energy; the symbol is J. (4-211)

key a checklist of structural traits used to identify organisms. (5-259)

kilowatt hour a unit in which electrical energy, as consumed by a household or for a building, is measured; the symbol is kW·h. (4-211)

kingdom the largest group into which a living thing is classified; the five kingdoms are: Monera, Fungi, Protista, Plantae, and Animalia. (5-252)

larva the juvenile form of certain animals; for example, the caterpillar is the larva of a butterfly. (5-230)

law (scientific) a general statement that sums up the conclusions of many experiments. (1-31)

Law of Conservation of Mass in a chemical reaction, the total mass of the reactants is always equal to the total mass of the products. (1-31)

Law of Conservation of Energy energy is never created or destroyed by ordinary means. (3-153)

lens the part of a microscope that changes the path of light in order to magnify an image. (SB5-359)

life cycle the stages of development that an organism goes through in its life. (5-223)

light energy the form of electromagnetic energy that is visible to the human eye. (1-37)

line graph a graph that has a line drawn through the points plotted on it. A line graph shows changes in variables in relation to each other. (SB4-354)

liverworts type of small plants with no vascular tissues; in the phylum Bryophyta. (5-266)

load (electric) the part of a circuit that changes electrical energy into some other form of energy. (4-200)

lubricant a material (often oil or grease) used to reduce friction; the lubricant makes two surfaces slide over one another more readily. (2-47)

mammals vertebrates that have a covering of hair, have mammary (milk-producing) glands to nourish their young, and are endotherms; in the class Mammalia. (5-279)

manipulated variable the condition that is controlled by the experimenter in an investigation. (SB4-356)

manometer a U-shaped tube containing a liquid, used to measure pressure. (2-98)

mass the amount of matter in an object. (1-3)

matter anything that has mass and occupies space. (1-1)

mechanical energy the energy possessed by any system with moving parts. (1-37)

mechanical mixture a mixture in which the different parts can be seen. (1-6)

medusa the free-floating, umbrella-like form of some cnidarians, with tentacles trailing downwards; for example, a jellyfish. (5-273)

meniscus the curved surface of a liquid in a container. (SB3-348)

mercury barometer a device containing a column of mercury, used to measure air pressure. (2-87)

metal an element that is shiny, bendable, and a good conductor of electricity. (1-11)

metamorphosis a dramatic change in an organism's appearance and habits during its life cycle. (5-229)

meteorologist one who studies the atmosphere and weather. (2-95)

metre the base SI unit for the measurement of length; the symbol is m. (SB1-342)

millipedes arthropods with many segments, having two pairs of legs per segment; in the class Diplopoda. (5-276)

model a picture, diagram, or other means of representing something; for example, the particle theory is a mental picture of the nature of matter. (1-1)

molecule a particle formed when two or more atoms combine. (1-12)

molluscs soft-bodied animals, with well-developed systems; most molluscs have a shell; for example, scallops, snails, clams, and squids (whose shells are greatly reduced and are beneath a skin-like covering); in the phylum Mollusca. (5-275)

Monera [maw-NAE-ra] see Prokaryotae.

mosses type of small plants with no vascular tissues; in the phylum Bryophyta. (5-266)

multicellular organism consisting of more than one cell; cells are specialized to perform different functions. (5-253)

natural selection a process occurring in nature by which individuals that are not well adapted to their environment do not survive. Those that are well adapted do survive to reproduce more of their kind and, thus, are said to be ''selected''; over time, this process can lead to changes in species. (5-246)

neutral wire a wire that is not ''hot,'' a wire that is ''grounded''—connected to the ground. (4-213)

neutral (substance) a substance that is neither an acid nor a base. (1-40)

neutralization a reaction between an acid and a base. (1-44)

neutron a subatomic particle with no electric charge, found in the nucleus of the atom. (1-16)

newton the unit used to measure force; the symbol is N. For example, one newton is the force used to lift a mass of 100 g on Earth. (SB1-343)

nitrogen oxides gases formed when fossil fuels are burned at high temperatures in engines and furnaces. (6-307)

noise pollution excess noise that can cause damage to hearing and to various types of organisms. (6-295)

non-biodegradable unable to be broken down and used as food by organisms, as with, for example, many plastics. (6-298)

non-metal an element that does not have the properties of a metal. (1-11)

non-renewable resources resources that are in a limited supply; for example, oil or coal. (3-157)

nuclear reaction a type of reaction that affects the nucleus of an atom, emitting particles and energy; occurs in the interior of stars, in atomic bombs, and in nuclear reactors. (1-31)

nucleolus [noo-KLEE-uh-lus] an organelle within the nucleus of a cell; involved with the transfer of substances containing ''information'' from the nucleus to the cytoplasm, where proteins are produced that make up the entire cell. (SB5-361)

nucleus [noo-klee-uhs] of an atom; the dense central core, consisting of protons and neutrons. (1-16)

nucleus of a cell; an organelle that contains the chromosomes carrying genes that are responsible for all inherited characteristics. Sometimes called the ''central control centre'' of the cell. (SB5-360)

open circuit an electric circuit in which the conducting path is broken or incomplete. (4-200)

organ structure made up of tissues performing the same function. For example, the liver is an organ. (SB5-365)

organelle a structure within a cell that performs a specific function; for example, nucleus, chloroplast, cell membrane, etc. (SB5-365)

organic chemistry the field of chemistry concerned with compounds of carbon. (1-10)

origin the point on a graph at which the x-axis and the y-axis meet. (SB4-356)

oxidizing substance a substance that causes combustion or contributes to the combustion of another material. (Intro-xii)

ozone an unstable molecule of oxygen that is present in a layer of Earth's upper atmosphere. (1-15)

parallel circuit an electric circuit that provides alternate conducting paths for electric current. (4-200)

particle theory a theory stating that all matter is made up of extremely small particles in constant motion. (1-12)

particulates [pahr-TIK-yuh-lats] tiny solid or liquid particles released when fossil fuels and some other materials are incompletely burned. (6-307)

passive solar heating the use of solar energy to heat structures directly. (3-158)

periodic table an organization of all the elements, in a manner that show some major properties and similarities. (SB2-346)

permanent magnet a piece of hard steel alloy (usually straight or U-shaped) that stays magnetized for a long time. (4-188)

phosphate a type of compound that always contains the elements phosphorus and oxygen, and another element or elements. (6-295)

photoelectric effect an effect in which light, shining on metal, causes electrons to be emitted from the surface of the metal, producing an electric current. (4-196)

physical change a change in matter in which no new type of matter is produced; for example, freezing, melting, and boiling. (1-23)

piezoelectric effect an electric current produced by sound, due to the changing pressure of sound waves on certain crystals. (4-198)

Plantae [PLAN-tih] one of the five kingdoms of living things; multicellular organisms that make their own food by photosynthesis. (5-254)

poisonous causing immediate and serious harmful effects to humans or other animals. (Intro-xii)

pollution any change in the chemical, biological, or physical characteristics of the soil, air, or water that can be harmful to living organisms. (6-295)

polymorphism the existence of several distinct forms within the same species. (5-231)

polyp [PAWL-ip] the attached form of a cnidarian, with a cylindrical shape and tentacles facing upwards; for example, a sea anemone. (5-273)

poriferan [pawr-IF-uhr-an] an animal with no true tissues; the sponges; in the phylum Porifera. (5-272)

potential energy energy that is stored; for example, water behind a dam has potential energy, and a stretched elastic band has potential energy. (1-37)

pressure the force acting on a certain area of surface. (2-87)

primary sewage treatment the first level of waste water treatment, where materials are removed by filters and sedimentation. (6-314)

Principle of Heat Transfer in a perfectly insulated system, when two substances at different temperatures are mixed, the heat released by the hotter substance equals the heat gained by the colder substance. (3-152)

product (chemical) any substance produced in a chemical reaction. (1-27)

Prokaryotae [proh-kar-ee-OH-tih] (also **Monera**) one of the five kingdoms of living things; it contains the bacteria and blue-green bacteria. (5-252)

properties characteristics used to help describe or identify substances. (1-3)

protein a type of compound produced by living organisms, containing the elements carbon, hydrogen, oxygen, nitrogen, and often sulphur and other elements. (1-11)

Protista [proh-TIS-ta] one of the five kingdoms of living things; it contains many unicellular organisms with nuclei. (5-253)

proton the subatomic particle with a positive electric charge, found in the nucleus of atoms. (1-16)

pump a device for moving a fluid. (2-52)

pure substance a substance that contains only one kind of matter; elements and compounds are pure substances. (1-9)

R-2000 the program that aims to improve insulation in all new buildings (R refers to the resistance to heat transfer, and 2000 refers to the target year for all new buildings to have improved insulation). (3-148)

radiant energy energy transformed by means of radiation. Some examples of radiant energy (mainly from the Sun) are radio waves, heat rays, visible and ultraviolet light, X rays, and gamma rays. (3-136)

radiation the transfer of energy in a wave-like form. (3-136)

reactant any substance used up in a chemical reaction. (1-27)

reaction rate the speed of a chemical reaction. (1-47)

reactive undergoing a vigorous chemical reaction. (Intro-xii)

recover one of the ''Four Rs'' of waste management; to reclaim either waste material or energy, so it can be put to another use. (6-302)

recycle one of the ''Four Rs'' of waste management; to collect waste of a certain type in order to break it down and rebuild it into other products. (6-302)

reduce one of the ''Four Rs'' of waste management; to decrease the amount of waste produced. (6-301)

renewable resources resources that can be replaced; for example, wood. (3-157)

reptiles vertebrates, such as snakes, turtles, or crocodiles that have dry, scaly skin and lay eggs with a protective covering; in the class Reptilia. (5-279)

residue material that is held back by a filter, or left in place following distillation. (6-324)

resistance a measure of how much a material resists the passage of electric charges. (4-184)

resistor a device with a certain resistance to the passage of electric charges; used in electronic devices. (4-184)

responding variable the condition that changes in an investigation because of changes to the manipulated variable. (SB4-356)

reuse one of the ''Four Rs'' of waste management; to use things either for the same purpose a second time, or for a new purpose. (6-301)

rheostat a variable resistor; used to control devices such as electric motors and dimmer switches. (4-184)

rhizome [RIH-zohm] underground stem. (5-268)

roundworms slender, pointed worms with a two-ended digestive system; in the phylum Nematoda. (5-274)

RSI value the resistance to heat transfer of a material of a specific thickness. (3-144)

salt a substance produced when an acid and a base react; for example, table salt (sodium chloride). (1-44)

sanitary landfill site a supervised solid waste disposal site at which incoming waste is compacted and covered with soil. (6-298)

science a search for explanations; knowledge as to the structure and patterns that explain the nature of the world around us, based on curiosity, objectivity, and experimentation. (Intro-xii)

scientific notation a useful way of expressing very large and very small numbers; for example, the mass of the Earth is 6×10^{24} kg; the mass of a proton is 1.7×10^{-24} kg. (SB1-343)

secondary sewage treatment the level of waste water treatment that includes breaking down wastes by bacteria. (6-314)

segmented worms worms in the phylum Annelida. (5-275)

septic tank a large underground container for waste water and sewage. (6-313)

series circuit an electric circuit with one conducting path and no branches. (4-200)

sewage liquid and solid waste, usually carried away in underground pipes. (6-313)

sexual dimorphism a significant difference in the appearance of the sexes within a species. (5-230)

sexual reproduction a method of producing offspring by combining special reproductive cells (gametes) from two individuals. (5-233)

SI abbreviation for the *Système international d'unités;* that is, the metric system. (SB1-342)

solar cell a device that converts light energy into electrical energy. (4-196)

solar energy radiant energy that reaches Earth from the Sun. (3-157)

solid waste household garbage and other forms of waste that accumulate in large quantities. (6-298)

solute the substance in a solution that is present in the lesser quantity. (2-82)

solvent the substance in a solution that is present in the greater quantity. (2-82)

sound energy the energy of vibrations that we hear as sound. (1-37)

source (electric) the part of an electric circuit that changes some other form of energy into electrical energy. (4-200)

species a kind of organism; organisms that are very similar to one another; they usually reproduce only among themselves. (5-227)

specific heat capacity the amount of heat transferred (absorbed or emitted) when the temperature of 1.0 kg of a substance changes by 1.0°C. (3-152)

sphygmomanometer [sfig-moh-ma-NAHM-e-tuhr] an instrument used for measuring blood pressure. (2-112)

split-ring commutator a contact on the armature of a motor; a split cylinder that acts as a switching (control) device. (4-190)

sponges simple animals in the phylum Porifera. (5-272)

spore tiny reproductive structure of some groups of organisms, such as ferns, bryophytes, and fungi. (5-268)

states of matter the forms in which matter can be found: solid, liquid, and gas. (1-3)

static electricity the build-up of electric charges on an object. (4-159)

subatomic particles particles that make up all atoms— protons, neutrons, and electrons. (1-16)

subsystem one of the working parts of a complex machine; for example, the crank on a pencil sharpener is a subsystem of the sharpener. (4-185)

sulphur oxides colourless gases with a pungent odour, produced when certain fossil fuels, especially coal, are burned. (6-307)

switch device that can open or close an electric circuit; used as a control in an electric circuit. (4-200)

system (technological) a complex machine made up of simpler parts, each of which is made up of one or more simple machines. (4-185)

system (biological) specialized organs that together perform a body function, such as digestion (by the digestive system). (SB5-359)

technology techniques or inventions that solve practical problems; usually these are based on applying scientific knowledge. (Intro-xii)

temperature a measure of the average energy of the particles that make up a substance. (3-123)

tertiary sewage treatment the third level of waste water treatment, one in which specialized chemical and physical processes are designed to remove certain pollutants still remaining in waste water after primary and secondary treatment. (6-315)

theory a hypothesis that has been supported over a long period of time by many experimental results. (1-12)

thermal the rising part of a convection current in the atmosphere. (3-133)

thermal conductivity a measure of the ability of a material to conduct heat. (3-126)

thermal energy the total energy of all the particles of a substance. (3-123)

thermal pollution pollution of water by the discharge into it of heated waste from industrial or power plants. (6-295)

thermocouple device consisting of a junction of two different metals; it produces an electric current when the junction is heated. (4-193)

tissue group of cells specialized to perform a certain function. For example, nerve tissue and muscle tissue are each made up of cells specialized in different ways. (SB5-365)

toxic poisonous. (6-295)

trait a characteristic of a living organism; in the study of inheritance, the term refers to those characteristics inherited from parents. (5-234)

turbid referring to a liquid that appears cloudy because of the presence in it of undissolved solids. (6-322)

turbulence swirling motions within a fluid. (2-66)

unicellular consisting of a single cell. (5-252)

universal indicator a mixture of several acid-base indicators. (1-43)

urbanization the process by which human population becomes concentrated in cities. (6-5)

vacuole within a cell, an organelle surrounded by a membrane and filled with a particular substance. (SB5-362)

valve a device that controls the flow of a fluid. (2-109)

variable any changeable factor that may influence the outcome of a scientific test. (SB4-356)

variable resistor a device that can be used to control the amount of current in an electric circuit. (4-184)

vascular tissue a network of conducting vessels that transport water and dissolved minerals in plants. (5-267)

vertebrate [VUHRT-ibret] a member of the largest group of chordate animals, having a bony spinal column or ''backbone'' protecting the nerve cord. (5-279)

vertical axis the number line on a graph that runs from the bottom to the top. The vertical axis shows the responding variable. It is often called the y-axis. (SB4-356)

viscosity the resistance to flow of a liquid (measured as a rate of flow). (2-65)

volt unit used to measure voltage; the symbol is V. (4-178)

voltage a measure of the amount of electrical energy supplied to each unit of electric charge. (4-178)

voltmeter a device used to measure voltage. (4-178)

volume the amount of space occupied by a substance. (1-3)

waste a by-product considered to be useless or worthless. (6-286)

watt unit used to measure electrical power; the symbol is W. (4-211)

watt second a unit of electrical energy equivalent to one joule. (4-211)

wet mount a glass slide prepared for viewing through a microscope, using water, the object to be viewed, and a plastic cover slip. (SB5-363)

word equation a way of representing a chemical reaction, showing what is used (the reactants) and what is produced (the products). (1-27)

zygote a fertilized cell that results from the union of gametes and that can develop into a new organism. (5-233)

Index

Credits

Key to abbreviations: r = right, l = left, c = centre, t = top, b = bottom.

Illustration

Suzanne Bohay: xiii, 63, 64, 73, 76(t), 77, 83, 86(tr & b), 89, 97, 101, 103, 104, 106, 107, 109, 111, 300, 309(r), 310, 311, 315, 320, 331, 339. Compeer Typographic Services: 105. Pat Cupples: 23, 28, 30(b), 40, 46(b), 47, 49, 55, 74, 86(tl), 88, 112, 141, 152, 153, 185, 211, 233(t). Sam Daniel: 148, 149, 150, 158, 159, 170, 171, 173, 174, 177, 179, 180, 181, 183, 184, 187, 189, 190, 191, 192, 194, 195, 197, 199, 200, 201, 202, 203, 204, 205, 207, 217, 219. Norman Eyolfson: 140, 230, 231, 245, 249, 258, 263. Carlos Freire: 62, 67, 70, 76(b), 79, 82, 105, 110, 247(b), 296, 313(b), 329, 351(t), 352, 365. Full Spectrum Art: 14, 15, 19, 24, 25, 29, 30(t), 48, 163, 165, 214, 305(t), 306, 307, 324, 325, 326, 348, 349, 350, 351(c), 354, 355, 356, 357, 358, 361, 362, 363. Celia Godkin: 240, 241, 251, 252, 253, 254, 260, 269, 271, 272. Don Hobbs/Angry Cow: 70. William Kimber: 6, 7, 11, 27, 80, 81, 90, 92, 93, 95, 98, 99, 108, 115, 116, 117, 120, 121, 124, 125, 127, 132, 133, 134, 135, 146, 147, 164, 193, 215, 224, 225, 228, 229, 232, 233(b), 283. Paul McCusker: 17, 20–21, 122, 123, 136, 137, 138, 139, 157. Russell Moody: 129, 131, 142, 145, 161. Julian Mulock: 85, 261, 274, 278, 279, 284, 285, 330, 332–33. Loreta Senin: xix, 71, 176, 235, 236, 237, 246, 291, 298, 299, 305(b), 308, 313(t), 314, 318–19, 323, 337, 341. Henry Van Der Linde: 12, 13, 32, 33, 34, 35, 36, 38, 39, 43, 45, 46(t), 50, 52, 57. Peter Van Gulik: 226, 227, 239, 247(t), 256, 257, 281, 359.

Photography

F. Baumann: 97(t & bl). Bettmann Archive: 55(t-UPI), 175(b). Derek Bullard: 75, 94. The Calgary Herald: 319(Bill Herriot). Canada Packers: 316(stirring carrots). Canadian Forces Photo: 35. Canadian Petroleum Association: 113. Canadian Red Cross Society: 4. Canapress Photo Service: 16(tl), 28, 31, 70(b), 72(both), 81, 85(tl), 90, 108(Edmonton Journal), 119(tl-Uniphoto), 160(Hamilton Spectator), 161, 209, 219(Brian Gavriloff), 221(cl-Paul Drummond), 223(r-Richard Harrington), 250(Lynn Ball), 264(l-Ben Cropp), 269(t-Terry Self), 273(br-Ben Cropp), 293(b), 296, 302(c), 303(r-Larry MacDougal), 308. Consumer and Corporate Affairs Canada: 3, 42(br). CP Rail Corporate Archives: 1(dynamite), 37. Dr. Aubrey Crich: xviii(tr). R.H. Czerneda: 338(man in front of plant). Ducks Unlimited Canada: 118(r). From Nicole Dhombres, Les Savants en Revolution, 1789–1799: 9(t). Educational Images: xviii(cl-Matthew R. Gilligan), 221(tr & bl-Gary S. Zumwalt), 248(Charles Belinky), 266,(t-Charles Belinky), 273(t & bl-Gary S. Zumwalt), 274(t-Vernon Oswald; b-William Marquardt), 275(bc-Alex Kerstitch), 277(crab & shrimp-Alex Kerstitch; millipede & spider-Charles Belinky; centipede-John R. Macgregor). Environment Canada: 309. Federation of Ontario Naturalists: 268. First Light: 231(Peter McLeod). Fitness Slide Library: 59(runner). Ford Motor Company: 66. S.T. Fox: 9(b). Fundamental Photographs: 87(tl), 97(br), 198(t-Kip Peticolis). Glenbow Archive, Calgary: 290, 303(l-Byron May Company, Edmonton). Gordon Gore: 5(all), 22(r), 165, 168, 175(t), 178, 188, 190, 196(t), 198(b), 213(r). Grant Heilman Photography: 82, 228(plantain-Runk & Schoenberger), 234(John Colwell), 264(r-Runk & Schoenberger), 317(both; washing potatoes by Larry Lefever). Al Hirsch: 216(l). IBM Canada: 53(tr). The Image Bank Canada: xvi(tr-G.V. Faint), xvi(b-Herbert W. Hesselman), 56(jello-Obremski; painting-Alvin Uptis), 100(deep sea-Armando F. Jenik), 128(Leo Mason), 222(br-Charles C. Place), 254(Harold Sund). Imperial Oil Limited: 2(plastics), 102(b). Industry, Science & Trade Canada: 2(steel), 41(Bruce Paton), 119(c), 134(John de Visser), 140(Mike Beedell). Bill Ivy: 56(log), 59(tub), 118(l), 119(bl), 169, 221(bc), 242, 270(c), 275(t), 276(r), 277(beetle & wasp), 280(bird), 295. Lockwood Survey Corporation Limited: 291. L.T. Webster/Libra Photographics: xvii(tl), 1(milk, hydrogen peroxide, antacid), 6, 13, 26(potato, coffee), 39, 42, 43, 50, 53(b), 56(ice), 144(tr), 151, 154, 166–167, 182, 200, 208, 210(all except hairdryer), 243, 327, 364. Robert B. Mansour: 2(fabric), 16(tr), 22(l), 26(frost), 59(forklift), 61, 87(tr), 91(b), 102(tr), 119(tl), 172. Marine Atlantic: 78(b). Steve Mash: 20–21. Metropolitan Toronto Library Board: 18(Mendeleev), 78(t), 238. Miller Comstock: 239(all-H. Armstrong Roberts), 324(H. Armstrong Roberts), 334(H. Armstrong Roberts), 335(Lambert). NASA: xv(r), 56(rocket), 119(tc), 136, 147, 156, 196(b), 304. National Capital Commission: 1(Parliament Bldg), 51. Niels Bohr Library/American Institute of Physics: 18(all except Mendeleev). Birgitte Nielsen: xv(l), xvi(tl), xvii(tr), xvii(bl), 16(b), 63, 69, 74, 87(b), 135, 138, 301(both), 316(carrots & chips). Ontario Hydro: 57, 119(br), 158, 163, 186(b), 210(hairdryer), 292(t), 310. Ontario Ministry of Agriculture and Food: 47, 86, 115, 221(tl), 265(wheat). Ontario Ministry of Energy: 2(wood), 144(b), 145, 213(l). Ontario Ministry of the Environment: 59(irrigation). Ontario Ministry of Natural Resources: xviii(br), 222(tr-Erika Thinn), 228(clover), 232(deer, weasels, rabbit), 269(b), 270(l), 277(butterfly-R. Presgrave), 279(all except lamprey), 280(snake), 287(tr, tl, b), 320(R. Kaplan), 322. Ontario Place: 1(fireworks), 27. Ontario Science Centre: 275(br), 287(cr), 302(t). Arlene Penner: 276(l). Radio Shack/Intertan Canada Ltd: 206. James D. Rising: 60(b). Reptile Breeding Foundation: 265(snake-John Mitchell). Royal Ontario Museum: 2(crystal). Saskatchewan Ministry of Energy and Mines: 148. Shell Canada/Petroleum Resource Communications Foundation: 70(t). SSC Photocentre: 300(Tom Boshler), 328. Take Stock Photography Inc.: 59(street light repair-Bilodeau & Preston), 91(t-Mike Keller), 102(tl-Ray Ooms), 112(David Bentley), 287(cl-Rick Rudnicki), 288(Agnus McNee), 292(b-Bilodeau & Preston), 293(t-John Dean). Taurus Photos: 100(scuba school), 345(David Phillips). Trans-Alta Utilities: 186(t), 293(cr). Travel Alberta: xvii(br), 59(mountains), 144(tl). Verena Tunnicliffe: 96, 114. TV Ontario: xviii(bl), 59(bike), 228(dandelion), 277(grasshopper-Noel Keenan). Valan Photos: xviii(tl-Herman H. Giethoorn), 65(Aubrey Diem), 221(tc-Francis Lepine); cl-Arthur Strange; br-R.C. Simpson), 222(tl-Fred Bavendam; bl-Paul Janosi), 223(l-Valerie Wilkinson), 232(seals-Fred Bruemer), 265(panda-R.C. Simpson; rhino-Kennon Cooke, 266(b-Herman H. Giethoorn), 267(J.A. Wilkinson), 270(r-Karen D. Rooney), 275(bl-J.A. Wilkinson), 276(c-Paul Janosi), 277(mites-Thomas Kitchin), 279(l-Paul Janosi), 280(seal-Francis Lepine; bat-Wayne Lankinen; gorilla-Kennon Cooke). John Wiley & Sons Canada: 30, 44(b), 53(tl), 54, 126, 162, 216(r), 282, 302(b), 338. World Wildlife Fund: 280(turtle).

Special thanks to Philip Baer, science teacher at Annette Public School, and all students who participated in our photo sessions.